AFRICAN DINOSAURS
UNEARTHED

LIFE OF THE PAST
JAMES O. FARLOW, EDITOR

AFRICAN

DINOSAURS

UNEARTHED

THE TENDAGURU EXPEDITIONS

Gerhard Maier

INDIANA
University Press
Bloomington & Indianapolis

THIS BOOK IS A PUBLICATION OF

Indiana University Press
601 North Morton Street
Bloomington, IN 47404-3797 USA

http://iupress.indiana.edu

Telephone orders   800-842-6796
Fax orders   812-855-7931
Orders by e-mail   iuporder@indiana.edu

Library of Congress Cataloging-in-Publication Data

Maier, Gerhard, date
African dinosaurs unearthed : the Tendaguru expeditions / Gerhard Maier.
p. cm. — (Life of the past)
Includes bibliographical references and index.
ISBN 0-253-34214-7 (alk. paper)
1. Animals, Fossil—Tanzania. 2. Paleontology—Mesozoic. 3. Scientific expeditions—Tanzania—History—20th century. I. Title. II. Series.
QE731 .M25 2003
560'.9678—dc21
2002015732

1 2 3 4 5 08 07 06 05 04 03

TO COLLEEN

Who understood how important it was.

CONTENTS

Photographs follow pages 82 and 178.

FOREWORD

More than two dozen principalities on a plain bordering a northern sea unite to form a country that rapidly became prosperous. People in the new country, seeking adventure and knowledge, begin to explore a region located on a continent in the distant tropics. A man searching for minerals there finds pieces of giant bones near a "steep hill." The amazing discovery comes to the attention of a professor who lives in the northern country. He has seen huge bones taken from the ground and assembled into giant skeletons in another country beyond an ocean to the west. The professor travels over land and sea to the distant hill in the tropics and finds huge quantities of giant bones scattered over the ground. Very excited, he returns to his country and tells the director of a famous museum about the hill with all the bones. The director speaks to princes, to community leaders, and to captains of industry. They all agree that an expedition should be sent to the faraway hill in the tropics.

In such manner began the greatest intercontinental expedition ever to collect dinosaurs. The discovery was made in 1907 at Tendaguru in what is now southeastern Tanzania, thirty-six years after the unification of Germany. The centenary of the discovery is now approaching. The story of that expedition and the British expedition that followed may be told in numbers: funding, people, ledgers, specimens, casts, boxes, and weights. These details are important, for they address the scope and results of the excavations. But with only that information, the story would remain incomplete—more fundamental is the fire that caused people to dedicate themselves to its success.

It was a question of national honor. German researchers who participated in the expedition later died in the defense of their colony. A dinosaur, *Dryosaurus lettow-vorbecki*, was named in homage to the officer who directed its defense. Has any other dinosaur ever been named after a general? Many years later, old men could be found in the Tanzanian *bundu* (bush) who would proudly say, "Once I was a German soldier." After a war and a depression, a German paleontologist returned to "Steep Hill" to be warmly welcomed by men who had worked there in the quarries. A leader of British excavations humbly confessed his ignorance of bone morphology and requested instruction that never came; another worked until he died. How often it was said of a European that he labored in the tropical lowlands of east Africa "until his health was broken." And how many families, German, English, and Tanzanian, were separated by months and years of fatigue and loneliness because of the giant bones of Tendaguru.

For Germany, the half-century from 1909 included nearly thirteen years of totalitarian government, ten years of disastrous war, and ten years of economic turmoil. Museums and collections were destroyed; the scientific community and their families were decimated by death and emigration. Yet the period also saw the cleaning and preservation of Tendaguru fossils through an effort equal to half that invested in the field work; the construction of five dinosaur mounts, including one that remains the largest reconstructed dinosaur skeleton in the world; and the publication of the expedition's monographs. Paleontological research rose from the ashes.

Much has been written about good government, and history amply demonstrates that it is humanity's greatest challenge. Similarly, the heroic efforts at Tendaguru are worthy of the attention of those who would collect dinosaurs. They were inspired by love of country and a love of knowledge. To keep that vision clean, it was necessary to compensate as much as possible for an array of inevitable human failings. At Tendaguru, success followed the clear identification of individual responsibilities, and the encouragement of collegiality and consultation. In a multicultural setting, multilingual participants were essential. So was adequate nourishment.

In modern laboratories, lined with expensive technological devices, researchers still probe the contents of casts made so long ago at Tendaguru. Their vision is enormously extended by their instruments. They see and understand the implications of microscopic spores and pollen, of microlaminae of bone deposited in osteons, of mineral particles, and even of isotopes of atoms. The detail on such small scales seems to collapse the separation in time between the researchers and the archaic materials they study to zero.

On a human scale, missionaries, health workers, and teachers continue to labor in the service of others in the region of the faraway hill in the tropics. Across the *bundu,* the dry season prevails. Leafless forests are compressed between a burned-grass floor and a vault of electric blue. The bite of the tsetse fly is powerful, and that of the mosquito, lethal. Nearby, but 150 million years away in time, forests of flowerless trees line a subtropical coast. Their sparse crowns decorated with sinuous, reptilian fronds are slowly being stripped by giraffoid giants. On tidal wetlands between the forest and the sea, huge, dismembered bones and broken, knife-like teeth lie baking beneath a younger dry-season sun. The stench of decay, of both vegetation and flesh, wafts through the salty air.

The apparent rupture is artificial, for the two scenes are bound together in a continuum of space and time. The continuum lies at the root of the mystery of our existence within a world that vastly exceeds human dimensions and lifespans. A conviction that the multidimensionality of space-time is understandable and that knowledge is useful, opens the world to exploration. The effort will be costly, but the cost will be eclipsed by what is gained. The epic of the great expedition is now lost in time. It will never be repeated, for only the birth of a country could have sustained it. Its memory, however, lingers in those who cherish it, and in the fortunate few who can say, "Once, I too climbed Tendaguru Hill."

In the book that follows, the author expertly and compassionately describes the saga that began with Bernhard Sattler's discovery of immense bones in the *bundu.*

Dale Russell
December 2002

PREFACE

The region surrounding a small hill in southern Tanzania, known as Tendaguru, became famous almost a century ago. An intensive search for the remains of enormous Jurassic dinosaurs took place there between 1907 and 1931. Many hundreds of Africans were organized—first by German, then by British paleontologists—to unearth over two hundred tonnes of bones. The expeditions were multidisciplinary in nature, examining the geology and stratigraphy of the area and of other regions of the country.

German fieldwork was terminated prior to the First World War, but excavation resumed under the British. Preparation and study of the fossil material continued for decades in Germany, as scientific monographs were published and five skeletons were mounted in Berlin. Other institutions revisited the site, hoping to answer different questions about dinosaurs and their world, as the science of vertebrate paleontology evolved.

This comprehensive historical reconstruction introduces the people and their accomplishments. Field and museum campaigns that spanned almost a century are recounted against the background of worldwide political events that occasionally had a devastating impact. Scientific results are not a major theme. Research findings are available for all to read, but historical documentation of the day-to-day work of excavating, preparing, and mounting the finds is less accessible. Language, handwriting styles, geographic dispersal in museum archives—all make a thorough overview a time-consuming task. It is hoped that others will find the tale equally engrossing, and will fill in the gaps and correct the inevitable errors and misinterpretations, which are the author's alone.

The tradition of generosity and hospitality extended by the Museum für Naturkunde, Berlin over the course of two decades is acknowledged with the deepest gratitude. Hermann Jaeger and Wolf-Dieter Heinrich enthusiastically supported every effort to tell the story of Tendaguru. Archivists at the Natural History Museum, London, and the University of Tübingen provided access to materials, without which this book could not have been written. A debt is owed to many researchers who responded to numerous difficult questions.

AFRICAN DINOSAURS
UNEARTHED

-1-

1907

Something Curious in the African Bush

Cooling rains were returning to the interior of eastern Africa, repeating a cycle stretching back countless millennia. Soon the parched landscape would turn green with vegetation, and trails in the isolated hinterland would be less easily traveled. Of course, trails were sparse in German East Africa (Deutsch Ostafrika) in 1906, especially in the far south.[1] It was here, in Germany's prized colony on that vast continent, that an object of Bernhard Wilhelm Sattler's curiosity would involve thousands of Africans, as well as Germans and Englishmen, in monumental labors for decades to come.

Sattler, seasoned by long years of experience in Africa, was in charge of a garnet mine operated by the Lindi Prospecting Company (Lindi Schürfgesellschaft). He was traveling to the mine, south of the Mbemkuru River, when he noticed an enormous bone weathering out of the path near the base of a hill. In the language of the local Wamwera people the hill was known as Tendaguru, or "steep hill."

Subsequent accounts, likely apocryphal, relate that Sattler stumbled over this object on the path, and that he had his bearers carry fragments back to the nearest port, Lindi, a four-day march. Whatever the circumstances, he was sufficiently impressed by the great size of the remains that he forwarded a report and sketch to Wilhelm Arning, the director of his firm in Hannover, Germany. Arning, formerly a military surgeon in the colony, appreciated Sattler's conscientiousness. Arning's medical training and Sattler's drafting skills enabled Arning to recognize bones of prodigious size in Sattler's renderings as soon as the report arrived in Germany in early 1907. A chain of events was set into motion with Sattler's recognition of something unusual in the remote African bush. It would connect an ever-growing group of commercial and scientific men who shared a common interest in the state of Germany's overseas colonies.

Arning informed the Commission for the Geographical Investigation of the Protectorates (Kommission für die landeskundliche Erforschung der Schutzgebiete) in Berlin of Sattler's report. The commission had been established in 1904, at the suggestion of famed geographer Dr. Hans Meyer, who was its current director. Its mandate was to support research in the colonies by dispatching scientific and technical specialists. Meyer was absent when Arning's call for action arrived, and the commission recommended that Sattler should send some of the bones to Germany first. Arning, however, felt that it was irresponsible to encourage unscientific excavation, and continued to pressure the commission.

Germany was a latecomer to the nineteenth-century European competition for colonies. Influential groups quickly organized to lobby the German government in support of colonial development. Like other European nations, Germany sought territory in Africa. Persian and Arab traders had dominated the eastern coast of that vast continent between the ninth and sixteenth centuries A.D. Portuguese merchants enjoyed a period of success there during the sixteenth century. By the eighteenth century, Arab traders again controlled the coastal towns and monopolized the market in spices, slaves, ivory, and other commodities. In 1884, German adventurers, representing an ambitious colonial association, signed treaties with African tribal leaders in the interior. The legitimacy of these treaties may have been questionable, but the association was granted a charter over the newly claimed territory in 1885. Its agents formed a private company to administer the region, setting up coastal trading posts and plantations.

Agreements with Germany's colonial neighbors Portugal and Britain defined new geographic boundaries in East Africa. Inevitably, local Omani Arabs, loyal to the sultan of Zanzibar, resented this interference in their domain and rebelled violently. German chancellor Otto von Bismarck ordered the riots to be suppressed with military force, and reluctantly declared the area a protectorate of the Imperial Government in 1891. As von Bismarck's largest territorial claim in Africa, German East Africa extended over 946,500 square kilometers, an area almost twice the size of Imperial Germany.

Governors, who were often professional soldiers, were appointed to develop a bureaucracy responsible for customs, taxation, postal services, law and order, and other matters. A new currency was introduced and a bank opened. Police and military forces were established and began suppressing the slave trade. Yet the number of men responsible for governing this extensive region was small.

Uprisings and bloody punitive expeditions blighted the ensuing 15 years, as power was consolidated in the hands of Europeans. Military and civil administrative districts were expanded from the narrow coastal strip to the interior. Often only tenuous authority was exercised from widely scattered fortified garrisons. Local administration in the far-flung reaches devolved upon trusted Arabs. Communication was improved when telegraph and telephone lines were strung and the British laid an underwater cable to Zanzibar. Roads, bridges, ferries, and docks were built to upgrade transportation within the colony. Steamers plied the oceans between Africa and Europe. A northern railway was laid down and a central line was begun. Hospitals, schools, rest houses, and a prestigious agricultural research station were built. A meteorological service and veterinary station developed. Newspapers and a brewery were founded. Overseas investment financed plantations where German settlers grew crops that included sisal, cotton, rubber, and coffee. Plantation labor was supplied by the indigenous population.

Although changes were imposed unevenly and at different times within the districts of the protectorate, their impact undoubtedly caused the most radical and far-reaching transformation of East African cultures since the arrival of the Arabs. Brutal excesses characterized the first two decades of Germany's struggle to secure its authority over the country. Events in the homeland were soon to signal a change in approach.

A general election was held in Germany in 1907, with colonial policy a primary issue. Due to a growing number of scandals in its overseas possessions—another savage revolt in German East Africa and the Herero War in German Southwest Africa (Deutsch Südwestafrika, currently Namibia)—opposition parties demanded reform. One result was the establishment of an Imperial Colonial Office that was independent of the Foreign Ministry (Auswärtiges Amt). The secretary of state for the colonies (Staatssekretär des Reichskolonialamts) had authority over the governors of all colonies, who in turn supervised officers of the civilian or military districts.

How the economic development of the colony should proceed was hotly contested in German East Africa in 1907. In this era a range of improvements would be implemented that would allow the colonial power to effectively extract the maximum benefit from the foreign possession. Pressure groups in Germany lobbied the Reichstag to encourage European settlement, arguing that African labor on plantations would drive the colony's economy by providing European markets with new raw materials. German East Africa's governor, Baron Albrecht von Rechenberg, felt that such a policy would only result in further exploitation and alienation of the native population. He advocated cultivation of cash crops by Africans for trade with nations bordering the Indian Ocean.

The exhortations on overseas development delivered by His Excellency Dr. Bernhard Dernburg, secretary of state for the colonies, in January 1907 convinced the commercial councilor (Kommerzienrat) Heinrich Otto, the wealthy owner of a textile plant, to invest in German East Africa. Dernburg was about to leave for Africa, to confer with the governor, tour several districts, and recommend changes, and Otto was to accompany him. Otto and the consul Albert Schwarz, owner of a Stuttgart bank, formed a company to investigate commercial possibilities such as growing cotton, raising cattle, and operating steamships on Lake Victoria. Hoping that his geological expertise would prove useful in identifying coal deposits and valuable minerals, Otto invited Professor Dr. Eberhard Fraas, a Stuttgart paleontologist, to join them as a scientific advisor.

Eberhard Fraas was born in Stuttgart on June 26, 1862. He was the second son of Oskar Fraas, conservator (Konservator) at the geological-paleontological department of the Natural History Collection (Naturalienkabinett) in Stuttgart. In 1899, the institution was renamed the Royal Natural History Collection (Königliche Naturalienkabinett).

Eberhard inherited his father's passion for the earth sciences. The younger Fraas's academic training focused on geology, petrography, mineralogy, paleontology, and zoology. Classes at universities in Leipzig and Munich introduced him to Germany's eminent paleontologists, among them Karl Alfred von Zittel. Fraas received his Ph.D. at Munich in 1886.

Map A.
AFRICA

EQUATOR

TANZANIA

N

0   1000 km

Map B.
German East Africa
Redrawn from Wenig, 1920, <u>Kriegs-Safari</u>

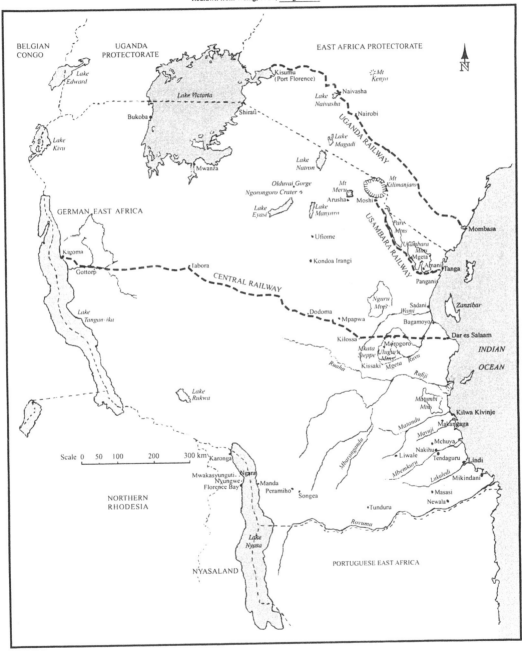

Eberhard succeeded his father at the Stuttgart museum in 1894. The title of Professor, with its attendant prestige in German society, was also conferred upon him that year. His enthusiasm for scientific research took him throughout Europe, and then to Egypt in 1897.

In 1901, Fraas visited America at the invitation of Henry Fairfield Osborn. Osborn, curator of the Department of Vertebrate Paleontology at New York's famous American Museum of Natural History, was one of America's preeminent and influential paleontologists, and had once studied German in Fraas's homeland. Fraas was especially enthusiastic about visiting Eastern museums and dinosaur localities in the American West, opportunities of which he had dreamed for years.

After viewing the New York collections, Fraas traveled to Princeton, Washington, Chicago, New Haven, and Ann Arbor. He then rejoined Osborn, and the two men boarded a train to Colorado. From Denver, they made several excursions through the mountains beyond Canon City, and continued to Utah and Wyoming. These western states were the source of a fabulous variety and abundance of Jurassic dinosaurs, whose resting places were visited by Fraas and Osborn.

Fraas was privileged to participate in the American Museum's excavation at Bone Cabin Quarry in Wyoming that year, alongside the crew of Walter Granger, Peter Kaisen, and George Olsen. This tremendously productive locality had been worked annually since its discovery in 1898, and would eventually yield over 81 tonnes of dinosaur remains.[2] According to the AMNH Annual Report of 1901, the best collection to date was removed that year. Fraas reveled in life in the field, where a sleeping bag and raincoat provided the only protection from the elements. He also acquired skills that in later years would prove invaluable, namely expertise in excavating huge fossil remains.

Leaving Osborn, Fraas took part in a "special expedition" of the United States Geological Survey, led by Nelson Horatio Darton. Its members would collect geological information for Osborn's comprehensive monograph, *The Titanotheres*. Legendary collector John Bell Hatcher of the Carnegie Museum accompanied them to the Oligocene badlands of South Dakota, where they spent at least three grueling weeks. Traveling through the Black Hills by horse and wagon, they encountered quicksand, and Hatcher and Fraas had a narrow escape while crossing a river in flood. Fraas then headed west to Yellowstone Park in Montana, and upon his return to New York, Osborn presented him with sauropod limbs from Wyoming.

In Germany, Eberhard Fraas was awarded the Knight's Cross First Class (Ritterkreuz I Klasse). German Southwest Africa was his destination in 1904, and Egypt in 1906. Eberhard's elder brother Victor secured funding that allowed Eberhard to undertake a lengthy camel trek to the Fayum Oasis, 70 to 80 kilometers southwest of Cairo. He was accompanied by Richard Markgraf, an impoverished fossil collector, who had enriched museum collections in Stuttgart, Frankfurt, Munich, and New York for over a decade. Markgraf had uncovered remarkable specimens in Egypt, including an early Oligocene primate that would lead to a subsequent connection with Walter Granger of the American

Museum of Natural History. At the Fayum they found remains of the large herbivorous mammal *Arsinoitherium* and the early whale *Basilosaurus*. Now, at
Otto's invitation, he was off to East Africa. Wilhelm Arning, hearing that the
renowned paleontologist would soon be in Dar es Salaam, placed both Bernhard Sattler and laborers belonging to the Lindi Prospecting Company at his
disposal gratis.

Fraas was granted five months' unpaid leave from the Royal Natural History
Collection. Otto, Schwarz, and Fraas departed Stuttgart independently of Dernburg, whose party had been delayed, on June 1, 1907. After a stormy passage,
they disembarked on June 21 at Dar es Salaam, Arabic for "Haven of Peace."
The capital city was growing rapidly at the time of their arrival. Construction of
a new railway spurred many changes. An influx of technical staff and their families, along with hordes of laborers, stimulated local businesses. By 1900, it was
home to approximately 20,000 people.[3] At the time of Fraas's visit, six or seven
hundred were Europeans, out of a total of 2,792 Europeans in the colony.[4]

In the 1860s, the sultan of Zanzibar had established a post at a sheltered East
African harbor. The settlement, known as Dar es Salaam, had languished in the
humid equatorial heat, and interest shifted north to the rival port of Bagamoyo.
German traders in the 1880s became embroiled in a series of rebellions and
made little substantial progress. A decade later, the seat of German government
was transferred to Dar es Salaam from Bagamoyo. Government buildings were
erected and an extensive botanic garden was planted. By the time of Fraas's arrival in 1907, the spires of the Catholic cathedral and the Lutheran church were
familiar landmarks for steamers entering the harbor. A lighthouse, a floating
dock, and electric cranes allowed heavy cargo to be safely shipped and unloaded. Stationary engines in the rail yards supplied electricity for the town.

Otto, Schwarz, and Fraas probably moved around the capital by rickshaw,
the most common form of transport. Horse-drawn carriages were used by only
the most senior officials. Greek- and Syrian-run hotels were available for travelers, but the Stuttgart trio likely checked into the Kaiserhof, built the year before
and featuring hot and cold water and accommodation for thirty guests. Well
laid out streets and comfortable residential dwellings characterized the European district. The original native quarter had shifted to a new location, for
which an orderly grid of roads was surveyed. Between these two districts flourished an Indian bazaar, a labyrinth of shops and African dwellings. Otto's group
may have purchased provisions here.

Their intention was to leave immediately for Lake Victoria to establish a shipping company. These plans were altered when the governor recommended that
they tour the central regions, along the line of railway construction from Dar es
Salaam to Morogoro. Otto hoped to establish a cotton plantation in the interior, to provide a reliable supply for his factory in Germany. On June 26, 1907,
Eberhard Fraas celebrated his 45th birthday, and the group departed Dar es
Salaam by rail.

Work on the meter-gauge Central Railway (Mittellandbahn) had begun two
years earlier. With the railhead at Dar es Salaam, the line would eventually traverse the country and terminate at Kigoma on Lake Tanganyika, 1,244 kilome-

ters to the west. Trade around the great lake and throughout the interior would therefore have access to an ocean port. In 1907, however, the end of steel was at Ngerengere, 149 kilometers west of Dar es Salaam, and Fraas happily collected fossils en route from Jurassic limestones exposed by the railway cut. An engineer named Kinkelin had gathered ammonite specimens at a stone quarry near Pendambili, at kilometer 128, during the construction of the rail line. Fraas relocated the quarry. At Ngerengere, the travelers went on to experience firsthand the time-honored form of travel in Africa, the safari.

Fraas declared the three-day trek on foot to Morogoro to be a "quite healthy though also strenuous roving life."[5] They reached the town, at the northern foot of the Uluguru Mountains, where Fraas inspected local mica mines. Northwest of Kilossa, Otto became ill and had to be rushed to a nearby mission station, where he remained under care for a week. Returning to the rail line, they reboarded at Ngerengere, and their train rolled back into Dar es Salaam on July 23. What had been envisaged as a week-long excursion had lasted almost a month and had broken the health of one of its participants.

Otto was advised to return to Germany immediately. Schwarz and Fraas, though not unscathed by their 28-day adventure, boarded a passenger liner to Mombasa. Mombasa was the major port of Britain's East Africa Protectorate and the railhead of the Uganda Railway. Known to critics as the Lunatic Line, it stretched inland 966 kilometers to Kisumu (Port Florence) on Lake Victoria. Devastating rains, tsetse flies, disease outbreaks, and man-eating lions had bedeviled the construction program. Fraas and Schwarz rode the line to Kisumu, a five-day journey across savannah teeming with wildlife.

When pressing business interests forced Schwarz to return to Germany, Fraas carried on alone, boarding a lake steamer. In the course of his travels, he nearly circumnavigated Lake Victoria, the world's third largest lake. When he returned to Mombasa, the scheduled steamship to Dar es Salaam was long overdue, but Fraas made excellent use of the delay, scouring Jurassic-age exposures for fossils.

His official duties for Otto and his partner discharged, Fraas eagerly read several letters that had arrived during his first safari. Dr. Hans Meyer of Leipzig urged him to investigate Sattler's finds. Arning provided additional details and instructions for contacting Sattler, and Otto had left him the necessary equipment and supplies from their earlier reconnaissance. He determined to do so.

A coastal steamer slipped south through the Indian Ocean, conveying Fraas from Dar es Salaam to Lindi. It dropped anchor on August 30. In 1905, the little town had hosted a population of 3,500.[6] Spread along a creek, it possessed a postcard look from the sea—endless sandy beaches lined with tall palms, and Arab dhows moored in the harbor. The seat of government was in the boma, a fort with ornately carved Arabic doors, and there was also a small medical dispensary. A shed located behind the fort served as the main customs office (Hauptzollamt), and the nearby market hall was overgrown with vivid bougainvillea flowers. Fraas was dismayed to discover that he had somehow become separated from his servant and baggage, but the local administration acted quickly to provide him with replacement gear, bearers, and guides for an expedition to Sattler's fossil site at Tendaguru.

Doubts still assailed the paleontologist. He too, had suffered badly from dysentery on the two-month inland safari and still felt very weak. What was the significance of the purported discovery of bones? How could he be certain that the remains were as large as claimed, or even fossilized? What if they were too badly weathered to collect, or, worse yet, were not even bones?

On August 31, the day after his arrival in Lindi, he found himself marching to Tendaguru. District Administrator ten Brink and the portly Dr. Wolff, chief surgeon of the Defense Force (Oberarzt der Schutztruppe), accompanied him. His escort was a well-equipped caravan of 60 bearers and African soldiers (askaris).

The column marched north along Lindi Bay and ascended a limestone ridge of Eocene age known as the Kitulo Heights. Night was approaching, so the first camp was pitched. Fraas's excitement was tempered by the realization that heavy undergrowth made geological observations extremely difficult. Further hindered by poor health, he was restricted to examining exposures along the trail and telegraph line, and was carried in a litter at times.

The next day they crossed the two-hundred-meter summit of the Kitulo Heights and descended steeply into an open plain known as Yangwani. Their second camp was erected at the settlement of Namudi. They traversed heavily forested plateaus, through seemingly endless tracts of thorny acacias, gnarled baobab trees, and tangled grasses infested with swarms of mosquitoes and tsetse flies. Daytime temperatures climbed to 35°C. Annual rains normally rejuvenated the arid terrain from December to May, so the safari was marching through the heat of the dry season.

Fraas recorded geological observations, stopping occasionally to check the exposures for fossils. By the time they reached the Mikadi Plateau, the vegetation had been transformed into extensive stands of bamboo. For hours, the column marched through desolated settlements, abandoned during the Maji-Maji Rebellion that had raged through the district.

Two summers before, southeast German East Africa had been convulsed by a fierce rejection of German rule. The complex causes are debated to this day. One factor was the institution of compulsory labor on rubber and cotton plantations. Harsh abuse and taxes aggravated the already unjust system that was enforced by German overseers and their Arab and African proxies. Throughout the summer of 1905, unrest grew among the people of the Matumbi Hills near Kilwa. A cult claimed that anyone partaking of magic water (maji) would be rendered immune to bullets. As a defiant gesture, a small number of Africans uprooted one symbol of their oppression, cotton.

The rebellion spread rapidly. In Lindi District, missions and mining projects were destroyed. To the north, cotton crops were torched. Trading posts were burned, missionaries were shot, and German troops were ambushed. Panic ensued among the European population, which suddenly felt isolated and vulnerable.

Kaiser Wilhelm II dispatched two naval cruisers and a complement of marines from the German protectorates in China and New Guinea. Three military columns took the offensive. Reprisals were brutal: rebel leaders were hanged and a scorched-earth policy was adopted. Crops and villages were burned and food supplies and cattle were seized. Deprived of operational support, the remaining warriors waged guerrilla warfare. It was not until the end of January 1906 that full control was reestablished in Lindi District. A year later, all resistance ceased.

The toll on the African population was appalling. Starvation was widespread. Chronic malnutrition led to an increase in infant mortality and decreased fertility in women. Tsetse flies infested the abandoned countryside, adding the threat of sleeping sickness to other diseases ravaging already weakened survivors. The net result was a major depopulation of southeast German East Africa. Casualty estimates ranged from an official contemporary figure of 75,000 to more recent estimates of 120,000, or even 200,000.[7] News of the uprising reached Germany and sparked questions in the Reichstag. Secretary of State Dernburg's visit to the colony was one consequence of the war, as were Governor von Rechenberg's attempts at reform.

It was through this devastated landscape that Professor Fraas and his party advanced. On the evening of the fifth day of marching, they reached the hill known as Tendaguru. It was situated on the border of a plateau that fell westward to the Mbemkuru River. To the east the terrain rolled gently to the base of the Likonde and Noto Plateaus. This circuitous route had been chosen over a more direct one, to spare Fraas the rigorous ascent of the Likonde Plateau.

Upon his arrival at Tendaguru, Fraas was suffering gravely from amoebic dysentery. But his pain and exhaustion were immediately forgotten in the elation and relief of discovery:

> The joy that inspired me as I caught sight of the immense bone fragments for the first time, and immediately recognized them correctly as dinosaur remains, can only be appreciated by someone who himself is a scientist, and who has reached a long-desired goal after privation and hardships. There now lay, weathered out and washed, in soft sandstones, the prodigious parts—foot bones of more than one meter's length, finger elements, claws, and vertebrae—and [they] spoke in an eloquent language of the extinct primeval world.[8]

A punishing journey through desiccated thorn scrub and waterless streambeds in the glaring African sun had been rewarded with unimagined success. Bone lay strewn everywhere, but darkness arrived all too quickly: "Full of plans and thoughts about the course of the investigation, I spent a restless night until the morning invited the commencement of excavation."[9]

Although Tendaguru reminded him of Bone Cabin Quarry in Wyoming with respect to the size and abundance of bones, it did not boast the exposed strata of the American site. The overall extent of bone was not immediately obvious, since the region was cloaked in heavy undergrowth. Lacking exposures, Fraas selected a spot where bones were concentrated on the surface and set his men excavating. Complete skeletal parts or perhaps even entire skeletons were needed to identify the types of dinosaurs represented at the site. Digging would allow him to interpret the sequence of rock layers.

Disappointingly, he was not equipped to undertake the lengthy and large-scale investigation that Tendaguru demanded. About 30 laborers, out of the 60 men on the safari, were supplied with small hoes. After several hours of effort, the inexperienced crew had penetrated only a few centimeters into the hard ground. Frustration is obvious in Fraas's early report: "I thought uneasily about the number of weeks and months I could well fry in the sun here, until a few cubic meters would be dug up."[10]

Encouragingly, District Administrator ten Brink quickly fashioned a comfortable camp in the bush. Tents were pitched on the southern flank of the hill. A spacious and airy bamboo hut was constructed, complete with chairs and tables. Overhead, the black, white, and red flag of Imperial Germany waved in the breeze. Unknown to Fraas, W. B. Sattler, the European who first reported the fossils, was rushing to the site to assist.

Bernhard Wilhelm Sattler, son of an artist, was born in Cronberg near Frankfurt am Main on August 17, 1873. At age 16, he joined a Berlin firm as a business apprentice. During his compulsory one-year military service he contracted a severe lung infection, and at age 21 traveled to South Africa to recover his health. In Johannesburg, he trained as a pharmacist, and soon obtained a position with a mining company as a chemical analyst.

Sattler was a member of De la Rey's commandos, the last force to yield to the British in the bitter Boer War. Returning to Schweinfurt, Sattler befriended Dr. Wilhelm Arning, and was promptly hired to work for the Lindi Prospecting Company, a firm established in 1903. Sattler took up his position as mining engineer (Bergingenieur) in Germany's East African colony around 1904.

A polyglot, Sattler spoke German, English, Italian, Afrikaans, Swahili, and several African dialects. Lindi District became his second home—the location of his greatest accomplishments. During the Maji-Maji Rebellion, he led a volunteer corps against the Wamwera and Wangindo tribes, and was awarded the Royal Prussian Order of the Crown, Fourth Class with Swords (Königliche Preussische Kronenorden, IV Klasse mit Schwertern) in 1907. Despite the cruelty of the conflict, Sattler provided food for starving women and children after the rebellion. He is reported to have had a special understanding of and sympathy for Africans. In turn, they are said to have respected and trusted him. To fellow Europeans he was an old Africa hand with an inexhaustible store of hard-won knowledge about the country and its peoples. He was renowned for his willingness to assist anyone in need. These qualities, combined with a practical interest in scientific matters, proved invaluable at Tendaguru.

Notified of Fraas's presence in the colony, the 34-year–old Sattler and his team of Africans arrived at Tendaguru one day after Fraas, by forced marches. Sattler's men were skilled miners equipped with German-made picks and hoes, so progress accelerated markedly. By the evening of Sattler's first day on site, fragmented hind limbs, pieces of a pelvis, and a partial vertebral column were exposed. All were lying in association in a pit half a meter deep. Fraas designated this find as Skeleton A. It was subsequently pronounced a type specimen, based on two caudal vertebrae, partial ribs, an ischium, a femur, a tibia, and an astragalus.

Massive fires were lit to remove the tangled thicket that obscured the landscape. Tendaguru revealed itself as overwhelmingly rich—a veritable graveyard of extinct giants. Three more partial skeletons came to light after burning, but Fraas's failing health allowed him no more than a few brief orientation excursions.

Approximately four hundred meters south of Skeleton A lay two femora, each 1.4 meters long. They spanned a path and formed a slight rise. These bones had first aroused Sattler's curiosity. Another badly eroded pelvis was uncovered some 250 meters southeast of Skeleton A. It was cleaned, photographed, measured, and sketched, since Fraas realized that the limited means at his disposal would never allow him to transport such a heavy and fractured specimen to the coast. Quite close to this pelvis was an entire right hind leg. Much sturdier, it was packed out in pieces and labeled Skeleton B.

It must have been frustrating for Fraas to have come so far, at such cost in time and health, only to superficially sample this fossil treasure trove. In all, he noted about 20 femora, but was forced to be selective about what he could bring home. An intact femur could tip the scales at 115 to 160 kilograms, while a single articular end represented a 42-kilogram load for a bearer.[11]

Fraas was only too aware of the time and resources required to collect such enormous dinosaurs—his experience in North America made them abundantly clear. Luckily, certain surface remains were worth saving, so he had a number of isolated foot elements and vertebrae collected.

As Fraas's illness worsened, Bernhard Sattler stepped in and ordered his crew to pack the collection. Altogether, a column of about 90 men carried more than 30 bulky loads back to Lindi.[12] The paleontologist was far too weak now to search for fossils or even make detailed geological observations. He was forced to return to the coast in short, painful stages, accompanied by Dr. Wolff. They arrived in Lindi on September 16, having spent approximately one week at Tendaguru. The collection was repacked into crates and bundles for shipment to Stuttgart.

A coastal boat bore Fraas to Dar es Salaam. Back in Stuttgart on October 13, he resumed his work at the Museum three days later. Fraas had spent four and a half months in Africa.

-2-

1908

A Matter of National Honor

In Stuttgart, museum authorities recommended that Bernhard Sattler be recognized for his substantial assistance. In January 1908, he was awarded the Knight's Cross Second Class of the Royal Württemberg Order of Frederick (Ritterkreuz II Klasse des Königlich Württembergische Friedrich-Ordens). A fortuitous chain of events had brought the fossils to light: Sattler's scientific interest and energy, the conscientiousness of various colonial officials, and the enthusiasm, expertise, and American experience of Fraas.

In early 1908, the precious African cargo reached Stuttgart, although numerous vertebrae and foot bones had somehow been accidentally left at Tendaguru. Fraas energetically publicized the discovery as newspapers spread the word across Germany. Within weeks, articles appeared in popular German science and natural history magazines. Fraas spoke to learned societies and, in 1908, published a scientific description of the finds. In a special volume of the prestigious journal *Palaeontographica,* he described and illustrated the geology and fossils of the site. Two species of a new and enormous type of dinosaur were established—*Gigantosaurus robustus* and *Gigantosaurus africanus.* Fraas believed they were similar to the North American Jurassic sauropods *Diplodocus* and *Camarasaurus.*

Although dinosaurs on several continents had been described by 1907, most of those found in Europe could not compare with the sheer size of the American sauropods. In Africa, only Madagascar, South Africa, and the French Sahara had yielded dinosaurs. In that era, dinosaurs were universally considered slow and unintelligent reptiles. Interpretations of the lifestyle of sauropod dinosaurs had varied since their discovery. In Britain, Richard Owen was convinced that they had lived exclusively in water. In the United States, Marsh and Cope initially proposed a terrestrial, giraffe-like existence, but later adopted Owen's arguments, as did Osborn and Matthew. One of the few dissenters at the time of Tendaguru's discovery was Riggs, who maintained that a gigantic form he had described, *Brachiosaurus,* was adapted to life on land. Skeletons had been restored with limbs held vertically beneath the body, but this too, was disputed by the American Oliver Hay and the Germans Gustav Tornier and Richard Sternfeld, who both argued for a lizard-like pose.

Fraas believed that the sediments of Tendaguru were of Upper Cretaceous age, and postulated that a land bridge had existed between North America and Africa during the Upper Jurassic. Tendaguru dinosaurs, he suggested, represented a relict fauna that had survived after North American forms became extinct. Fraas recognized two strata at Tendaguru, the lower of marine origin, and the upper of continental origin.

The invertebrate fossils collected during his excursions in the northern part of Germany's colony and the East Africa Protectorate were forwarded to Dr. Edgar Dacqué in Munich for description. Those from Tendaguru were sent to paleontologist Dr. Erich Krenkel, also in Munich. These two researchers erected 21 new species of ammonites and about a dozen bivalve species, and drew paleogeographic conclusions. Dacqué believed that the East African invertebrate fauna had shifted from an Indian to a European character between Oxfordian and Kimmeridgian times. Krenkel argued that Fraas's interpretation of the Cretaceous in East Africa was incorrect, as the ammonites could be dated to the Lower Cretaceous.

Fraas assigned the fossils to the sole preparator in the Museum's geological department, Max Böck. Born on February 27, 1877, in Herbertingen, about 40 kilometers north of Lake Constance, Böck was a stonemason and sculptor by trade. Hired as an assistant preparator at Stuttgart in mid-April 1907, the 30-year-old Böck was initially trained in paleontological preparation by his predecessor. He was later taught by Bernhard Hauff in Holzmaden, a locale famous for its beautifully preserved ichthyosaurs. Conservators at the Museum and workers at fossil excavation sites alike considered Böck an outspoken but tireless master craftsman.

Böck mounted a *Diplodocus* femur from Bone Cabin Quarry next to a composite right hind limb from Tendaguru. This limb was assembled from the phalanges, astragalus, metatarsals, tibia, and fibula of Skeleton B (*Gigantosaurus robustus*), combined with a *Gigantosaurus africanus* femur from Skeleton A. A *Gigantosaurus africanus* ischium from Skeleton A was placed beside the limb. Finally, there loomed a complete *Diplodocus* left hind limb, also from Bone Cabin Quarry. At least one more mount existed, a right hind limb from *Gigantosaurus robustus*. Only the lowermost portion of the femur, a mere 1.5 meters high, was preserved.

News had spread to the English-speaking scientific world by December 1908, with a notice in *Nature*. At a lecture to the Berlin Society for Earth Sciences (Gesellschaft für Erdkunde zu Berlin), perhaps the second oldest geographic association in the world, Fraas vigorously championed further investigation at Tendaguru.

There was concern among members of the museum community that foreigners would become interested, though it was felt that competition at Tendaguru should not be refused, for two reasons. Firstly, to maintain good political relations, any external application to excavate should be approved, and secondly, they viewed scientific research as international in scope.

Nonetheless, pressure built to place a German stamp on the enterprise. Fraas's experience in the United States had introduced him to the scale on which Americans were able to operate, thanks to funding by philanthropic millionaires such as J. P. Morgan, George Peabody, and Andrew Carnegie. A significant discovery on German soil obliged German scientists to demonstrate to the world that they were equally capable of developing such a tremendous resource. In Fraas's words,

> we have at Tendaguru a locality for dinosaurs that hardly falls short of the rich American sites. However, I maintain that our German science is honor-bound to see that this work is begun as soon as possible and supported by the required resources.[1]

Fraas's institute in Stuttgart possessed neither the personnel nor the finances to adequately investigate the site. Rather than approach other German museums to launch a cooperative venture, Fraas turned to a long-time friend, Professor Wilhelm von Branca. Between 1901 and 1907, they had co-authored a series of papers explaining geological features of volcanic origin in southern Germany. By 1907, von Branca was a highly regarded scientist with a distinguished career. More importantly, he was also the director of the Geological-Paleontological Institute and Museum of the Royal Friedrich-Wilhelm University of Berlin (Geologisch-Paläontologisches Institut und Museum der Königliche Friedrich-Wilhelm Universität zu Berlin). In later years, von Branca acknowledged that Fraas's broader vision of German science had presented an unparalleled opportunity to the capital, Berlin.

Karl Wilhelm Franz Branco, who later became known as Wilhelm von Branca, was born in Potsdam, near Berlin, on September 9, 1844, and raised in Silesia. His interest in natural history drew him to the University of Heidelberg and culminated in a Ph.D. in 1876. He did postdoctoral work in Rome, Berlin, Strassburg, and Munich. It was in Munich that he, like Fraas, studied stratigraphy and paleontology under Karl Alfred von Zittel.

Von Branca held lecturing positions at universities and government ministries in Berlin, Aachen, Königsberg, Tübingen, and Hohenheim. In 1898, he accepted a post at the University of Berlin as lecturer in geology and paleontology, and became director of the Geological-Paleontological Institute and Museum. As the years passed, numerous honors were bestowed upon him in recognition of his academic achievements. Contemporaries described him as being of noble and distinguished bearing, but approachable.

Once again, fortune favored the Tendaguru fossils. Fraas's energetic promotion found its mark in a personal acquaintance whose reputation, position, and connections could advance the investigation. Like Fraas, von Branca instantly recognized the scientific importance of the remains.

In 1907, Germany was a young and ambitious nation, having been unified by von Bismarck, the Iron Chancellor, only 36 years previously. Presently under the rule of the headstrong Kaiser Wilhelm II, Germany prided itself on its achievements. These ranged from music and philosophy to technical, industrial, and military breakthroughs. Industrial and commercial growth was rapid after 1871, fed by the immense reparations paid by the French after their defeat in the Franco-Prussian War.

Convinced that national honor was at stake, the 63-year-old von Branca was determined to commence excavations with the backing of his institute. Collections of "curiosities and petrefactions" had accumulated in Berlin since the 1600s. The Academy of Sciences (Akademie der Wissenschaften), established in 1700, held specimens, as did private men of means. Natural history had enjoyed widespread popularity among all classes of people in Britain and Europe from the mid-1800s onward. With the expansion of global trade and the onset of the imperial era, few parts of the world remained unvisited by military, trading, or scientific expeditions. With the founding of Berlin University in 1810, these scattered groups of fossils found a new home in a wing of what had once been a royal palace.

By the 1870s, three separate museums filled two-thirds of the university, and it became obvious that the palace was not designed to adequately store and display natural history specimens. Therefore, in the fall of 1883, plans were laid for a new natural history museum and teaching institute.

Museum architecture in the latter half of the 1800s was often monumental, reflecting the prestige that immense collections brought to cities like London, Paris, and New York. Designs often incorporated glass atria reminiscent of railway stations, the characteristic manifestation of progress in the Victorian era. Such an edifice would soon grace Berlin on the site of the old Royal Iron Foundry (Königliche Eisengiesserei) at Invalidenstrasse 43. The new museum would occupy the central position in a grouping of three structures of the same architectural style. On the west stood the National Geological Institute and Mining Academy (Geologische Landesanstalt und Bergakademie). To the east stood the Royal Agricultural College (Königliche Landwirtschaftliche Hochschule).

In September 1886, it was decreed that the facility would be known as the Museum of Natural History (Museum für Naturkunde). It had three stories and a basement. Cut tuffstone outer walls supported by iron girders formed the main frontage, facing Invalidenstrasse. The rear portions of the structure were clad in economical fired brick. A sandstone facade at the main entrance bore reliefs of German scientists who were influential in establishing programs at the university.

The main square block enclosed the Lichthof, an atrium with a glass roof supported by iron girders. Front galleries had vaulted ceilings supported by polished diorite and syenite columns. A rectangular block extended transversely from both ends of the main square block. The transverse blocks each had two rectangular wings extending northward. Each of the four wings was separated from the next wing by a courtyard, to allow ample natural lighting. The entire edifice was heated with steam and hot water from four boilers, and initially illuminated by gas jets.

New teaching institutes were founded: a Geological-Paleontological Institute and a Mineralogical-Petrographic Institute. The Zoological Institute formed the third division of the Museum. Each division was self-contained and autonomous, with its own staff.

Construction was nearing completion in 1886, and some of the University collections were moved and installed in 1887 and 1888.[2] Dedication ceremonies took place on December 2, 1889, in the presence of the Kaiser. It was not until 1892 that the Museum was officially turned over to its director. Construction costs ran to 4,170,000 marks.[3] The Berlin public had access to a relatively unified natural history display for the first time.

Von Branca's directorship put a vigorous personal stamp on both the public and academic sides of the 10-year-old Museum. He had successfully applied for state funding amounting to 60,000 marks in 1901, to purchase fossil mammal remains. In 1908, the distinguished director was again fighting for support. He was convinced that the enterprise at Tendaguru had to be conducted on a massive scale if it was to be successful: "Admittedly, it was obvious that very large resources would be required, if something truly good were to be accomplished."[4]

In his opening gambit, von Branca drafted letters to the secretary of state of the Imperial Colonial Office. The secretary pleaded a lack of resources, but recommended that the authorities in German East Africa ensure the site remained undisturbed. In June 1908, the colonial department passed the request to state authorities, the Prussian Ministry of Culture (Preussisches Kultusministerium). Circumstances had changed since von Branca's earlier monetary successes: "From the outset it was, at that time, out of the question to obtain even a small portion of the necessary funds from the state, since Prussia was then forced to exercise the greatest thrift."[5]

Though the German economy had performed strongly after 1895, a brief but serious crisis lasted from July 1907 until December 1908. Several hundred thousand industrial workers were unemployed. In large cities, thousands were put to work shoveling snow over the winter of 1908–1909. It is no wonder that little money was available from the public sector to excavate dinosaur bones in Africa. It must have been disheartening to be stymied so soon in the campaign, but von Branca was undaunted: "thus my only remaining option was to go public with an attempt to awaken on the private side the necessary enthusiasm to obtain the required resources."[6]

Von Branca's determination, eloquence, and credentials soon produced a startling reversal of fortune. Well connected, he was able to operate at the level of German society that was both moneyed and influential. Professional, academic, political, and business success was highly respected in Wilhelmine Germany, and the director knew individuals of note in these fields. Among his acquaintances was Dr. David von Hansemann, a renowned pathologist at the Rudolf Virchow Hospital and a private lecturer in pathology and anatomy at the University of Berlin. Von Branca sent his head curator to speak with von Hansemann, who had performed wonders raising funds for other expeditions.

Von Hansemann sprang into action on several fronts. A representative of the Ministry of Culture conceded to the pathologist that a one-time grant of as much as 80,000 marks was theoretically possible. Von Hansemann considered such a grant unlikely, concluding that the ministry did not appreciate the significance of the Tendaguru discovery. But he felt it was important that the enormous dinosaur remains and, possibly, fossil mammals be collected. Such finds would draw international attention to the Berlin Museum, attracting paleontologists from around the world and convincing the ministry to support further excavation. Von Hansemann suggested applying to private and governmental agencies simultaneously. The Africa Funds disbursed by the Imperial Colonial Office were one possibility, since the Africa specialist Dr. Hans Meyer had great influence there.

Fundraising in the private sector for a noncommercial venture could only succeed if men of substance backed the scheme. Von Hansemann astutely proposed to establish a committee of such men, with royalty at the head. He approached the brother of Johann Albrecht, duke of Mecklenburg and regent of Braunschweig, to determine if the 50-year–old aristocrat would act as its president. The widely traveled Johann Albrecht was president of the German Colonial Society (Deutsche Kolonialgesellschaft). Its 39,000 members vigorously promoted the benefits of colonialism in 389 branches throughout Germany and 25 overseas offices as far flung as London, Chicago, and Japan. Von Hansemann knew that when he met the duke in person, he would be asked the identities of the other committee members, so he rapidly set up meetings with candidates in scientific and commercial fields.

The Tendaguru Committee ultimately included Wilhelm von Branca; Anton Reichenow, of the Zoological Museum; Dr. August Brauer, director of the Zoological Museum and overall director of the Museum of Natural History; Ernst Vanhöffen, curator at the Zoological Museum; Dr. David von Hansemann; and Gustav Tornier, curator at the Zoological Museum and president of the Society of Friends of Natural Science of Berlin (Gesellschaft Naturforschender Freunde zu Berlin). Members of this august natural history society, founded in 1773, pledged the first 10,000 marks.[7]

Committee members contacted many well-placed citizens and the campaign snowballed. Though only a modest number responded to the monetary appeal, individual donations were substantial. In a short time, a remarkable sum was gathered for the German Tendaguru Expedition (Deutsche Tendaguru Expedition).

Responses flowed in from many influential and respected jurists, financiers, industrialists, physicians, and academics, and at least three other aristocrats. In addition to His Highness Johann Albrecht, donors included such prosperous citizens as Fritz Baedeker, son of the guidebook publisher; Arthur von Gwinner, director of the Deutsche Bank; the wife of steel magnate Friedrich Krupp; Arnold and Wilhelm von Siemens, sons of the founder of the mammoth electrical concern; Adolph Woermann of Hamburg, founder of a major shipping firm; three members of the powerful banking dynasty of von Mendelssohn, including Paul, grandson of composer Felix von Mendelssohn-Bartholdy; and Dr. Anton von Rieppel, director of the enormous MAN industrial concern. With the financial campaign underway, von Branca obtained the necessary approval of government ministries.

Von Branca and many others had labored mightily to overcome countless hurdles. Another concern was leadership and staffing for the Expedition. Von Branca had a candidate in mind as the scientific representative—Dr. Werner Janensch, curator of the Geological-Paleontological Institute and Museum.

Werner Ernst Martin Janensch was born November 11, 1878, in Herzberg, about 90 kilometers south of Berlin. The elder Janensch, a circuit judge, passed away while Werner was still a schoolboy. Geology and paleontology stimulated Werner's interests at the Universities of Marburg and Strassburg. A Ph.D. was conferred upon him in 1901 at Strassburg.

Wilhelm von Branca employed the 22-year-old Janensch as first collections assistant (Erster Assistent) in 1901. Janensch's scientific investigations spanned a broad range of topics, from Triassic-age ammonites of Germany, to Eocene-age snakes of Italy and Egypt, to Quaternary-age glyptodonts of Patagonia.

Janensch's appointment stipulated that he not become qualified for the status of academic lecturer (habilitiert), but concentrate exclusively on research and curatorial duties. On April 1, 1907, after five and a half years in the capital city, Janensch was appointed curator.

It is clear that von Branca thought very highly of his young protégé. He assigned the 30-year-old curator to lead an expedition to distant Africa. Unlike the widely traveled Fraas, Janensch had never left Germany. In the brief time allotted to him, the reserved and modest curator planned and prepared for Africa. Naturally, he was not alone, as efforts were underway to attract veterans of overseas fieldwork, who would organize and direct the logistics.

Janensch gave full credit to his fellow museum assistant, Wilhelm Herrmann, who possessed the practical experience Janensch lacked. Herrmann, an engineer, was selected to accompany Janensch because of his expertise in retrieving an extensive assemblage of fossil mammal remains under demanding physical conditions. In 1903, Herrmann had recovered Pleistocene mammals from Tarija in Argentina and Bolivia. A member of the German Pilcomayo Expedition to Bolivia in 1906–1907, he returned with ethnological, archaeological, paleontological, and zoological treasures. Supplies for the upcoming African venture were ordered and crated by Herrmann in the winter of 1908–1909.

The European discoverer of Tendaguru, Bernhard Wilhelm Sattler, had switched employers during his holiday in Germany. The German East Africa Company (Deutsch Ostafrikanische Gesellschaft), impressed with his extensive knowledge of the colony, had engaged his services. Upon hearing of plans for an expedition to Tendaguru, Sattler applied for a leave of absence.

In a meeting with Janensch in December 1908, Sattler agreed to accompany the Berlin scientists to Africa for at least the first two weeks in the field. Sattler outlined several conditions for his participation: his regular salary of six hundred marks per month for the time spent in Africa, travel expenses for himself and a servant from Dar es Salaam to Lindi and back, and a daily stipend of 12 rupees while in Lindi District to cover the cost of his porters.[8] Sattler promised to recruit African porters in advance, if Janensch would do the same through the offices of the German East Africa Company in Lindi. This was essential, since local plantations would require many Africans at the same time the Berlin group expected to be in Lindi. An early cost estimate for a six-month campaign ranged from 18,000 marks for one European to 50,000 marks for two Europeans and a hundred African laborers.

Fritz Linder was recruited in German East Africa with an offer of six months' employment at 150 marks per month. Linder had settled in the colony at the turn of the century, first serving in the administration in Lindi. Like Sattler, he had made a great impression during the Maji-Maji Rebellion, and later managed two plantations at Mikindani, south of Lindi. Popular with Africans, Linder was able to attract almost two thousand workers to his plantations. He was unable to direct operations at Tendaguru full-time, but was willing to act as a labor agent to supply contract workers.

Von Branca asked the Imperial Colonial Office to second NCO (Unteroffizier) Czeczatka of the German East African Defense Force, who had served on the Central African Expedition of 1907–1908, to the Expedition for six to eight months. Unfortunately, he was unavailable. The Imperial Colonial Office suggested a Swede named Knudsen, who had accompanied Professor K. Weule of Leipzig on an ethnographical investigation around Lindi in 1906.

A contact at *Kolonie und Heimat,* a publication featuring colonial matters, sent Janensch a list of other reliable men who had served in Africa. Among them was ten Brink, formerly the district administrator of Lindi who had escorted Fraas to Tendaguru, who was presently living in Strassburg.

There was no shortage of volunteers once the Expedition was publicized in newspapers throughout Germany. In January and February 1909, at least six men applied to join. Some, like Dr. Paul Vageler, a soil scientist, were highly qualified. There were also a multilingual mining engineer, a professional photographer, a mechanical engineer, a chemist trained as a geologist, and a 26-year-old Viennese who lived with his parents and cautioned that he was "not musical."

Colonial experts across Germany were consulted regarding everything from supplies to staffing to travel conditions in German East Africa. The list of requirements must have appeared infinite, as von Branca envisioned a two-year project. The remote location of the site called for a high degree of self-sufficiency. If an item was forgotten or defective, a four-day trek would be required to replace it. Extremes of climate and the rigors of transport by ship and human bearer demanded durable equipment of limited size and weight. Wherever possible, bulky items were ordered from suppliers in Hamburg, to save the expense of shipping them from Berlin.

In response to von Branca's plea, the German East Africa Line (Deutsche Ost-Afrika-Linie), a steamship company founded in Hamburg by Expedition patron Adolf Woermann, granted a 50 percent reduction on the cost of transporting Expedition freight back from Lindi to Germany, on the condition that individual shipments not exceed 10 cubic meters in volume. This reduction would be valid for the life of the venture, a significant concession. The company also agreed to a 50 percent fare reduction for Janensch and Herrmann. The German East Africa Company, a major player in the commerce of the colony, offered its extensive services in Africa, from expediting Museum business to providing free accommodation.

Arthur von Gwinner, whose bank financed international ventures, inquired about the availability of motor vehicles in the African colony. They were extremely rare, but a few were used by the builders of the railway. Von Gwinner speculated that a truck and trailer might be leased or purchased to haul supplies and fossils at Tendaguru.

Two medium-sized tents, complete with furnishings, were chosen for shelter and storage. A major outfitter of tropical gear to royalty and expeditions, Robert Reichelt of Berlin, was commissioned to manufacture them. A third tent, without any extra features, was reserved for a third crewmember. A large awning was ordered to protect fossils in open quarries. Accessories included camp beds, mattresses, camel-hair blankets, horsehair pillows, folding chairs and tables, hammocks, mosquito nets, and sundry other camping equipment.

Food supplies would be carried to the site, and supplemented with locally available items. Several German corporations approached the Museum upon hearing of the Expedition. Carl Bödicker & Co., the Hamburg firm that had provisioned the German Antarctica Expedition of 1901–1903, offered drinks and tinned food. All goods would be shipped from Germany in convenient porter loads weighing 28 to 31 kilograms. Maggi Corporation, known for its soup flavorings, offered to supply tinned soup extracts at no cost. Wooden cases contained spices, bouillon cubes, tins of mushroom, potato, pea, vegetable, and barley soup, tapioca, and oatmeal porridge.

Richter and Nolle, dealers in tropical and overseas equipment, approached the Museum for a contract. Excavating implements such as shovels, hoes, hammers, and chisels were ordered. A portable forge was selected, along with tools to maintain and repair digging equipment. Carpentry tools were included so that wooden handles could be replaced and crates assembled. Plaster of Paris and gum arabic, supplied by Theodor Wilckens in Hamburg, were stockpiled for collecting and preserving the fossils. An ample quantity of rope was laid in to package the bones.

Scientific instruments, including thermometers, barometers, a barograph, and compasses, were added to the growing inventory. Temperatures and altitudes could thus be recorded. Supplementing this apparatus was camera gear to document the fieldwork. Voigtländer & Sohn of Braunschweig generously loaned two 13 x 18-centimeter plate cameras with a complement of lenses and tripods, a 9 x 12-centimeter mirror reflex camera, and loading cassettes. Prism binoculars and an eight-power telescope completed the package. The total value of this outfit was over 2,500 marks.

A wide range of scientific literature was packed. As replacement items or fresh materials were required, they would be ordered from within the colony whenever possible, or from Germany.

Janensch corresponded with Eberhard Fraas about bone preservation at Tendaguru and suitable consolidants. When they finally met in Stuttgart in March 1909, the elder scientist was able to offer valuable advice based on his own extensive travels in the tropics. Janensch examined Tendaguru material and discussed the geology of the area.

By early spring 1909, preparations were complete, but at the last moment, another personnel change at the Museum upset the carefully laid plans. Engineer Herrmann departed suddenly for yet another assignment in South America. The man chosen to replace him was Edwin Hennig. Hennig later wrote, "I had the good fortune, just at the last moment, to take part as companion and helper in the duties of the Tendaguru Expedition."[9]

Edwin Georg Eugen Hennig was born in Berlin on April 27, 1882, the third of five children. Edwin's father, a merchant, died when Edwin was only 10 years old, creating considerable financial hardship for the family.

In 1902, Edwin enrolled in a broad selection of subjects at Freiburg University. Philosophy, the natural sciences, anthropology—his interests were catholic. He attended lectures at the University of Berlin, where von Branca and Otto Jaekel were among his tutors. In his free hours he fenced and even drew up plans for a flying machine.

In 1905, Hennig earned his doctorate under Jaekel. Von Branca hired his former student as third assistant (Dritter Assistent) at the Museum of Natural History in 1906. Three years later there came sudden and startling news: his assignment to the Tendaguru Expedition, at a salary of two hundred marks per month.[10] Edwin received the call on March 4, barely a week before he was expected to leave. He had just enough time for a medical examination, and one of his last official acts prior to departure was to draft a document absolving the Museum of all liability for his ill health or death while he participated in the Expedition.

Crates were shipped, personal affairs placed in order, and farewells made. Janensch left first, on March 8. Hennig's train steamed slowly out of Berlin's Anhalter Bahnhof at 8:00 A.M. on Thursday, March 11, 1909. The departure must have felt abrupt and more than a little rushed to him: "When I received the summons to participate in the Expedition, the ship from Hamburg was already underway. I only had a few days to meet it in Marseilles."[11] A tremendous undertaking was about to begin.

-3-

1909

A Cemetery of Giants

Hennig joined Janensch in Marseilles, and on March 13, they boarded the *Feldmarschall* of the German East Africa Line. The 5,400-tonne Imperial Mail steamship, built in 1903, was a recent addition to the fleet. Bernhard Sattler awaited them in the Italian port of Naples.

Hennig began a course of malaria prophylaxis as the ship passed through the Red Sea, with a dose of 3/4 grain of quinine every four days. A polio vaccination from the ship's doctor completed his medical preparations. On the afternoon of April 2, 1909, the *Feldmarschall* pulled into Dar es Salaam harbor after a 23-day voyage.

Janensch was introduced to Herr Voertmann, "an important personage" who represented the German East Africa Company.[1] The firm had been granted a charter to trade in German East Africa before the Imperial Government assumed control, and was one of the most influential commercial concerns in the country.

The Expedition members were received at Government House by Governor Albrecht von Rechenberg. Von Rechenberg, 48, was a highly capable administrator, fluent in Swahili, Arabic, and Gujerati as well as English, Spanish, French, and Russian. The Museum scientists would later note the considerable support he extended to the Expedition.

Two servants were hired, and Hennig chose a man named Ali as his personal attendant. Ali had the benefit of much experience with Europeans, gained during the years he had spent aboard a German naval vessel. The *Feldmarschall* weighed anchor again, and entered Lindi Bay on April 6. Janensch was delighted with the setting: a small town nestled among swaying coconut palms. He observed that "Soon after the ship's arrival . . . the Europeans [of Lindi] appear on board in the usual way, in order to once more taste a glass of beer on ice."[2]

E. P. W. Schulze of the German East Africa Company informed Hennig and Janensch that the Gebrüder Kritikos hotel was full, and he could only offer them two small rooms in the boma. The Berliners decided to stay on the ship for the night. Life ashore promised tantalizing new experiences: "Out of the distance, as if issuing from the depths, the rumble of a lion pressed my ear after I lay down to rest."[3]

The *Feldmarschall* weighed anchor before sunrise the next morning to avoid being stranded by the ebbing tide. C. W. Besser, outgoing representative of the German East Africa Company, was introduced to the Museum party. Janensch and Hennig were warmly welcomed at the firm's Usagarahaus, where they dined as boarders. As newcomers to Africa, they were regaled with tales of lions fearlessly strolling into the market at Lindi, and the perils of malaria and blackwater fever.

The ensuing days were filled with the business of the Expedition. Masses of equipment required customs clearance. While Sattler was in Germany, his African assistant had assembled bearers. Sixty of them had been waiting since the end of March and were now restless to be on their way. For ease of transport, equipment was divided into individual loads of about 30 kilograms, and additional supplies were purchased for camp members and excavation crews.

On the morning of April 9, 40 bearers departed Lindi for Tendaguru to prepare a campsite. Roughly 50 more were released by the district office to assist the Expedition, making a total of 162 individuals.[4] Another source states that 160 men joined them, which puts the total at 200 men.[5]

Janensch and Hennig met with District Administrator Wendt, whom they found "sympathetic and amiable."[6] Civil and military districts of the era were supervised by district administrators. Local administration was delegated to liwales, who were appointed from the population in major coastal towns. Around the towns, Arab and African akidas were paid officials who were empowered to collect taxes and mete out corporal punishment over a wider area. Finally, jumbes were headmen of individual villages.

On Easter Sunday, Hennig remarked that two months ago he could not have dreamed that he would be spending the holiday in Africa. The evening before their departure for the field, Sattler arranged a sendoff celebration at the Museum campsite on the beach. A marvelous scene unfolded as the European population of Lindi made itself comfortable on chairs and crates, amidst the rustle of palms and flicker of candles.

Crowds of locals assembled. The bearers, mainly Wamwera people, performed their rarely seen battle dances. Choruses of "Maji-maji" rose and they recited verses to Bwana Saturo (Sattler) and Bwana Fotumani (Voertmann), both of whom were in Lindi during the 1905 troubles. Then Wangoni dancers took over, with even greater energy. Wayao tribesmen added more variety. The setting was electrifying:

> The harsh glare of the flickering fire played in wonderful competition with the rays of the moon on the ghostly black forms that turned in a circle to the beat of the drums and strange songs—forms that first disappeared into the deep shadows of the palms, then were splashed by the sudden lights. The silvery bay peeped in between the dark palm trunks and far out on the high sea summer lightning flashed, only now and then sending a distant, hollow rumble into the noise [on the beach].[7]

As April 12 dawned, camp was struck and Janensch and Hennig set out with Boheti bin Amrani as their guide. Part Arab, Boheti owned a small plot of land near Lindi. He would prove invaluable in the years ahead. Sattler organized the bearers and trailed after the vanguard.

They marched north along the sandy shore of Lindi Bay, in the direction of the telegraph line. The rainy season had not ended and strong downpours were common in the afternoons. This made the ascent of the Kitulo Heights treacherous. Millipedes, 30 centimeters long and as thick as a thumb, infested their first camp in great numbers. Janensch observed that the water used for cooking and washing was the color of cocoa with milk, but lunch was welcomed all the same.

The proposed route north to the Likonde Plateau was revised the next morning due to the poor condition of the paths. Sattler chose the road westward to the Noto Plateau, but the valley was wet and streams were in spate. Narrow winding paths were flanked by three-meter-tall grass. Eventually the Expedition was moving upward again onto the Nkanga Plateau. Here they managed to link up with the network of wide, well-maintained roads that spread throughout the district.

At Nkanga Creek, it took two hours to clear a campsite in the dense bush. The jumbe explained that a spot had not been prepared in advance because lions had terrified the inhabitants of the area. Rifles were loaded and Hennig was taught how to fire them.

These worries appeared groundless the next morning, but the Europeans carried their rifles themselves for the first half-hour of marching. The caravan ascended the Noto Plateau, traveling along one of the main roads maintained by the colonial government. Animal tracks were common: leopard, wild pig, and antelope.

Hennig was impressed with the bearers. He noticed that they ate only one large meal of mtama or millet porridge a day. Barefoot and naked to the waist, young and old alike sang while carrying heavy loads on head or shoulders at a pace that matched that of the unladen Germans. After three to five hours on a rudimentary trail that could be stony and flanked by thorn brush, bamboo, and tall grass, they dropped their loads and immediately prepared camp.

A thorn hedge encircled the next village, and posts bore the skulls of rebels executed during the Maji-Maji Rebellion. The Expedition exchanged silver coins and cloth for chickens, eggs, and flour.

The government road was abandoned as they marched down the steep plateau walls. Despite the striking contrast between bright red soil and deep green vegetation, Hennig was frustrated by the lack of accessible exposures. Their destination, Tendaguru Hill, was pointed out to them in the rolling country that fell westward to the Mbemkuru River.

The terrain became forested the next day, as the Mchinjiri River, still swollen with rain, was forded. At 10:15 A.M. on April 16, 1909, they reached their long-sought goal: Tendaguru. Five days had passed since their departure from Lindi. Hennig noticed the first exposed dinosaur bones a quarter hour before their arrival.

The chaos of loads, crates, tools, the swarming throng of bearers and locals arriving either for work or to greet the newcomers gradually dissolved, the three tents were pitched next to one another, and thus we took possession of the spot that remained the center of the excavation efforts for the duration of the Expedition, due to the richness of the finds.[8]

Much had been accomplished by the advance crews. A wide area had been cleared a few hundred meters to the south of the hill, perhaps a 15-minute walk away. To the east stood a long dwelling hut. A food storage shed had been constructed and a kitchen shelter had been framed. Slender tree trunks formed the main supports of bamboo walls, and bark strips held together a roof of long grass. Now tents were erected, with large canvas tarpaulins stretched over them as a sunshade. Cots were assembled under mosquito nets and personal gear was stowed inside.

Janensch and Hennig walked around the site, delighted with the quantity of surface bones. Although it rained again that evening, nothing could dampen the celebrations that took place. An immense amount of planning and effort by many individuals had brought the Expedition to its starting point at last.

Sattler guided the paleontologists to ten bone localities to the north and south on the following morning. A system of rewards was instituted to encourage the locals, who soon pointed out good finds. The preliminary assessment was most positive:

> A true cemetery of dinosaur remains—saw at least ten specimens, mainly limb bones, a few vertebrae; a new large bone exposed in $1/2$ hour. . . . Prospects good, work for a long time.[9]

> Everywhere one sees bone fragments lying about, so that I am filled with joyful hope of good success.[10]

The Expedition was to collect more of the immense dinosaur remains sampled by Fraas. The geology of the region was to be unraveled. Marine invertebrates were required to date, establish the sequence of, and correlate the rock outcrops. In addition, modest accumulations of modern mammal, reptile, insect, and plant specimens were to be obtained.

On April 20, the excavations commenced at a locality five minutes south of camp. Two years previously Fraas had found several caudal vertebrae of *Gigantosaurus* here. A long trench, Quarry I (Graben I), was cut into the north face of a gentle rise of ground, in the hopes of reaching a layer of bone that was untouched by erosion. Within a short time a large limb bone was uncovered, then several articulated vertebrae. By April 22, the collection consisted of a tibia, two incomplete femora, a radius, an ulna, forefoot elements in articulation, ribs, a sacrum, and a scapula. Some of these remains were incorporated in Fraas's Skeleton A, and others were designated Skeleton C. The bones appeared about 20 meters from the edge of the trench, at a depth of 1.5 meters.[11]

At the end of five days of quarry work, about 15 men had dug a trench 50 meters long and one to two meters deep. They exposed another stretch 40 meters long to a depth of half a meter.[12] A third site was excavated and yielded more sauropod bones 1.5 meters below the surface.

Numerous specimens could be found quickly with limited, shallow digging in an area where a concentration of bone was weathering out at the surface. Alternatively, deeper trenches could be sunk in areas thought to contain bone. Deeper trenches required moving more earth, but the finds were often much better preserved.

Sixty local men had been engaged by April 23, for the purposes of camp and excavation duties. This grew to 70 and 80 in the first two months.[13] A small village was forming at the foot of the hill, as wives and children arrived. Some of the men were trained as excavators and others were assigned to the support tasks. Loads of equipment and food supplies waited in Lindi for transport to Tendaguru. Sturdier hoes were ordered from Dar es Salaam and Tanga, since those being used were suited to finer work. Paths had to be cut and the campsite had to be expanded. Besides the storage huts for excavation equipment, a shed for tinned food, and a "bone hut" were erected under Sattler's watchful eye. The bone huts protected recently excavated specimens and loads packaged for transport to the coast. A round grass-roofed shelter with open sides served as a workplace for the Germans.

Sattler arranged the laborers' huts according to tribal affiliation to avoid conflicts. They were built in long orderly rows near the tents of the Germans, with sufficient space left between dwellings to prevent the spread of fire.

Both paleontologists were more than pleased with the caliber of the quarry crews:

> the people work more quickly than I had expected, but also at the same time quite carefully; they also have a good eye for what is bone and what is rock. Some of them understand how to expose with the greatest care and exactitude bones as difficult as ribs.[14]

> Supervision at the excavations is almost superfluous; the people work very carefully with well-applied African patience. We can also rely on Boheti.[15]

Many quality finds were reported regularly. Hennig enthusiastically reported,

> By all means start to build a new museum. It appears that we must level the entire hill, since there is hardly a spot without bone remains.[16]

In time, the track to Lindi assumed the appearance of a wasteland, as holes and mounds of earth stippled either side of it. So many trenches extended across it in the vicinity of Sattler's original finds that the trail had to be diverted on several occasions.

Expedition crews regularly patronized a hut containing a small selection of items for sale. In his field notes Hennig marveled, "Thus I also became a shopkeeper here."[17] Becoming increasingly involved with the day-to-day routines of Expedition life, Hennig added physician to his occupations by bandaging injuries, mainly hoe wounds to the feet. He later treated a victim of snakebite.

The health of the workers was generally good. Complaints usually consisted of toothaches or pains in the head or chest. Rats gnawed the feet of the sleeping Africans so seriously that they had to be bandaged at morning roll call. To control the spread of disease, anyone found guilty of unhygienic practices within the confines of the settlement was fined or punished. When disease did break out the sick individual was isolated.

Hennig, Janensch, and Sattler climbed Tendaguru Hill most evenings for the cool breezes and quiet. A path wound up its east flank, terminating at a round pavilion with a bench. Many peaceful hours would be spent watching the beautiful sunsets from this location in the years to come. On April 27, Hennig's 27th birthday, the three went up the hill to celebrate and to dedicate the little pavilion. The setting obviously struck a chord with Hennig:

> The second glimpse into the virgin land of which we are absolute monarchs, the sunset behind the mountains of the horizon, the clear air, the strange silhouettes of the foliage against the evening sky, the musical cries of nocturnal monkeys, guinea fowl, and the campfire of our black assistants, and the merging of everything in the silent moonlight provided enough atmosphere for a pleasant memory of the sadly short day to be retained.[18]

His field notes, however, indicate that most of his thoughts were far away with family back in Germany.

C. W. Besser, of the German East Africa Company, arrived on April 28. He was about to retire after eight years in Africa and offered his services for six weeks prior to his return to Europe. Janensch hired him on the same terms as Sattler. This was a fortuitous development, since Sattler's leave had expired. The two Berliners, though learning quickly, still did not speak enough Swahili to direct such a complex operation.

Hardly a single aspect of the undertaking had not benefited from Sattler's experienced hand. Without his knowledge of the country and the languages and customs of its inhabitants, the two Berliners would have been able to achieve but a fraction of what had been accomplished.

Hennig and Boheti were invited to accompany Sattler as far as Namwiranye to search for the contact between the ostensibly Cretaceous sediments of Tendaguru and the much older granitic rocks at Namwiranye. Reaching the village on May 2, the Germans inspected graphite and garnet mines that Sattler had established for the Lindi Prospecting Company. Since 1900, various Indian firms had collected garnets exposed on the surface, or worked eroded gneiss deposits in the far south of the colony. They flooded the market with the semiprecious stones, and many companies went bankrupt. After 1909, production was to dwindle rapidly and the industry would fall stagnant.

On May 2, Hennig found himself unaccompanied by Europeans for the first time. Sattler was returning to Dar es Salaam to take up his post with the German East Africa Company at Morogoro. He would be in charge of establishing plantations and the company's mining interests in the area.

Map C.
Main Area Investigated by Tendaguru Expedition
Redrawn from Hennig, 1914,
Beiträge zur Geologie und Stratigraphie Deutsch-Ostafrikas, <u>Archiv für Biontologie</u>, 3:3

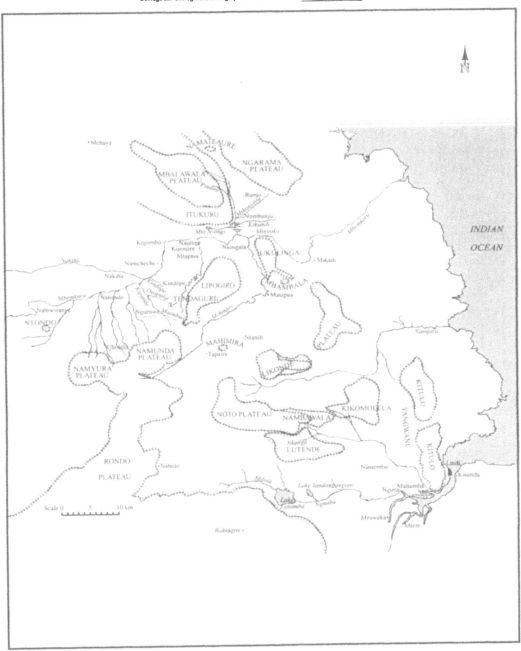

Boheti led Hennig to the village of Ubolelo, at the foot of the Namyura Plateau, to investigate reports of bone; the deposits proved worthy of future excavation. When the group arrived back at Tendaguru on May 4 after five days' absence, they found Janensch confined to his tent with malarial fever. This surprised them, as Janensch had taken quinine regularly since arriving in Lindi, and the mosquito-breeding season was finished.

The size of the project was increasing. Most fossil finds were reported by locals, who were rewarded with bonuses. Villagers in the surrounding countryside were also trained to identify bone and were paid to search for it. From time to time lines of workers were sent out, walking a short distance apart to survey a region for more finds.

As the excavations stretched over an ever-widening area, Janensch and Hennig could not visit every site each day, so they split the trips between themselves. Once the pits were more than two hours apart, it became impossible to inspect each one daily and skilled locals were assigned this responsibility. The Germans would decide where to dig and when to stop. They oversaw the preparation of the most important discoveries, numbered specimens, and entered all records in a catalogue.

The work day began at 6:00 A.M., when an overseer would use two sticks to beat a large drum made of goat hide stretched over one end of a section of hollow log. Workers assembled in front of the tool shed, where excavation and preparation implements were distributed. Those who did not live within the camp left their homes in the early hours to reach Tendaguru by sunrise. Initially, names were called out from a roster, but as the ranks swelled, this became impractical. The sick were required to appear to offer an explanation or send someone to report their absence. Parties were assigned to specific quarries and columns of 15 to 25 men marched off under the direction of a native supervisor.

When a column reached a designated site, the area was cleared of brush and other obstacles. The sturdiest members of the crew removed overburden, rocks, and roots with heavy two-handed hoes and shovels. When the bone-bearing layer was reached, another group replaced the first. These men would carefully shift stone and earth using hoes without handles, sitting among the bones to work. When a more delicate touch was required, they switched to knives, hammers, and chisels of various sizes. Should a trench require extension or walls require cutting back step-wise to prevent collapse, the first group of men would be recalled to remove matrix piled up at the edge of the pit.

The rock consisted of friable gray and red sandy marl, a mix of sand and clay rich in calcium carbonate. It was easily worked with a hoe and normally separated nicely from the bone, but was extensively fractured, due to its chemical composition and the variable climatic conditions. Soil became desiccated during the dry season, its surface further affected by numerous brush fires. Grass and tree roots also disturbed it. Consequently, bones on or near the surface were badly split and shattered. Fractures in shallowly buried elements were often filled with marl or lime deposited by groundwater. Bones found at greater depths exhibited less infilling. Some clay ironstone was present but few fossils were found within concretions. The contents of a few pits were covered in a thin crust of lime or hard, iron-rich clay.

Some bones exhibited crushing or distortion due to the weight of overlying sediments. Generally, vertebrae were poorly preserved due to their complex shape. This was especially true for cervicals, with their wafer-thin processes. Limb bones, although split into large pieces, remained relatively intact. Often the articular ends suffered the most. When limb shafts eroded, the surface spalled off in layers.

Once exposed in the quarry, bones were left standing on pedestals as the surrounding rock was cut away. Surfaces were carefully cleaned and several dilute coats of gum arabic were applied. This water-soluble glue was obtained from the processed resin of various species of acacia trees found in Africa and Australia. Local tree gums were tried as a cheaper alternative but proved unsuitable except as a last resort. Tins of this mixture, as well as brushes, were carried to each site. Loosely woven cotton strips a few centimeters long were dipped in a concentrated adhesive solution and pressed into every contour of the bone.

Larger specimens were given coatings of plaster of Paris with wire mesh inserted between the layers. Greased iron rods stiffened and supported the largest plaster jackets. The quality of the plaster from Dar es Salaam was poor, so a telegram was sent to Berlin requesting more of the European product. It quickly became obvious that this essential material was prohibitively expensive. When purchased in Germany it had to be soldered into tin cases to keep moisture out. Moreover, shipping from Hamburg to Lindi, followed by caravan transport to Tendaguru, inflated the cost dramatically, from 2 marks to an unacceptable 33 and even 45 marks per 50 kilograms.[19]

Trial and error produced a suitable local substitute. Reddish brown clay, used by the locals to seal the walls of their huts, was found in abundant quantities on the Namunda Plateau to the south. When finely ground and mixed with a strong gum arabic solution, it could be spread over bones. Coarse fibers were obtained by breaking up coir ropes, and adding them to the clay-glue compound produced a hard shell. Bones were coated with this concoction and sometimes partly banded with plaster reinforced with wire mesh. Only the most irreplaceable skeletal elements were totally encased in a plaster of Paris jacket.

Once this coating had dried, the pedestal of rock upon which the bone sat was undercut and flipped over. Excess matrix in the jacket was removed and the underside of the bone was exposed to receive several layers of glue. Another shell of clay-glue-coir was applied, so that the jacket protected the entire specimen.

A uniform cataloguing system was established for the localities and specimens. Articulated skeletal elements were assigned lower- or upper-case letters or a combination of the two. Quarries containing the remains of numerous individuals, or isolated bones, received Roman numerals along with the abbreviations of site names or preparators' names. Individual elements within these quarries were denoted by consecutive Arabic numbers. Any isolated discoveries were simply given sequential Arabic numbers. Each fossil was dutifully entered in a catalogue that was maintained in duplicate.

Atop each bone, Janensch or Hennig placed a small piece of split bamboo, upon which the appropriate catalogue designation was written in pencil. Another specialized work group of scribes followed them, transferring the markings on the bamboo labels to the bones with brushes and waterproof India ink. Paper labels could not be used since they were susceptible to damage by moisture, rats, and insects. In the case of fragmented elements, colored lines were drawn across the breaks to aid in their eventual reconstruction. The Expedition leaders also sketched individual bones and drew maps of the quarries to illustrate the positions of the parts of the skeleton. Two or three African scribes also kept worker attendance records, marked packing crates, and maintained an inventory of excavation tools. They were required to record the quantity and sale of grain in camp as well.

Once the field jackets were completed and labeled, they were carried to the storage sheds at Tendaguru Hill. In camp, a group of men was further organized according to specialization. Materials had to be procured, so some cut bamboo in the bush. Others sawed it up, while still others felled trees. Roughly a dozen men then built bamboo cases and packed the field jackets into them.

Locally available wood was too brittle for crates that faced a long, rough journey. It was also not feasible to carry empty wooden crates to the site from Lindi. Once loaded with specimens they would have made an impossibly awkward and heavy load for humans. Motorized transport was unavailable. Pack animals could not long survive in the tsetse fly–plagued region. There was no river or rail route. Instead, bamboo stems were cut into 70-centimeter lengths. They were bound together with wire and coir rope, forming a flat mat. A field jacket was packed in grass and set onto the bamboo mat, which was then simply rolled into a flexible but sturdy and portable cylinder around the jacket and tied up with more wire or rope. A circular piece of wood was wired to each open end of the cylinder and labeled. The Africans dubbed these cylinders "bamboo corsets."

Occasionally, jackets containing very fragile bones were enclosed in lumber sawn from indigenous hardwoods. Bamboo was nailed to the outside of the planks. Exceptionally heavy loads were given greater rigidity by adding bent iron rods to the bamboo. Other protective measures were also devised. The fruit of the baobab tree was enclosed in a hard-shelled gourd that could be sawed open and emptied. Small bones would be wrapped in soft grass or raw cotton wool and placed inside. A high degree of improvisation and innovation was required to overcome technical and logistical problems in this remote setting.

Camp maintenance and miscellaneous support activities occupied yet more locals. One or two experienced smiths sharpened and repaired the preparation tools. As the labor force grew, 10 to 20 young boys hauled water for workers in the quarries and the Europeans in camp.

At 2:00 or 2:30 in the afternoon, the big drum signaled the end of another working day. Laborers returned to camp to rest during the hottest hours. Their wives prepared the first meal of the day. Occasionally women or children would bring a small meal to the men at the more distant excavation sites.

Hennig reassured his mother that the quality of his diet was outstanding. The large stock of tinned items allowed for a varied selection. Breakfast consisted of cocoa, coffee, or tea, along with a choice of Quaker Oats, eggs, bread, butter, and jam or cheese. There was no shortage of drinks, from lemonade, soda water, ginger ale, and raspberry juice to wine, beer, champagne, and other liquors. All local water was filtered and then boiled. At 4:00 in the afternoon, there was tea or coffee. Evening meals often consisted of chicken and rice, or guinea fowl, pigeons, bushbuck, or antelope, when available. Hunting became more of a necessity than a sport as the population of the camp grew. Wild game could also stretch the reserve of tinned foods, which took three months to arrive after an order was dispatched to Hamburg. Lemons and bananas were available locally and oranges could be ordered from Lindi. Hennig's appetite was so robust that Janensch and Besser could only watch in amazement as he ate.

Among Hennig's duties were the time-consuming and often contentious shauri meetings. These sessions were held each afternoon to resolve problems and complaints from the work force. Supervisors reported new finds and outlined the day's progress. Ingeniously, they would cut branches or twigs to the length of a newly uncovered bone and bring this back to camp as a measurement for the Europeans. Where Janensch was the more reserved of the two Berliners, preferring to deal with the excavation business and correspondence, Hennig was energetic and extroverted.

Hennig also learned the routines of accounting and wage payment from Besser. On Saturdays workers received a one-rupee advance for food purchases, while payday fell at the end of each month. Wages were scaled according to the level of skill a job required. The monthly rate for heavy laborers was established at 5 rupees, while preparators and supervisors earned 6 to 7. This was equivalent to 12 U.S. cents per day, or 9.5 to 12.5 German marks.[20] Hennig quoted figures of 9 rupees for laborers and 10 to 11 for the supervisors and preparators, which presumably included the weekly advance.[21] Pay was withheld for days lost to illness.

Although plantation jobs offered higher wages of 12 to 15 rupees per month, there was no difficulty in obtaining workers in the immediate area of Tendaguru, since there were no nearby plantations. The rates paid by the Expedition therefore remained unchanged. It is possible that the German Tendaguru Expedition was in competition for labor with commercial operations on the coast. Five large plantations had been established around Lindi after 1905. By 1910, plantation managers were compelled to send recruiters as far inland as Songea. Although about ten thousand African refugees returned to Lindi District in 1909, the Maji-Maji Rebellion had had a devastating effect on the male population that provided the bulk of workers.[22]

Despite the lack of wage raises, there was never a shortage of volunteers. Numerous applicants were turned away each month. This steady supply of men may have been partly due to the lack of alternatives other than plantation jobs, for in 1897 a head tax had been instituted, payable in April and October of every year.

Absenteeism at Tendaguru was very low—only two or three men failed to appear on any given day, even when the work force numbered in the hundreds. In contrast, plantation overseers in the northern regions of the colony could expect only two-thirds of their workers on any given day, and listened skeptically to Janensch and Hennig's reports.

Copper money was kept in small, unlocked cases in the Germans' tents. It remained unguarded during the day, since theft was virtually unknown. Couriers for the transport of the payroll from the coast were carefully chosen, but the added precaution proved unnecessary. The Germans had also set up a "bank" into which many workers deposited their wages. As the camp population grew, keeping the accounts demanded much of Hennig's time.

While the work at Tendaguru was not always easy, discipline was rarely harsh. Dismissal was the most serious punishment, though District Administrator Wendt had forwarded the official regulations governing disciplinary action against Africans. Corporal punishment could be meted out by employers only for breaches of discipline: a maximum of 15 lashes with a hippopotamus-hide whip or 10 strokes with a cane. A list of offenders was to be forwarded to the district office every quarter, and one survives from the German Tendaguru Expedition.

Entertainment was limited. There was a plentiful supply of reading material, which Hennig devoured. The Germans often played chess and Hennig painted. On Sundays, Hennig worked on two writing projects that he never described beyond their somewhat cryptic titles: *Leben* (Life) and *Jesus*.

Large campfires were lit at night since darkness fell early and quickly, a consequence of the proximity to the equator. In June the sun went down around 5:15 in the afternoon. By 6:00, after barely three-quarters of an hour of twilight, night had fallen. On most days, the Germans retired to their tents by 8:00 or 9:00 at night.

As the work force grew and the dry season progressed, the camp's water supply became a concern. Hennig had been searching for additional sources for some time. In early May, he dug into the bed of the Kitukituki Stream. Further along, at a small grotto in the walls of the streambed, he found a spring that seemed likely to flow continuously. Though it was a short distance from camp, Hennig only discovered it almost a month after his arrival, so overgrown was the region.

Temperatures rarely dropped below 28°C once the sun was up, and averaged 32° to 36° in the shade in the afternoon. An attack of fever or dysentery under these circumstances was particularly trying. Bouts of malarial or tick-induced recurring fevers were serious. Dysentery was commonplace. Hennig remarked that enforced idleness in a tent while feverish made him feel like a piece of bread growing crisp and crackling in an oven. Whereas Janensch and Besser often enjoyed afternoon naps in camp, Hennig left for nearby excavations around 1:00 or 2:00 to stand down the workers for the day.

Living conditions were not luxurious, but a little ingenuity could go a long way to increase comfort. The open thatch of the hut roofs made doing paperwork indoors bearable. The legs of tables in the work and dining huts stood in small metal pans filled with water or oil to discourage termites and other insects from crawling up them. Nonetheless, insects could make mealtimes a trial. Of greater concern were the tsetse flies and ticks that could cause debilitating fevers, and the biting insects and scorpions that could inflict painful wounds. Mercifully, army ants only occasionally wreaked havoc in camp.

On May 7, Fraas's original diggings were considerably enlarged by another trench adjacent to and at right angles to the first. It was designated Quarry II, under the direction of Boheti. Within a few weeks, the site would be transformed into a trench 115 meters long. On May 8, Hennig thought he discovered fossilized bird bones, a claim that was not verified subsequently. This illustrated how difficult it could be to identify bones in the field before they were fully exposed. May 20 saw the start of another trench, Quarry III, as an extension to the others. It too reached a impressive length, about 63 meters. The last few days of the month were spent measuring, sketching, and photographing the quarries.

As the mild days of May ended, the camp, which was already inhabited by a diverse array of humans and animals, was enlarged by yet more unique visitors. Two baby serval cats, found in the bush, were brought to Tendaguru, and shortly thereafter, a pair of guinea fowl. The menagerie expanded when Janensch purchased a small monkey as a pet. A turtle or tortoise was set free, as it never ceased trying to escape. By this time, there were 13 chicks in camp, hatched by hens sent as gifts by local jumbes.

On June 2, it was time to say goodbye to the second German who had generously offered his expertise in the remote setting, C. W. Besser. He was honored with an evening of camaraderie atop Tendaguru Hill. From their vantage, the Germans watched the magic of distant flames in the blackness. The first large-scale torching of dry grass by natives was taking place in the valley of the Mbemkuru.

The first European visitor who was not connected in any way with the Expedition arrived on June 23. Kaiser, who was a planter in the Lukuledi Valley to the southeast and a representative of the Lindi Prospecting Company, proved to be very interested in the fieldwork.

A new site south of the hill was designated Quarry IV and was placed under the supervision of Godfrey, who commenced digging on June 4. Quarry V was started June 5 as the responsibility of Seliman Kawinga.

Salim, a trusted preparator, exposed a scapula, a forearm, and numerous ribs of what appeared to be a large sauropod to the northeast of Tendaguru. Work on this specimen began in earnest around June 22; officially Skeleton D, it was whimsically dubbed "Salimosaurus." A second scapula and humerus, a 1.80-meter-wide sacrum, and 29 caudal vertebrae lay exposed in a few weeks.

On the last day of the month, June 30, Hennig noticed some sandy shales at a work site he was inspecting. He spent some time splitting the rocks, thinking they might yield fossil fish. To the immense delight of both himself and Janensch, he discovered what appeared to be the lower jaw of a tiny fossil mammal. This was a find of great paleontological significance, as virtually no mammal remains were known from that geological era in all of Africa. The surrounding area was scoured without further finds.

The growth of the work force is indicated by Hennig's statement that the month-end payment of wages had occupied him from 2:00 to 6:00 P.M. nonstop. Furthermore, volunteers continued to appear daily, as the locals finished harvesting their crops. This was fortuitous, as District Administrator Wendt reminded Hennig that Africans were not to be diverted from planting and harvesting. Also on June 30 five Wayao men arrived, claiming they had traveled for a month to reach Tendaguru. When asked who sent them, they replied that they had simply heard of the project through word of mouth. Hennig originally refused to hire them, but when he heard that local Wamwera would not allow the Wayao into their settlements or sell them food, he relented.

In early July, a letter was sent to all the jumbes in the area, offering to buy food in preparation for the coming months, when supplies would inevitably run out and prices would rise. The precautionary measure proved highly successful, as carriers arrived with rice and mtama (sorghum) throughout the ensuing days.

The local harvest of 1909 was adequate for the Expedition's needs, for the population of the camp varied from 150 to 180, not including the workers' families. Consequently, there was little demand for the grain that had been purchased. Cereal crops were procured locally to avoid the prohibitive cost of transporting them from the coast. Sattler estimated that over the two-year period envisaged for the Expedition, it would cost at least three thousand marks just to pay the porters to transport food loads.

There was some difficulty with the jumbe of Nanundo, about two hours to the southwest. He forbade the people in his district to sell rice to the Germans at Tendaguru. When it was established that local villagers had lodged numerous complaints against him, he was sent to Lindi for questioning by the district office.On July 10, Hennig encountered a trio of Germans on a game-hunting safari, and introduced himself to Themistokles and Margarethe von Eckenbrecher and Lieutenant Colonel (Oberleutnant) F. von Porembsky. The von Eckenbrechers had settled in German Southwest Africa but left because of the turmoil of the Herero Rebellion. Upon their return to Germany in mid-December 1910, Margarethe wrote a book about their half-year safari, *Im Dichten Pori,* and described her visit to Tendaguru with more enthusiasm than accuracy:

> On the plateau sat a bungalow, a kind of verandah; located all around and constructed of bamboo were the museum, the supply and grain sheds, and the huts of the laborers. The German flag waved from the tents. A true small colony had formed.
>
> The two gentlemen on Tendaguru were absolutely not as fossilized as their finds. Champagne was even imbibed. The first drop naturally was intended for the bones, which provided the opportunity for five whites to be standing together in inner Africa.[23]

The rainy season left the landscape cloaked in thick stands of bamboo and grass, up to three or four meters tall. Throughout June, July, and August, the area around camp was put to the torch. A 10-meter-wide strip was cleared around camp as a firebreak, a safety measure to reduce danger from nearby wildfires, be they natural or man-made. Deliberate burning also became a tool in the hunt for fossils, though this could not take place until the heat of the dry season's afternoon. Dew on the undergrowth was heavy enough to completely soak clothing until about nine o'clock in the morning.

When fires burned, waves of flame a meter or more high coursed through the forest. Grass often disappeared completely, but the lowest tree branches and leaves were only singed. The foliage would change color and drop off. Stretches of a kilometer or more would succumb, but green grass or heavy forest stopped the fires' advance. Some fawns, snakes, snails, and insects perished, but neither people nor larger animals were endangered. Flocks of birds were attracted to the clouds of smoke, seeking the flying insects that fled the heat. Raptors soon appeared to prey upon small birds and escaping rats and hares.

For days following a burning, crews would return from work blackened with ash. Signs of life sprang up almost immediately. By the morning after the burn, spiders had already respun their webs and antelope tracks could be spotted. Within one week, new buds appeared in the crowns of trees.

Work expanded as more and more sites were discovered. In early July, bones were discovered at Mtapaia, one and three-quarters hours from Tendaguru. On July 19, Quarry VIII was opened just north of camp. It yielded a string of articulated caudal vertebrae and a two-meter-long pubis in the ensuing weeks, and by September enormous ribs, dorsal vertebrae, three limb bones, and a huge cervical vertebra had emerged.

Hennig became increasingly busy with personnel matters—the count now stood at about 130 men.[24] He was also anxious to fulfill the other mandates of the Expedition. His request to go on safari, a desire he had harbored for the last month, was granted by Janensch. Hennig's impatience and frustration are obvious in his journal entry: "In more than a quarter year we two have not accomplished as much geologically as Prof. Fraas did in one week."[25]

Janensch, the dedicated curator, likely wished to repay von Branca's trust, not to mention the investment of considerable funds. As Expedition leader, he must have felt the Museum's expectations of tangible results. Collecting, studying, and describing were the elements of his profession. Hennig, less certain of his future career, may have wished to gain a broad base of practical scientific experience. The opportunities for independent study, be it geological or anthropological, were legion and not to be missed due to camp logistics. There were, of course, outlets for his abundant energy. Virtually every Sunday morning Hennig was hunting game. He used these daily outings to increase his understanding of the geology of the bone-bearing horizons.

Sixteen men left camp with Hennig on July 22. They headed east to Matapua, some six hours distant. Lions were purportedly in the area, but as they left the village the next day, Hennig remained in high spirits: "A totally lion-free night; pity!"[26] Their march headed northward to Pindiro, three and a half hours away. Hennig had expected to find fossils in the Pindiro Valley but instead discovered finely layered shales, which were folded sharply by the forces of the earth. Hunting was unsuccessful, so he was forced to purchase food. Conscious that he or others might be passing this way again, he paid generously to ensure a friendly reception for future safaris.

Hennig and a guide hiked up the Itukuru Plateau. He was struggling to correlate the fragments of his many observations into a coherent picture of the regional geology. On July 27 he and his entourage returned to Tendaguru, having been absent for six days.

Two days later, the camp was in an uproar. Lions had been heard all around the night before, but this was not considered an extraordinary event deep in the African bush. As a precaution, Hennig and Janensch brought a rifle and pistol into the field. Suddenly at nine o'clock, a breathless messenger appeared at the quarry with heart-stopping news. A lioness had attacked a woman filling water containers at the spring. The terrified victim had her robe torn away but the big cat fled at the screams of the other women.

Janensch and Hennig dashed back to camp for their rifles, gathering a few workers along the way. A motley array of weapons was brandished, including sticks and the spears of the Wangoni. A noisy group raced to the stream and searched the area, but there was no trace of the lioness. The robe was recovered and returned to its shaken owner. That evening the last detail to carry water was provided with an armed escort. Hennig wrote a letter to his mother upon his return to camp, no doubt relating the day's dramatic events. More roaring was heard that evening.

Weapons were taken into the field again the next morning. Judging by the nightly concert and pugmarks, the district was being visited by an entire family of lions. Hennig sounded cautious but not panicked: "It did happen that the rifles stood behind the chairs during nightly chess games by lantern, or that besides knife and fork, the Browning pistol also belonged to the cutlery of supper: lions *ante murus!*"[27] The locals really had cause for fear, as they were comparatively defenseless. Since the Maji-Maji Rebellion all firearms had been confiscated from the Africans and even bows, arrows, and spears were officially forbidden. A limited number of men were designated as hunters and were supplied with small amounts of powder and shot. There were perhaps 150 muzzleloaders in native hands in all of Lindi District.[28] If a lion threatened a village the government sent traps and a soldier to capture or shoot it. Only six hours away, the village of Matapua had been besieged by a man-eater during the first two months of the Expedition's presence at Tendaguru. Ten people fell victim in two months.[29] The great cats even prowled the streets of Dar es Salaam, where traps were set.

Hennig held a shauri on August 2, and sentenced an accused thief to dismissal and corporal punishment. The latter proved a difficult decision, as his diary shows: "a sentence which would be abhorrent sadism to me at home; but here somewhat more understandable even if not willingly applied by us."[30] Hennig had assumed the role of judge because Janensch's knowledge of Swahili was still weak. However, Janensch left partway through the proceedings, giving an impression of indecisiveness and disinterest. With the growing size of the work force, Hennig felt that issues of this sort had to be dealt with quickly and consistently. In his opinion, this inability to reach a decision was a part of Janensch's personality, and he filled almost three pages of his field notebook with examples of Janensch's behavior that obviously irritated him. After further thought, Hennig came to a balanced view of the matter:

> as calm temperaments we get along well with one another. . . . I also in no way want to diminish his great merits, only wish him livelier energy. Our undertaking is so unusually magnificent that the smallest thing cannot be left undone and—delayed can so easily become canceled here![31]

One man was deliberate and contemplative, the other energetic and incisive.

In mid-August, Hennig was on his way to the northern quarries when he met a German family named Christens. Theirs was an adventurous outing, as two family members were girls only seven and eight years old. The Christenses were en route to the Mbemkuru River, scouting for land suited to cotton cultivation. Charmed by the little girls, Hennig promised to have a large dinosaur exposed for them to ride when they passed by the next time.

Hennig was spending more time each morning on geological work. An understanding of the sequence of strata was crucial if they were to identify changes in the fauna of the stratigraphic intervals represented at Tendaguru. He was confident he could trace three distinct layers of dinosaur-bearing sediments, and a general slope of the faulted landscape from south-southeast to north-northwest. Janensch agreed, but was less certain about the third, deepest dinosaur horizon.

On August 17, Janensch's servant located bones in dense bush to the east of Tendaguru, and the site became Quarry IX. Mohammadi Keranje was placed in charge of the specimen, which was very well preserved. He was already supervisor of Skeletons F, G, and H, the latter having been dubbed "Mohammadisaurus." Hennig felt the discovery in Quarry IX might again be a new type of animal, perhaps the fourth genus of dinosaur uncovered to date. The new quarry was a rich one: about eight limb bones and numerous vertebrae. With a work force of 170 by late August, quarries everywhere were yielding quality material:

> at the vertebral column near camp a dermal plate besides the scapula, on the path to Lindi a second very nice sacrum from a small form [of animal]. . . . Strangely, it appears that most often two skeletons (fortunately of different types) are mixed together. Also new cervical vertebra at Skeleton A.[32]

At the end of August, Janensch's monthly report to von Branca announced that work on Fraas's *Gigantosaurus robustus* had ended after a thorough search. A nearly complete manus had been recovered. With an eye to future displays, Janensch singled out the hind limbs and pelvis of the first quarry of the season as a potential mount. This report and others appeared in the journal of the society that backed the Expedition, the *Sitzungsberichte der Gesellschaft Naturforschender Freunde zu Berlin*:

> Bones of a very large dinosaur have now also appeared . . . among them a mighty femur of ca. 1.70 m. length and great massiveness [in Quarry II]. It is lying so deep that the complete uncovering of it could only be accomplished yesterday after a

three-week effort. Close to camp, an articulated series of caudal vertebrae 3 m. in length was found in a trench, as well as a number of other bones. Northeast of camp we discovered a series of vertebrae with ribs of unusual strength and great length (2 m!) not far beneath the surface.[33]

Skeleton I was placed under the direction of Salim Tombali on September 2. Its proximity to the village of Mtapaia earned it the appellation "Mtapaiasaurus." Four days later, Skeleton L, or "Wangonisaurus," was probed, an imposing animal. At the end of the month, it included 10 dorsal vertebrae, huge ribs, and four immense cervical vertebrae. Seliman Kawinga was placed in charge of Quarry XII on September 10.

Another find, the eventual significance of which could not even be imagined, was inspected by both paleontologists on the afternoon of September 21. The site was located in the exposed walls of the Kitukituki Stream. Skeleton M, under the hand of preparator Nyororo, was opened September 24. Its unofficial name was "Nyororosaurus."

One afternoon Hennig followed Janensch's example and had his tent set up under a wooden framework with a grass roof. This made the high temperatures more bearable and gave him more workspace and privacy. He confessed in a letter to his mother that more space was needed because they had allowed camp standards to slip badly since Besser's departure. The table in the kitchen pavilion was now so cluttered with newspapers, bottles, bones, letters, and instruments that it had become impossible to find a place to write.

From time to time Boheti was shown magazine photographs of inventions in Europe. It did not surprise Hennig that the intelligent head supervisor (Oberaufseher) expressed considerable interest, and wished to travel to the distant homeland of the Germans. Hennig considered Boheti a highly suitable candidate to oversee the preparation of Tendaguru fossils in Berlin. Janensch never lost sight of a critical component of the Expedition. He gave full credit to the Africans for exceptional efforts under trying conditions: "Some of our people have developed into excellent preparators and achieve more here than a white man could, who, for example, would find it impossible to uncover bones for a longer time out of the ground of a sun-burned quarry, the way our people do with skill and care."[34]

Many of the Africans could identify skeletal elements as accurately as the German paleontologists. The locals easily distinguished cervical from caudal vertebrae, or fossilized wood from limb bones, and independently discovered invertebrate fossils. When a copy of Karl Alfred von Zittel's classic work on paleontology arrived from Berlin, quarry supervisors and preparators showed great interest in the illustrations, asking to see them repeatedly.

In speaking to locals, Hennig gathered conflicting views on a topic that was puzzling him. Some told of a mythical animal called Majimwi or Ma'imi or Mazimwi, depending on the speaker's language. When asked about the rich deposits of bones strewn about, however, most Africans maintained that before the Expedition arrived, they never identified the fossils at Tendaguru as the remains of ancient animals.

In a letter home, Hennig cautioned his brother Richard about submitting excerpts from his Tendaguru letters to Berlin newspapers. This could only be done with the direct approval of von Branca or Janensch, in order to avoid the inaccuracies that such preliminary reports would inevitably perpetuate. On several occasions, nonetheless, portions of his letters were reprinted in dailies such as the *Vossische Zeitung, Berliner Tageblatt,* and *Berliner Lokal-Anzeiger.* Another popular journal, *Umschau,* offered twenty marks per page for brief accounts, which Hennig refused. The issue was becoming sensitive, as von Branca brought it to Janensch's attention. Janensch defended his colleague, assuring the director that Hennig had repeatedly asked his brother to desist.

There was trouble in camp on the evening of September 15, when a fire broke out. It spread quickly, consuming several huts. Supervisors tore down surrounding dwellings to prevent a calamity. Fortunately, no one was hurt, and despite accusations that the Wamwera were responsible, workers began singing as they cleaned up the smoldering ruins. Hennig remarked, "their cheerfulness is fireproof."[35]

By the 17th, both Germans were energetically removing specimens and packing them out of the quarries, an activity that greatly cheered Hennig. He now felt that the precious finds could be retrieved safely before the rains arrived. Several hundred bearer loads were stored in the bone magazine. The Expedition operated under the assumption that all work at Tendaguru would cease during the rainy season, from January until about mid-April, when quarries were in danger of flooding and the area became unhealthy due to malaria-carrying mosquitoes.

Plans for this period were never firm and there was much speculation about a stay in the region of the Great Lakes or the mountains in the northern half of the colony. A professor in Berlin recommended that Janensch visit the Seychelles or other islands in the Indian Ocean, like Madagascar. Fossilized remains of giant tortoises had been found on the Seychelles, and since German warships allowed their crews liberty ashore, living expenses should be low.

Eventually Hennig and Janensch decided to investigate the geology of the Usagara and Usambara Mountains to the northwest of Dar es Salaam, and visit Sattler at his mica mines near Morogoro. The Great Lakes were too far away to reach on foot in a reasonable time, and the expense of a safari would be prohibitive. Likewise, a steamship voyage to Madagascar or the Seychelles was considered an unwise use of Expedition funds.

The number of laborers was reduced as men departed to plant their crops. At least a dozen were packing fossils. Hennig and Janensch hoped that the first bone shipment could be sent with the next steamer out of Lindi. If bearers could be supplied from the coast, the work of skilled laborers at Tendaguru would not be interrupted.

More discoveries were made close to the original skeletons of *Gigantosaurus.* One locality near Skeleton D featured all four limbs complete with toes. It was designated Skeleton P, or "Nteregosaurus oedipus." A preparator named Hizza was responsible for the beast, which was tentatively identified as the sauropod *Gigantosaurus robustus.*

October 11 marked the commencement of labor on the Kitukituki streambed site. During rainy seasons, the rushing water had cut a bed six to seven meters deep. Massive bone fragments were found in the dry stream and others were still embedded in the walls, including a humerus that was standing nearly vertically. Other elements included a femur, a tibia, a fibula or radius, and a pubis. In time, this find, known as Skeleton S, assumed the appearance and proportions of a siege. It was placed under the supervision of the redoubtable Boheti bin Amrani, the most trusted and skilled of all the overseers. The colloquial title bestowed upon this beast was "Blancocerosaurus." It was embedded in the middle dinosaur-bearing layer.

Janensch came upon a 2.08-meter-long femur northwest of Tendaguru. The site became Quarry XV, and was formally opened on October 18 by Seliman Kawinga. A concentration of fossil fish remains added to the variety of fossils from Tendaguru. As Hennig was stalking wild pigs near Boheti's Quarry S, he came across the vertebrae of a small theropod mixed with bones of a large sauropod. Theropod remains were not common, other than their shed teeth.

Boheti, hearing of interesting discoveries in Quarry IX, hurried over to excavate the spot where numerous teeth had appeared. He prepared a group of five found in their original position, but only came across the most fragmentary bits of skull. On his way back to Tendaguru from Mtapaia, Hennig came across a very delicate fossil bone, which he thought might belong to a bird limb.

The feverish activity of the past weeks caused Hennig concern. In October alone, work had commenced at five skeleton localities (O, P, R, S, and T) and at two quarries (XIII and XV). As customary, informal names were bestowed on the skeletons: "Selimanosaurus" for O, "Mtotosaurus" for Q, and "Abdallahsaurus" for R. Unless the bones could be pulled out of the field and stored safely under cover, a great deal of effort would be lost. As many as five hundred loads, representing the partial remains of perhaps 30 animals, were already awaiting transport to the coast, a task that had to be completed before the return of the annual rains.

The Expedition's original call to Lindi for Wangoni porters went unanswered, since the Africans were working on their farms. So Hennig must have breathed a sigh of relief when 80 bearers marched into camp on October 19. District Administrator Wendt had sent a request out to district officials. Bamboo "corsets," stacked like drums in the bone storage magazines, were allotted to carriers, along with field jackets and crates.

Bundles weighing as much as 30 kilograms were swung onto sturdy shoulders in the early morning of October 20. Larger packages were slung from poles hefted by a man at each end. Larger bones were packed out in separate, carefully numbered sections. A single humerus provided 14 men with a load each. Shoulder blades and ribs were almost always fragmented and could be separated into smaller packages.

Massive skeletal elements such as sacra often had naturally occurring breaks, but at other times they were painstakingly split into manageable sections. Occasionally a pole was attached to both sides of a heavier item and four or eight men were assigned to carry it, spelled off periodically by a relief crew of the same size.

A line formed and the first caravan of Tendaguru bounty wended its way out of camp. Accompanying the procession was Werner Janensch, who wanted to be certain the packing methods developed in the field would withstand rough handling. Janensch would arrange for the shipping of loads from Lindi to Germany.

October 16 marked the end of the Muslim month of Ramadan, and brought with it the resumption of daytime meals and drinks for the faithful. Boheti too had observed the daily fast for the previous month.

On October 30, 40 more men left the Expedition's employ, which reduced the labor force to a little over a hundred.[36] Janensch returned from his trek to Lindi, after an 18-day absence. In Lindi, the German East Africa Company generously donated warehouse space, which protected loads from rain and insects. The head office in Berlin had instructed its branches in Dar es Salaam and Lindi to offer all possible assistance to the Expedition. In practical terms, this was an enormous benefit. All services were provided tax-free and a tremendous range of help was available: forwarding mail; housing staff at no cost; procuring bearers; ordering, packing, and forwarding supplies; and expediting shipping formalities. Fully aware of how materially the Expedition was aided, Janensch urged von Branca to formally acknowledge the Museum's gratitude to the firm.

Initially, the wood for large shipping crates was ordered from Dar es Salaam, a time-consuming and costly process, as sailing vessels from as far away as Norway, Finland, and Russia carried lumber around the Cape of Good Hope. This raw material was unloaded at the government vocational school in Lindi, where students fashioned the planks into sturdy crates. The average capacity was about $3/4$ cubic meter. About seven or eight loads, averaging 25 kilograms apiece, were placed inside, and dry grass was stuffed tightly around the jackets or bamboo cylinders. At 300 to 350 kilograms per loaded crate, it was soon evident that this made too heavy and awkward a burden, so smaller containers were assembled. About $1/2$ cubic meter in volume, they were filled with five or six bearer loads. Janensch spent from October 24 to 26 packing, sealing, and labeling the cases.

Most coastal steamers did not berth directly in the harbor. Consequently, boxes had to be moved from Lindi to steamers anchored offshore in Lindi Bay. Human power was available, so porters waded out into the surf with the crates and heaved them aboard Arab dhows, or sailboats. These wooden craft left the coastal shallows and transferred the cargo to waiting European steamers. In Dar es Salaam, the freight was reloaded onto the large German East Africa Line vessels for the long journey back to Hamburg. At that point, a local expediter was hired to forward the crates to Berlin. Accounts differ concerning the final stage through Europe. The fossil harvest out of Africa proceeded to Berlin both by rail and on freight barges along inland waterways such as rivers and canals. The number of fossil loads rapidly accumulating at Tendaguru and Lindi prompted Janensch to have von Branca inquire whether the German East Africa Line would increase the current limit of roughly one to two tonnes per shipment. Otherwise, the transport would be prolonged.

A small crate cost 22 marks and a large one 30 marks by the time the wood reached the colony from Scandinavia.[37] To economize, Janensch suggested that the containers be disassembled in Berlin, the walls nailed together in a stack and shipped back to the trade school in Lindi, complete with nails. Students re-assembled and reused some cases five or six times. The pioneering shipment from Tendaguru was a success, and Janensch predicted the first 11 crates would reach Hamburg in early November.

Janensch supplied detailed instructions to the Museum regarding the handling of specimens. He cautioned that the crates should not all be opened simultaneously, as there would not be sufficient space to store the contents in Berlin. In addition, it would be impossible to retain the context of the bones unless there was enough room to place all the related specimens from each quarry next to one another. A selection of bones suitable for public display was listed, including the enormous two-meter-long humerus. Janensch also recommended that extremely large bones should not be reassembled, as this would make it awkward to move them about while measuring, describing, and figuring.

It was resolved that all transport caravans would depart Tendaguru on Mondays, which again boosted the local economy. Volunteers from the surrounding area gathered on Monday mornings, especially around the time when hut taxes were due. To make the best use of the trained staff, Expedition members were employed to transport bones only when there were supplies to be carried back from Lindi. Usually these transport columns had an overseer, unless they were composed purely of Tendaguru laborers. In such cases, they marched unescorted.

Inevitably, news of the Expedition's aims had spread rapidly in a settlement as small as Lindi. Europeans in outlying regions began reporting fossils they had found on their plantations or noticed on their travels. While Janensch was occupied with the business of crating and shipping in town, he was informed of a bone discovery on the north slope of the Makonde Plateau. Accordingly, he chose a route that took him back to Tendaguru via the valley of the Lukuledi River, the Makonde Plateau, and the Rondo Plateau.

Before departing Berlin, Janensch had been urged by a coal specialist at the Prussian Academy of Mines, Dr. Potonié, to be on the lookout for peat bogs. Prevailing scientific opinion held that major accumulations of peat could not form in the tropics, and Potonié hoped for contrary evidence from Africa. As Janensch followed the course of the Lukuledi River southwest of Lindi, he questioned local Europeans about the presence of any such bogs.

Reinhold Körner, who grew coffee and cotton at Narunyo and Mroweka, located three small bogs. One was found 45 minutes west of Mroweka Village, and two others on the Narunyo Plantation, about two hours from Mroweka. As Janensch continued west, he was led to a fourth small peat bog by an assistant on Kaiser's plantation near the village of Mtama. Kaiser had visited Tendaguru in June.

Earlier in the season, Janensch had assured von Branca that a third European at Tendaguru was an unnecessary expense. His opinion had changed, and arrangements were underway for another field worker from Germany. This did not please Hennig, as he had hoped for greater independence from daily routines and the opportunity to sharpen his geological skills. Hennig feared these more interesting tasks might be assigned to the newcomer. To both men, the most important task was surveying the surrounding topography in order to obtain elevations so that strata could be correlated and quarries placed correctly on maps. It made the most sense to hire a surveyor who was already in the colony.

On November 11, the two Berliners celebrated Janensch's 31st birthday with champagne, wine, caviar, and fresh pineapples atop Tendaguru Hill. They sat for hours under an immense, star-filled sky.

Efforts to clear quarries of their contents were redoubled in the following days, but more material continued to appear. Quarry II delivered a theropod femur, and Skeleton U or "Ligomasaurus," located on the trail to Kerani Ligoma, was opened.

Bötzow, a planter from the hinterland of Tanga, visited Tendaguru on November 16. He was on a hunting trip in the south and was so captivated by the Expedition that he stayed two nights in camp, taking notes and photographing the work in progress. Before he departed, he invited Janensch and Hennig to stay in an empty building on his farm during the rainy season in the south.

In late November, when almost two hundred loads had been sent out of Tendaguru, Hennig proclaimed that "not stones [a great weight] but rather bones fell from our heart."[38] Another 350 loads were packed and waiting, roughly half the total that had been brought into camp to that date. On November 27, plans were thrown into disarray when 60 bearers arrived unexpectedly. Adding to the urgency was a reminder from the Usagarahaus in Lindi that there were numerous loads ready for crating in the warehouse.

Hennig left early the next day with 62 carriers bearing bone loads and 17 hauling his tent and gear.[39] He ensured that their entrance into Lindi had a dramatic flair, with all men singing enthusiastically and a safari trumpet of antelope horn blowing regularly:

> Waited here so that we could arrive in an orderly procession. Now steeply downward. What beauty! The profusion of bare, finely cut palm branches through which the sea breeze and surging waves blow; among them the trim European houses, cattle, goats. Then the Indian quarter with strange, colorful costumes. Past the market hall, completely submerged in violet bougainvillea blossoms, which tempted one to paint, through the broad, straight, well-maintained streets to the Usagara House, under the rhythmic singing of the Wangoni.[40]

It had been seven and a half months since Hennig first walked out into the bush from this little enclave of civilization. Now he was back in the company of Europeans, staying at the Usagarahaus. Tents were replaced with solid walls and a ceiling, and khaki with white cotton suits. An agreeable evening was spent at the tennis court beneath the palms.

Hennig was not on vacation, however, and by the end of the next day had crated 165 loads.[41] A total of two hundred loads were readied for the long sea voyage to Berlin by the third of December.[42] He wrote more letters, and assembled supplies for the trip back to camp. A few idle moments were spent wandering along the seashore and through the town. The days passed all too quickly:

> What an exquisite splendor of colors. The most lovely [is] the old Arab tower in the dark foliage and violet flowers, among them a tree with delicate green and brilliant red; behind—the palms and the lovely, living sea. I would like to sit here for 8 days without work, gaze and paint![43]

Yet such a break in routine was not to be, and he was back at Tendaguru on December 10, after 13 days' absence.

On the last payday Janensch dismissed 70 men and halted fresh excavation work. It was too humid and rainy for the gum arabic and plaster to set. This meant they could concentrate on clearing specimens out of the pits and packing them. The recent rains had demonstrated how leaky the roofs of many buildings were, so a large number of men were sent to cut grass and repair the important bone warehouse. Delicate white blossoms had appeared on low runners along the paths. It made an attractive picture, though the tendrils could easily trip the inattentive.

In his regular report to von Branca, Janensch was proud to list the growing yield from Skeleton S, which now included many limb bones, ribs, coracoids, pubes, and poorly preserved dorsal vertebrae. A total of about four hundred loads had reached Lindi. A collection of insects, carefully packed in cigarette boxes, was packed for Professor Brauer of the Zoological Museum. There was also the depressing report, from District Administrator Wendt, that Mohammadi Keranje, an overseer valued for his excellent work on Quarries IX, XVII, and Skeletons F, G, and H, had been sentenced to 18 lashes and three months in chains for theft.

Janensch proposed to retain a number of men at the site throughout the rainy season. They would build a new camp for the Europeans, repair storage huts, clear potential excavation sites of undergrowth, and collect invertebrate fossils. Half of the first letter of credit was now expended, or about 20,000 marks. If another disbursement of 30,000 could be sent, Janensch would be able to hire about three hundred workers for the 1910 field season. By his calculations, employing 80 men cost an average of a thousand marks per month.

Dr. Wolff, who was making his regular medical inspection of the district, stopped at Tendaguru on December 13. Although there was not much to be seen this late in the year, he was shown the 2.1-meter-long humerus of the giant Skeleton S, from the Kitukituki Stream. Hennig was still frustrated with the number of bones remaining in the pits: "A powerful and lengthy thundershower. . . . Now the quarries and many uncollected bones under water. Also a punishment for hesitating too long! And a 'present' for Christmas."[44]

Even Christmas Eve did not bring Hennig any special pleasure. In his notes, he sounds subdued and even melancholy: "As much as I feel comfortable here, I have spent these days better in the past! Toward evening alone for a long time on Tendaguru at a time when presents are distributed at home and loving thoughts also fly here to me like the swallows that shoot swiftly over my head."[45]

Christmas Day did not start out much more cheerfully. Hennig and Janensch's sense of urgency was tempered by District Administrator Wendt, who cautioned the Museum men to reduce their bone transports to 10 to 20 loads, as it was becoming difficult to handle the 50 or more at a time that that were being sent so regularly to Lindi. There was also the matter of undesirable competition with plantation work. Wendt reminded Janensch that many laborers were now required to process commercial crops, as well as their own.

New Year's Eve was celebrated with champagne and goose liver pâté. Hennig was typically philosophical on the end of the year:

> I leave behind the old year with heartfelt thanks. Once again I was able to set a whole series of new sails, and have them filled with favorable wind, true beautiful sea wind. Everything went according to my wishes; many expectations have been greatly exceeded. That the production was low despite such strong collecting activity is perhaps not bad, though it weighs upon me.[46]

The new year began with much activity, since they planned to leave Tendaguru on the coming Friday. The 60 remaining preparators were paid extra to work through Sunday. Every day they strove to remove the last bones from the quarries, pack the sundry collections of invertebrate fossils, and label the last of the field jackets.

Careful thought was given to preparing for the next field season. An intensive prospecting campaign was launched, spanning several days. New finds were marked by hanging fist-sized shells of the pulmonate snail *Achatina* in trees near the sites. Bleached to a brilliant white by the sun, these were highly visible. Potential excavation sites for the following season would thus be identifiable at a time when high grass usually frustrated prospecting.

It was agreed that Boheti, a scribe, the bone hut manager, and about a dozen Wayaos would remain at Tendaguru over the rainy season. Their tasks would include guarding the tools, the grain stocks, the two hundred leftover loads in storage under waterproof cloth and oilpaper, and the chickens. They had another assignment as well. Almost nine months of close contact with a growing, bustling village had convinced the Germans to relocate themselves to more peaceful surroundings. They ordered the caretaker crew to prepare a campsite and buildings for the Germans atop Tendaguru Hill. Larger huts were also to be constructed to house the bone loads and tools. A grass-roofed shelter was ordered, to protect the site of Skeleton G, which could not be completely emptied of bones.

Finally, all was ready. On January 7, 1910, the two paleontologists and 40 to 50 porters left camp. The site had been home to Janensch and Hennig for eight and a half months. Both were infinitely richer in experience. Africa, with all of its fascinating people and animals and its strange new sights, smells, and sounds, was forever imprinted upon them. The rainy season was underway with daily showers. Despite softened paths along the Kitulo Ridge, they reached Lindi on January 9, 1910.

About 70 crates were awaiting shipment, containing another two hundred bone loads.[47] Hennig and Janensch were put up first at the Usagarahaus and then at the Greek hotel at two rupees per day, while packing the last loads from January 10 to 15. In the final report of the season, Janensch advised that the size and preservation of skeletons such as B, C, D, and I or J, P, and S meant that preparing them for display would be too time-consuming to be done immediately.

The long-awaited steamer *Sultan* finally appeared on January 25. In the early hours of January 29, 1910, their ship entered the harbor of Dar es Salaam. It had been close to 10 months since they last set foot in the Haven of Peace.

Much had been accomplished, considering that the two specialists from Germany had arrived in Africa with so little local knowledge. Under the direction of experienced and sympathetic veterans, a smoothly functioning camp operation had been established. A skilled and reliable work force of up to 180 individuals had been assembled and trained, and an excavation program of ever increasing magnitude instituted. The geological sequence of the region had been probed to some extent. During the first season virtually all sites were located within a two-kilometer radius of Tendaguru Hill. Three separate bone-bearing horizons were recognized. Almost all of the finds of 1909 came from the upper horizon. Including Fraas's finds of Skeletons A, B, and C, a total of 21 associated or articulated skeletons had been excavated: from A to U. Fifteen quarries and trenches had been opened: I to XV. Again, they were predominantly from the upper bone-bearing horizon. The majority of the dinosaurs were sauropods. Enough differences could be detected during the initial exposure to confirm that several different genera had been found. Proper identification could only proceed once the remains were fully prepared in the Museum, where detailed study and comparison with other dinosaur fossils were possible. There were also a theropod and remnants of a stegosaur. Other significant finds included a fossil mammal, as well as crocodile and fish bones.

By the end of the season, over 700 bone loads had been carried to the coast, according to preliminary reports. Of these, about 300 loads represented the remains of associated or articulated skeletons.[48] The figure of 700 may be an initial estimate made in the field. In another source published a few years later this number was revised to 566 loads carried to Lindi by 585 porters, a number that more closely corresponds to the total of 600 loads given in Janensch's report to von Branca.[49] Some 108 crates containing 22,000 kilograms of material had been shipped to far-off Berlin.[50] By the end of the year 45 crates had arrived in the German capital.[51]

Total expenditures for the first season were calculated at 37,553.98 marks, or 28,150.48 rupees.[52] These figures do not include the cost of major items purchased in Berlin prior to departure, nor items purchased in Germany and shipped to Tendaguru during the course of the year. This amounted to another

15,000 marks, to make a total outlay of about 50,000 to 60,000 marks for the first year of the Expedition.[53] The lion's share, 20,000 marks, went to wages for the Africans.

But numbers cannot be the sole measure of an undertaking's success. The interaction between Europeans and Africans appears to have generally been positive. Both groups learned something of the other. The paleontologists, especially, benefited from this contact. Little of value could have been accomplished by a handful of Europeans in the bush without local cooperation. The skilled efforts of many local people bore a priceless harvest. The scientific community was certainly enriched by the bounty flowing into Germany for study and description. The curiosity and conscientious attitude of people like Sattler and Fraas was amply repaid. The fundraising effort of von Branca and his committee was well rewarded, as was their faith in the significance of the site. Tendaguru was indeed the treasure trove all had hoped it would be.

-4-

1909–1910

Geology in the Rain

Dar es Salaam felt like a sophisticated metropolis after such lengthy toil in the remote interior. Janensch and Hennig heard all the recent news that evening at Schultz's Biergarten, marveling at the busy streets filled with pedestrians, cyclists, rickshaws, and donkeys.

Their goal for the upcoming months was to correlate geological deposits of two different Mesozoic periods in the colony by collecting fossils along the grade of the Central Railway. The rail line was a shallow trench that provided a rare profile of subsurface strata in a country that was poor in exposures, and the quarries that had been opened to supply ballast for the line were prime fossil sites.

On February 4, a rail journey brought them to Mikesse and Sattler's camp. Hennig created a geological profile, running from Ngerengere at Kilometer 149 to Kitugallo at Kilometer 138. He was searching for the contact between the Permian or Triassic Karoo deposits and Jurassic strata, collecting fossils to help date the outcrops.

The ammonite locality, discovered in 1907 by Kinkelin at Kilometer 123, was the next goal. It was completely overgrown, and Hennig needed three days of effort and the assistance of Salesi and Wilhelm to re-locate it.

On February 27, the men trekked back to Sattler's camp. The safari had lasted 16 days. Though the invertebrate fossils were valuable, the frustrations were colossal. Walking along the railway grade was convenient, but also unbearably hot, with sun-heated rails, blinding reflective ballast, and windless passages. Examining the sites off the rail line demanded bearers and tents. During these side trips, rain had confined the group to their tents for hours, soaking the ground and encouraging rampant vegetation.

Specimens and equipment were packed for the next stage of their program around mid-March. Janensch and Hennig departed for a village 30 kilometers west of Dar es Salaam, where engineers had reported fossils. At Kilometer 23, near Pugu, the rocks closely resembled the dinosaur-bearing strata at Tendaguru, as they had at Kilometers 114 and 116. However, the fossils mentioned by engineers did not derive from sediments of the expected geologic age.

Map D.
Area Investigated by Hennig During Rainy Seasons, 1909/1910, 1910/1911
Redrawn from Hennig, 1924, Der mittlere Jura im Hinterlande von Daressalaam, <u>Monographien zur Geologie und Palaeontologie</u>
Scale: 1:300,000

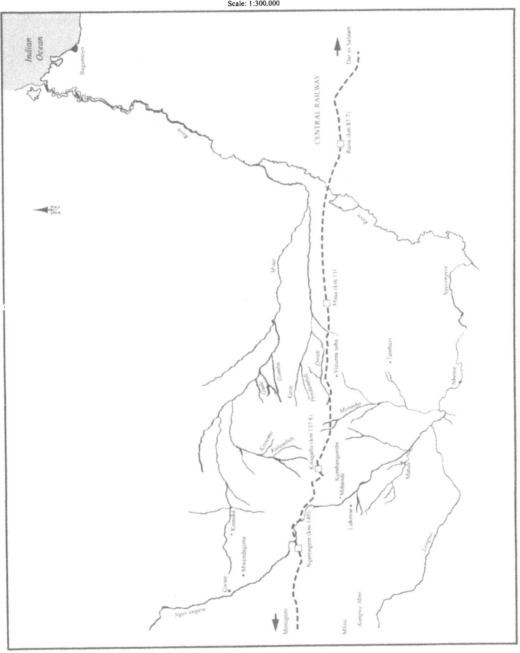

On March 21, the Germans set out independently for a 25-kilometer walk back to Dar es Salaam. They had spent some 46 days away from the capital. Hennig would later write that March was hopeless for fieldwork: "The impediment by vegetation and flooding or total soaking of all lowlands, particularly regions rich in clay soils, is hardly less than that of a wintry, snow-covered north German plain."[1]

Janensch and Hennig proceeded from Dar es Salaam to Tanga, a carefully laid out town that was home to a population of 5,689 in 1908, of which 141 were Europeans.[2] Besides the capital, Tanga was the only town in the colony to enjoy electric lighting. It also had a hospital, a newspaper, and dozens of European and Indian firms, and was the railhead of the Usambara Line (Nordbahn). Initiated in 1893, the line was not completed until 1912, following many financial and engineering setbacks. It would eventually stretch 352 kilometers inland to connect Tanga's excellent harbor with the rich agricultural district around Moshi at the foot of Mount Kilimanjaro.

They were joined at the Club in Tanga by Baron Walter von St. Paul-Illaire, the European who, in 1892, first brought the African violet from the surrounding Usambara Mountains to the world's attention.

The Amani Biological-Agricultural Research Institute was their final destination. Throughout the ensuing week, they marveled at the luxurious rain forest surrounding the 304-hectare institute. Hennig described the setting to his mother:

> The deep silence, only occasionally broken by cicadas or startled monkeys, and the deep shade seem to be as one, and a magical green light lies over everything, floods everything, only sprinkled here and there by dazzling sunlight. One wanders like a small worm in a tall cornfield and imagines that one is transported to the interior of a magic mountain.[3]

The Amani Biological-Agricultural Research Institute had been founded by Dr. Franz Stuhlmann in 1902 to conduct experiments in tropical agriculture. Spread over several ridges of the 1,400-meter summit, the facility was staffed by a dozen specialists.

After a brief halt in the capital, they came into sight of Lindi on April 19. A great deal of ground had been covered in the 12 weeks since their departure from Lindi. A strong foundation had been laid for continued fieldwork in the north during upcoming rainy seasons, but now it was essential to carry out plans for a second campaign at Tendaguru.

The warm surf still rolled ashore and the rain of the southwest monsoon still fell, delaying any thoughts of marching inland for a week. A fresh order of provisions from Carl Bödicker was waiting, along with equipment ordered from Berlin. Janensch deposited several thousand marks with the German East Africa Company to cover expenses. Von Branca had expressed concern, as only 25,000 marks remained in Expedition coffers. At two rupees per load from Tendaguru to Lindi, the cost just for bearers had been 1,200 marks in the first season.

A column swung out onto a familiar trail to Tendaguru on April 29, 1910. Despite the excitement of returning to the field, not all thoughts were fixed solely on earthly matters. Halley's Comet was streaking into the solar system, gradually becoming visible on the eastern horizon during the early hours of morning. Unspoiled by pollution or artificial light in 1910, the African heavens provided an awe-inspiring backdrop.

Tendaguru came into sight on May 1. During the previous field season, grains of corn and millet had fallen unnoticed as they were prepared for meals in the native village around the hill. Now a ragged field of these crops, three to four meters tall, waved in the breeze. At the six-meter-deep excavation site of Skeleton S, poles used to prevent the wall from collapsing had sprouted leaves and twigs.

Located at the base of the hill were a large new bone storage structure and a tool shed. The rainy season crew had erected a storage hut for tinned and other foods, dwellings for the cook and servants, and a kitchen, all partway up the south slope of Tendaguru Hill. Higher yet stood the barassa, a circular gazebo with a grass roof. Meals would be taken and administrative tasks completed here, on the site of the viewing pavilion of 1909.

Janensch and Hennig decided to relocate their tents well up the hill to isolate themselves from illnesses or fires in the laborers' camp. This ensured peaceful evenings despite a greatly expanded work force and allowed them to benefit from breezes that kept mosquitoes and other insects away. Hennig ordered an open-sided, grass-roofed shelter constructed over his tent. Janensch had a similar house built for his tent, with log and bamboo walls and a grass roof.

The village was segregated into five distinct districts according to tribal affiliation and was erected within a short walk from the tool shed, to which everyone was summoned by the morning drum. The huts were also closer to a water source.

Tribal migration and the plantation labor system had created a heterogeneous mix of Africans in the Tendaguru district.[4] Mwera or Wamwera constituted the majority of the population between the Mbemkuru and Lukuledi Rivers. Most commonly employed by the Expedition as preparators, they were considered the most peaceable of all the Africans. The Ndonde or Wandonde people had settled in the middle and upper reaches of the Mbemkuru. As peace-loving as the Wamwera, they were fewer in number, perhaps 6,000 compared with 40,000 Wamwera in Lindi District.

The three Bantu-speaking tribes of the region were quite different. Ngoni or Wangoni, descendants of the Zulus, had moved in from lands to the east of Lake Malawi by the middle of the nineteenth century. Renowned for their prowess in battle and their physical strength, they made up the bulk of the excavation teams at Tendaguru. They had a well-deserved reputation as hunters and warriors, and were in demand throughout the colony as bearers and plantation laborers. Perhaps 12,000 lived in the district, much reduced in number because they had proved to be such formidable opponents during Maji-Maji times.

Yao or Wayao people were also valued by Europeans for their strength and intelligence. Second only to the Wangoni as plantation labor, they were considered equally warlike. Some had fought on the side of the Germans during Maji-

Maji due to old animosities against the Wangoni. An estimated 45,000 lived in the region, having been encouraged by the Germans to leave Portuguese East Africa (presently Mozambique) and immigrate to German East Africa.

Makwa or Wamakwa tribes were most common around the Lukuledi Valley, with a total population of roughly 18,000. Like the other two Zulu-descended tribes, the Wamakwa were considered warlike. Their loyalty to the Germans was divided. Over the decades, the Bantu speakers had both intermarried and warred against one another. The Wangoni, Wayao, and Wamakwa fought not only among themselves but also with the peaceful Wamwera and Wandonde.

The African who had importuned Janensch and Hennig to open a shop was granted a concession at Tendaguru. He offered tobacco, salt, sugar, pots, cloth, and other items. Eventually his shop became so popular that it yielded a profit. The Germans were concerned, since the purpose of the store was to cover costs the Expedition incurred in purchasing these supplies, not to make life more expensive for its own employees. The arrangement eventually caused problems because it was not approved by the district office in Lindi.

Janensch's plea for funding to employ three hundred men had initially been met with some doubt, since this would require an additional 20,000 marks. Despite this pessimism, however, Dr. von Hansemann and the Tendaguru Committee had worked their monetary magic again, and the number of workers that could be hired increased substantially.[5]

Local interest in Tendaguru was high. By May 3, more than a hundred men were signed on and more were arriving daily.[6] Four days later the number stood at 175. More than two hundred men were put to work by the ninth, when further hiring was temporarily suspended.[7] A larger staff placed heavier responsibilities on Janensch, Hennig, and Boheti, since another European did not join them as originally planned.

With such a large, experienced staff at hand, extensive excavations could be begun at the very start of the season. Tremendous amounts of overburden had already been shifted, allowing promising sites from 1909 to be extended. The foresight of launching a bone-prospecting program over a wide area in late 1909 also paid off, since many widespread localities could be investigated immediately. Added to this were the discoveries that large numbers of former employees reported in outlying areas.

Throughout their second week at Tendaguru, the paleontologists had arisen early to enjoy the fabulous sight of Halley's Comet. It became particularly striking as it reached its closest point to Earth on May 18, 23 million kilometers distant. In Europe and America, there was exhilaration and hysteria. Astronomers detected minute traces of cyanogen gas in the tail of the comet and it was estimated that the Earth would pass through this stream of dust and plasma for six hours on the 18th. This news caused suicides in the United States. Reactions at Tendaguru were calmer, since the colonial government had reassured the Africans that no calamities would result. Hennig probed local opinion:

> I asked one of our supervisors whether anyone had become frightened in his home village, and a precious reply ensued: "No, it was of course forbidden (!), and

from the government edict we did not even know that we were supposed to be frightened of it."[8]

The vision was unquestionably awe-inspiring:

Glorious spectacle: Dark, clear night and the impressive shining apparition like a colossal exclamation mark over the black forests.[9]

The silent, profoundly dark tropical night hung down from the glittering starry heavens, but glowing in all its magnificence yet mystery that messenger from invisible distances of the universe stood in the firmament . . . outshining all other stars. . . . at last it spanned two thirds of the heavens on 19 May . . . like an immense searchlight the tail unfolded over our zenith. With the first shimmer of morning it rapidly shrank and before the sun peeped over the horizon, the fairy-tale-like vision was dissolved into nothingness.[10]

The comet disappeared as it hurtled around the sun at three thousand kilometers per minute.[11] Within a week it had vanished, heading toward the farthest reaches of its orbit: beyond Neptune, five and a quarter billion kilometers from the Sun.

Three weeks after Janensch and Hennig's arrival at Tendaguru, a crew was sent to resume digging at Skeleton S. This monster demanded extraordinary effort, since the bones were running into the steep wall of the Kitukituki Stream. Each meter of forward progress exposed an increasingly large volume of overburden that had to be cut back, a task that occupied a sizable group of men.

Bone fragments found at the northwest foot of the Likonde Plateau, eight hours distant, were brought to Tendaguru by the native discoverer. Ammonites were gathered from the marine *Trigonia* horizons that had been deposited between the dinosaur-bearing sediments, and scouts returning from Likonde confirmed reports of bone.

Tribesmen arrived from Masasi, five days to the southwest, with word of bones, so Hennig assigned some experienced hands to substantiate the claim. The result was disappointing, as the fossils were a weathered asbestos outcrop.

While rendering Skeleton S in oil paints on May 30, Hennig received word that a European was in camp. Hurrying back, he met Bishop Thomas Spreiter, a Dar es Salaam cleric. Spreiter had been in charge of the Benedictine mission along the Lukuledi in 1906, when it was burned to the ground by Maji-Maji rebels. He had picked up a dinosaur bone fragment at Mbate, north of the Mbalawala Plateau. Even more significant was a piece discovered near the village of Makangaga, a two-day trek west of the port of Kilwa Kivinje. This was truly cause for excitement. Dinosaur bone was in no way restricted to Tendaguru, but could be found a hundred kilometers to the north.

Around the time the grass was burned at Tendaguru, an experienced overseer was directed to assess Bishop Spreiter's reports of bones at Makangaga. Fossils were tracked down but the undergrowth was too wet to be reduced by fire.

The work force stood at three hundred in late July.[12] So that so many men could be occupied, it was important to burn the undergrowth. Hennig discussed the matter with an African supervisor:

Grass completely dry but it absolutely does not burn, and it makes me completely frantic and raving, when I see this tinder and my companion explains, as a calm matter of course, [that] before 12 it does not burn, it is of course still too cold (while the sweat covers my forehead and the sun glows). This physics is too complex for me.[13]

Fresh investigations were instigated at Skeleton G and at a discovery made by Janensch near Skeleton T. More vertebrae were soon exposed at the string of caudals that was Skeleton G.

Quarry XVII, another new locality, was placed under the direction of Mohammadi Keranje on July 1. The overseer presumably was none the worse for his three-month prison sentence of the previous year. The site was not far from the theropod, Skeleton E, and before long, a large humerus was exposed.

Quarries were now spread as far out as an hour to the north at Kindope, three hours to the southwest at Ubolelo, and two to three hours to the northeast at Mtapaia and Kijenjire. Tents were left standing at Ubolelo and Mtapaia so that Janensch and Hennig would have shelter should they decide to stay overnight. The optimistic tone of the Expedition was mixed with impatience: "in addition to numerous teeth, many vertebrae well preserved, a very desirable supplement. But where are the skulls?"[14]

By July 24, the efforts at Kindope were paying off. For the first time that season, Janensch recognized the small vertebrae that had originally weathered out of a low slope near the trail leading to the village. They came from a medium-sized stegosaur, so the locality was designated St. This quarry, in sediments from the second or middle dinosaur-bearing horizon, had been cut into the side of the hill, but till now had not rewarded its excavators. This had changed, as vertebrae, limb bones, pelvic elements, bony plates, and spines appeared. Several genera of dinosaurs were preserved in the quarry. On August 15, the jumble of disarticulated bones indicated no less than four stegosaurs. As many as 18 bones a day were showing up in St.[15] In late August, a one-meter-long stegosaur caudal spine and a sauropod femur 2.28 meters long were recovered.

Eighteen quarries were in progress at the end of July, with a work force of four hundred men.[16] After weeks of earth-moving, the bone level had been reached at Skeleton S. Boheti exposed ribs of startling dimensions: fully 2.5 meters long, 60 centimeters across the head, and 15 centimeters across the shaft. A giant was entombed here, and heroic efforts were required to unearth the skeleton. A few weeks later, the site produced four enormous cervical vertebrae and a 1.9-meter long scapula. Because the skeleton extended back into the wall of the watercourse, a woven mat of bamboo stems and tree trunks was laid against the quarry face to prevent falling rocks or slumping earth from injuring the workers or damaging the hard-won trophy. By late August, Janensch calculated that it would require as much as two months of shovel work before the bone layer was reached again. Two months later, the quarry was 10 to 12 meters deep.

Janensch had instructed that bones were to be left exposed in quarries for longer periods so that their position remained clear as more bones were uncovered. Only by preserving these relationships in the field did he feel that the articulation of elements like vertebrae and ribs could be understood. Quarry maps were sketched, but as the sites became more distant, it was impossible to fully record the position of their contents. Exposure naturally placed the bones at some risk, and delayed their removal. To compensate, Janensch occasionally ordered grass shelters to be thrown up over unfinished pits. This was also his plan for the upcoming rainy season.

Porter caravans had stockpiled such an accumulation of bones at Lindi that the German East Africa Company facilities were hard pressed to store any more. The last of the 1909 jackets had been transported out of Tendaguru, and 50 to 60 loads per week had been sent to the coast.[17] Janensch left for Lindi on August 10 to prepare the backlog for shipment. He had to consult with District Administrator Wendt on how heavy loads could be transported to Lindi. They could not be carried safely to the Noto Plateau due to the deep gullies that were encountered en route. Wendt was petitioned to improve the paths. A labor detail was assigned to cut a road that would connect Tendaguru with the government-maintained trail running across the Noto Plateau to Lindi. Deeper streambeds were bridged. Eventually, a vital crossing between the Noto and Lutende Plateaus was made accessible to the bulkiest loads when a bridge was thrown across Nkanga Creek.

Wendt's labor corps was not popular among the locals. Hennig was embarrassed to learn that many men fled into the bush to avoid the work, despite all disciplinary measures, which included imprisonment and flogging. Not all such roadwork was involuntary, for in other cases the Africans sought temporary positions to earn the annual hut tax. Throughout Lindi District, this system of road labor had created a comprehensive network of trails. They were kept free of foliage during the rainy season to allow year-round access to the interior. Trade caravans and military columns complete with artillery had easy passage.

In anticipation of the demand for packing material, Boheti had planted cotton. A supply was growing up around Tendaguru, and it was possible to gather baskets full of the bolls every few days for wrapping delicate fossils.

As the scope of the Expedition expanded, the available resources, both human and natural, were strained. Even an individual as energetic as Hennig was admitting to difficulties: "It is impossible to visit all the quarries."[18] Cereals, deliberately stockpiled in 1909 at a cost of 1,200 rupees, were rapidly depleted. Crops were poor in 1910, forcing the Expedition leaders to purchase whatever was available in the immediate environs. Initially less grain was requisitioned than in 1909, under the assumption that the fossil-collecting enterprise had a life span of two years. They were victims of their own success.

As the supplies dwindled, additional quantities were ordered at greater expense from the German East Africa Company in Lindi, now headed by E. P. W. Schulze. Demand outstripped supply to the point where the entire camp was eventually reduced to half rations on some days and no rations on a few days. Local Africans were more inclined to sell their corn harvest to Indian traders on the coast, who charged more, as the commodity became scarce.

Similarly, the water supply proved inadequate for the enlarged population. Women and children were compelled to make several trips a day to the springs.

Eventually these nearby sources were overtaxed and excursions to more distant points became necessary. Hennig was aware of the Expedition's impact on the area and made a special diplomatic tour, reminding himself to "make enquiries about the water conditions, since we are in strong competition with the locals and I wish to avoid small embitterments as much as possible: after all we are the intruders."[19]

Von Branca hinted that another German might excavate at Tendaguru after the Berlin team departed. Intriguingly, a man named Niedeck had visited Tendaguru before the arrival of the Berlin Expedition in 1909. He was a wealthy and well-traveled big-game hunter who intended to reopen excavations at his own expense. Unfortunately, nothing more is known of him.

On at least two occasions, Hennig denied a request from his brother Bruno for dinosaur bones. He explained that the Expedition was receiving a substantial reduction in freight costs from the shipping firm, and that the excavations were being privately financed. It would be embarrassing if sponsors discovered that fossils were being sent to Germany but were not going to the Museum. Hennig urged his family to visit the specimens that were prepared and mounted in the Museum, assuring them that a preparator named Borchert would be proud to tour them about.

On August 27, Janensch discovered the rarest of elements, the back of a skull complete with condyle. It appeared between Skeletons A and C. Hundreds of people had passed this spot along the main path through camp without seeing anything other than the cervical vertebrae in the pit.

Skeleton m, or "Issasaurus," after overseer Issa bin Salim, was just one of the dinosaurs of Kindope. It lay between the Lilahi and Lilombo Streams in the middle or second dinosaur horizon. A beautifully articulated series of 10 cervical, 9 dorsal, and 20 caudal vertebrae lay in the sun by September 1.[20] Ribs, hindlimb elements, and portions of the pelvis were also present.

A third team of men was set to work near Mtapaia on September 12. Most of the quarries near Kindope and Mtapaia were found in the middle dinosaur-bearing layer. There were also a few caudal vertebrae in the first or deepest bone layer. Not only giant sauropods were found at Tendaguru, but a 14-centimeter–long theropod tooth as well. New finds were reported almost daily: a new stegosaur in the Maimbwi Stream, another medium-sized sauropod reminiscent of the "Nyororosaurus" near the main road, and a series of articulated sauropod tail vertebrae at Locality no. Janensch sent word of another partial skull at Mtapaia and Hennig witnessed encouraging results as well: "At Salesi['s quarry] 2 scapulae, each 2 meters long with coracoids, a true wall of bone."[21] The number of laborers fell to 330 as men left to plant crops.

Moving the largest cervical vertebra from the immense sauropod known as Skeleton L was a challenge. Over 1.2 meters long, it was thoroughly encased in cloth and plaster, but a chunk of the jacket broke off. The weight was too much for the Wangoni assigned to it, so the crew was increased—first to 16, then to 24, divided into two teams.[22] Hennig accompanied the extraordinarily heavy package. The carriers skirted the Kitulo Heights and reached Lindi a day after the main caravan.

A total of 1,100 loads had been dispatched to the coast by September 11, 1910, including those from 1909 and 1910. An average of 10 jackets per day had been brought into camp during the last two months. By this point, about seven hundred loads, filling 146 crates and weighing 30,000 kilograms, had been sent to Berlin.[23] Another shipment, of 72 crates, left Lindi on October 26: a shipment destined for trouble.

Janensch repeatedly stressed how urgent it was for the Museum to return the crates to Lindi as quickly as possible. With the overwhelming stream of specimens arriving from Tendaguru, the containers were always in limited supply. He was also concerned that they might be delayed in quarantine because of plague in Lindi.

In mid-October, a piece of skull had appeared at the end of a series of cervical vertebrae at Skeleton S. Both Germans were elated when a nearly complete specimen, including lower jaws and teeth, was exposed. The occipital region resembled that of the Kindope skull, and while compression had displaced the individual elements a little, there was no damage. It was painstakingly plastered in two sections, and labeled *S66*. The jackets were placed in individual wooden boxes, which were in turn fitted into a larger crate. The plaster jacket of the Kindope skull, labeled *t1*, was shipped in another crate. Janensch cautioned von Branca not to unpack or prepare these delicate fossils until he and Hennig returned to Berlin.

By October 26, a Kindope quarry situated about four hundred meters to the south of St had been in progress for some weeks. The contents were wonderful—another type of dinosaur, and in unbelievable numbers: "Janensch with his excellent comprehension and memory for forms has recognized a very small ornithopod. I can only surrender and acknowledge that I personally (admittedly without any preparation) would not have performed such expert work."[24] The new quarry would be named Ig, for the small iguanodont-like dinosaurs that appeared in such abundance.

Boheti returned from Lindi with a plant press and O. C. Marsh's *The Dinosaurs of North America*. He also brought unsettling news of an outbreak of plague on the coast. Rats were the vectors of the bacillus and only quick action by Dr. Wolff had prevented a potentially disastrous spread to the interior. Lindi was the center of a great deal of porter traffic, so an outbreak could be rapid and wide-ranging. If it were transmitted to Tendaguru, so utterly dependent on the carriers and local labor, the consequences would be devastating. Two other doctors had been sent from Dar es Salaam to assist Wolff, and Boheti was delayed in Lindi for two weeks in order to satisfy the authorities that he had complied with the sanitation measures on his farm near town.

Janensch and Hennig ascended Tendaguru Hill with the aim of attempting a trigonometric survey to fix the many quarry locations more precisely. On most of his circuits Hennig recorded compass readings and tried to establish the elevations of geographic features for mapping purposes. The aneroid barometer that had been repaired in Berlin had rusted in the salt air at Lindi, and been sent home yet again. This was a setback, since the device was essential for measuring the elevations of geological strata. Janensch requested a distance meter for the prismatic compass and logarithmic tables to allow him to calculate readings taken while surveying.

Concern for the welfare of the workers grew, driving Hennig to great exertions to obtain provisions. There were no rations at all on November 8. Wild animals became less common in the area as a result of the growing village and the constant stream of women and children shuttling to and from water sources. Men were sent farther afield to buy up cereal grain in an attempt to remedy the shortage.

Several antelope were shot, though this only slightly eased the food crisis. After distributing meager rations on the 12th, Hennig immediately moved north for several days of hunting. On one trip he hoped for a glimpse of hippos at a little watering hole known as Mto Nyangi. This lovely spot, four or five hours from Tendaguru, was to be revisited time and again with great delight. The group retraced its path via Kindope and Mtapaia and inspected quarries. Hennig felt that one site, eventually designated Skeleton Aa, was promising enough to warrant the immediate start of earth removal. Salim was assigned this responsibility.

Late that month, Janensch was forced to expedite loads in Lindi once again. Over four hundred had accumulated, and the total number carried to Lindi since the inception of the Expedition in 1909 stood at 1,825. Now, though they did not know it, Hennig's shipment of October 26 was in peril.

The 72 crates, containing 487 loads, were aboard the steamer *Gertrud Woermann* in Marseilles harbor on November 21 when fire broke out on the ship. Rapid action saved the cargo, but the Tendaguru specimens were damaged by fire and water, some severely. The vessel reached Hamburg on December 10, where an assessment was made. In an insurance claim to the Transatlantic Cargo Insurance Company (Transatlantische Güterversicherungs-Gesellschaft), Hans Reck, a museum assistant, reported that 19 crates were 75 percent depreciated in value, 31 crates were 50 percent depreciated, and 17 crates were 25 percent depreciated. The insurance company paid over 1,200 marks in compensation.[25]

At the height of the season, the Expedition employed 420 men.[26] The number fell to 230 in November, to 200 at the start of December, and to 170 in the course of that month.[27] Food supplies were still dangerously low, with only half rations available some days.

Defense Force soldiers were bivouacked on Lipogiro Hill, and an officer invited Hennig and Janensch to their camp. Though the evening was enjoyable, Hennig felt uncomfortable as a sudden shower drove the party into a crowded tent: "Frankly I found it contemptible that we did not even get to know the blacks across from the two noncommissioned officers, and left them to eat by themselves. Caste spirit over racial awareness!"[28] Having worked with, lived beside, and relied upon Africans for more than a year and a half, Hennig had developed a very different relationship with them than had members of the traditional military hierarchy. Officers von Kornatzky and Winterer accepted an offer of a tour of Tendaguru and were met near Quarry X, where a dense accumulation of stegosaur phalanges was removed. They saw the large humerus of Skeleton t that afternoon.

Schulze, of the German East Africa Company in Lindi, stopped by on December 16, toured the quarries in Kindope and Skeleton t, and hunted. He had spent three years in Africa and was due for leave back to Germany. The two Berliners admired Schulze, who was deeply interested in all aspects of African human and natural history.

Around this time, the lobbying to officially protect the Tendaguru area from unauthorized excavation bore results. In December 1907, the district office in Lindi had informed the governor of the scientific significance of the site. The matter had been referred to the Imperial Colonial Office, which in June 1908 replied that although it did not possess the resources to support a program of excavation, the governor should pass legislation to control access by nonscientific parties.

The matter had become more pressing when the director of a museum in Pretoria, Union of South Africa, approached the German consul in Dar es Salaam. The director wished to know under what conditions his institute would be allowed to excavate at Tendaguru. In November and December 1910, the Imperial Colonial Office advised the German governor, von Rechenberg, to declare the area crown land. The secretary of state believed that this would provide a legal basis to reserve the right of investigation to German museums.

When the Prussian ministry responsible for museums and cultural affairs was asked what effect foreign competition would have on the scientific interests of German institutions, von Branca was consulted. He vigorously urged that all the paleontological resources of German East Africa be declared crown land and thereby reserved for German museums.

At Mtapaia the scapula, humerus, ribs, entire pelvis, tail, and neck of an incomplete but well-preserved large sauropod was exposed in the marine *Trigonia* layer in mid-December. The good fortune of seemingly endless excellent finds was tempered by the awareness that the fossils had to be removed before they were damaged by rain. Memories of the last-minute panic in 1909 drove Hennig to redouble his efforts.

Christmas Eve 1910 was, like that of 1909, a time for somber reflection for Hennig. News from Europe did little to ease his homesickness:

> Now it is blazing in German lands! We can only think homeward. Letter to mother about the recently communicated and to us completely . . . surprise extension of the Expedition for a further year by von Branca, in which I, however, hopefully no longer participate. Among all the reasons the unspoken one is perhaps the most fundamental, namely that I at least want to get to know ladies, in order to be able to progress to a choice . . . rather be carried away by passion in the not too distant future. Next Christmas I will moreover still be alone.[29]

Christmas dinner was interrupted by an irritating event. A thundershower earlier in the day had triggered the flight of winged ants. Swarming out of the ground, they were attracted to the lanterns and covered the Germans and their meal with shed wings. Disgruntled, Janensch and Hennig retreated to their tents.

Von Branca announced that the Tendaguru Committee was to meet again on January 27, 1911, and launch an appeal for a third field season. Hennig thought he was missing too many opportunities to advance his career in Germany while he sat in the African bush, and hoped to leave when the money was depleted.

The Expedition's funds would be reduced to 10,000 marks by May or June of 1911. For every additional 10,000 marks raised, the excavations could continue another two months. Janensch declared himself willing to carry on alone with a smaller crew and reluctantly asked for a replacement for Hennig. The Expedition leader warmly praised the tremendous energy and accomplishments of his co-worker and spoke of their compatibility under demanding conditions.

If more money was available, Hennig would be sent north to operate independently at Bishop Spreiter's sites at Makangaga, and Janensch would stay in the field until the end of 1911. Then there were reports of bones near Lourenço Marques in Portuguese East Africa, though Janensch felt that this prospect was less valuable than Makangaga. He resolved to leave Tendaguru immediately to personally assess the region before it was obscured by rapidly growing grass.

Despite the lack of celebration at the end of a year, Hennig was optimistic:

> 1911. The New Year! It shall bring me to my family again. . . . May I meet with all of them in good health. . . . For me, the question of the continuing career now becomes seriously vital, if I wish to expand and improve my life in a natural manner. The Expedition was a welcome springboard for me, hopefully it does not develop into a hindrance. My lust for adventure is stilled for the moment, which is not to say that I will not be hungry [for adventure] again tomorrow.[30]

Yet why enlarge the zone under investigation when the two Europeans were already near the limit of their abilities? Perhaps events in Berlin were adding pressure. Newspaper reports circulating in the homeland had criticized Museum personnel in the capital and in the colony. Why, it was demanded, had no skulls been found? It was charged that so many crates had arrived that there was no longer room to store them. Why had such a small area been searched in the field? The inaccurate allegations and innuendo were angrily refuted by Hans Reck:

> several nonsensical reports about the finds have already appeared in the daily press, originating from utterly nonauthoritative sides. In rebuttal, I state that all accounts that do not derive from a member of the Expedition itself and that are not covered by the name of a member, or that have not originated directly from the Geological-Paleontological Institute, are based on pure fantasy and can be nothing more than a distortion of a misinterpreted notice or remark that has been snapped up here or there by unauthorized persons.[31]

Janensch and Hennig were aware of the controversy through Hermann Stremme, another museum assistant, who informed them that supporters of the Tendaguru Expedition had expressed dissatisfaction with the finds on a recent Museum tour. This was scarcely credible, as there could be no question that the year was tremendously successful.

The quiet 32-year-old Janensch walked away from Tendaguru on January 2, 1911, after eight exhausting and rewarding months. What lay ahead in future seasons would dwarf that which had already been accomplished. Hennig accompanied him as far as Kindope. There was no emotional farewell: "alone I am not, where in my surroundings I know every black as an individual."[32] This was no vain boast, for Hennig took an exceptional interest in most everything around him, and genuinely enjoyed dealing with Africans. The daily round of discussions and his handling of the trying duties of a paymaster were testimony to his sympathy for people and ability to easily interact with diverse groups. He received a priceless education in return.

On his Sunday strolls through Tendaguru Village, he could see children playing in the sun while chickens scratched in the dust. Small boys might be making slingshots and flutes. Women would be pounding corn in mortars, collecting firewood, or hauling water. Men from the Makonde Plateau to the south might be carving ebony, a skill for which they were renowned. Single men might be out gathering firewood or hauling water. A centuries-old rhythm had developed at Tendaguru among distinct cultural groups.

Rain now fell daily. On January 10, Hennig gave some final instructions for fossil packing and left directions for the crew of one dozen men that would remain throughout the wet period. Hennig's confidence in the Africans was unassailable:

> Without their adaptability and skill, it would have been impossible with the presence of only two Europeans for the most part to develop the excavations to the degree that has occurred. Our black population is the most important factor in the external conditions under which the Expedition has operated.[33]

Hennig walked down the hill for the last time that season on January 11. He planned a nine-day geological safari to the coast. Once crates were packed and shipped, he would continue on a series of reconnaissance treks around Lindi. If von Branca's fundraising could extend the Expedition, Hennig would meet Sattler at Mikesse to again investigate the geology along the Central Railway. If not, he would return to Germany. Leaving Tendaguru was more poignant than he imagined: "everything shines and sparkles so dew- and spring-fresh, the distances are so pure and deep, the colors so rich and harmonious, that there really also is some farewell pain. How often, how often in my life will I yearn to return to this little place?"[34]

Traveling north, he reexamined the stratigraphy around Mtapaia. South again across the Mbemkuru toward Niongala, there were fossil localities to be viewed. In the valley of the Pindiro, Hennig re-located the outcrop of folded shales that had intrigued him in 1909. He altered course toward the coast, through an exhausting stretch of bamboo stands and heavy underbrush. He needed both hands to open a passage for himself and his porters, and was thus unable to use the newly repaired aneroid barometer.

Gastropods, ammonites, and bone were gathered, and elevations were calculated at Matapua. Although the fossils were not plentiful or diagnostic enough to sort out the stratigraphic sequence of the region, they had collected about 12 loads of samples by the time they reached Lindi.

Invited to dinner by Europeans in the town, the feisty Berliner revolted against the formal dress code: "In the insane heat this felt like suicide to me. . . . It was a joy to me to breach the influential spell of fashion slavery for once."[35]

A steady stream of porters were sent back to Tendaguru to retrieve more bone loads, and some of the many jackets stockpiled in Lindi were crated. Once order was brought to the Expedition's affairs, Hennig set out to review coastal geology on foot.

Throughout the latter half of the 1890s, several geologists from Germany and other nations had explored south of Mikindani. None was able to draw more than general conclusions about the age and relationships of the outcrops in the interior.

Hennig completed two one-day excursions, but was frustrated to find exposures covered in vegetation. Bearers were next sent southwest to Mtere on February 2. Hennig followed aboard a small boat on the Lukuledi River, a novel mode of transport for someone accustomed to dusty foot marches in the dry interior. Turning back toward the coast two days later, Hennig found deposits of Tertiary limestone, possibly an ancient Eocene coral reef, rather than the Cretaceous strata he was expecting.

After crossing Sudi Creek on February 6, he and his party arrived at Mbuo and carried on to Mikindani, a coastal town of 1,500.[36] Its deep, nearly landlocked harbor was the main outlet for locally grown sisal, rubber, and cotton. In addition to a substantial fort, the town also featured a postal and telegraph station. The Mikindani Beds could be examined in the vicinity. At the plantation of Fritz Linder Hennig spent a few hours preparing a geological profile. Linder had been contacted regarding the Tendaguru Expedition in 1908. Hennig boarded a steamer for Lindi on February 7, and the familiar shoreline soon came into view.

Hennig's final trip took him to Mroweka and Lichwaichwa aboard a small boat, in an attempt to understand the surface geology. He passed a succession of plantation properties flanking the Lukuledi. Filled with curiosity one evening, he persuaded his guides to recite sagas by firelight. Nature intruded abruptly:

> in a small grass hut the cook and the boys prepared unsuspected delights. Then a shout of alarm, immediately thereafter several; everyone leaped up: "Siafu!" Uncounted thousands of biting ants had attacked our peaceful camp, in the blink of an eye the kitchen and all tinned food was taken as booty, the boys driven back. . . . Oblivious to the fierce, very painful bites, the bearers grabbed torches, ripped the straw from the roof of a hut, and slowly passed through the rows of the advancing foe with firebrands, until all around our camp a wall of corpses and glowing ash was erected. . . . on departure the other morning only the tightly soldered provisions could be salvaged with some humor.[37]

On February 22, the group was hiking between tributaries that entered the Lukuledi as it flowed between the Rondo and Makonde Plateaus. The tired band returned to Lindi after one week on safari. An abundance of mosquitoes and the alarming number of cases of malaria in Lindi convinced Hennig to resume taking quinine. There had not been enough time to understand the exact relationship between Cretaceous and Tertiary deposits along the coast, but specimens had been collected.

What a season it had been. Over a period of about eight and a half months enough bones had been pulled from the earth to form 1,430 loads that were carried to Lindi by 1,634 porters.[38] Some were leftovers from 1909, but nonetheless the total number of crates from both seasons now stood at 418. Their contents weighed roughly 70,000 kilograms.[39] If the 1909 results are subtracted, then the 1910 season produced 310 cases containing 48,000 kilograms of fossils.

In an almost apologetic letter to von Branca in early December, Janensch offered reasons why the number of loads from the 1910 season was not substantially higher than in 1909, even though almost twice as many men had been employed. Fewer sites had yielded rich groupings of bones either at the surface or only shallowly buried. Other quarries, like Skeleton S, had absorbed tremendous manpower but yielded relatively few bones. Finally, quarries like St and Ig, which had also demanded large crews, yielded relatively fewer loads, but more bones. This occurred because the bones were so much smaller than the sauropod remains, and a number of bones could be taken out as a single load. Judging from published figures, this justification, though logical, was unnecessarily modest.

All jackets and crates from 1909 had been packed off to Europe. Some quarries of the previous year were now considered officially finished. Quarry XIV, a bonebed, had yielded a rich assortment, and Skeleton G, in the uppermost bone-bearing horizon, had produced additional dorsal and cervical vertebrae of this mid-sized sauropod. Skeleton S, though greatly enlarged in area, had still not been completely excavated despite major efforts.

Material from the middle bone-bearing horizon was of the best quality. There were 19 excavated localities within this layer, including St at Kindope, with its hundreds of stegosaur bones, and Ig, with its abundance of ornithopods.[40] Skeleton m, even further north, consisted of a beautifully articulated eight-meter-long series of 50 vertebrae and other elements.

The uppermost bone-bearing level produced the greatest number of bones. Altogether, 26 excavations had been undertaken in this horizon.[41] Among them were Skeleton k, which consisted of cervical, dorsal, and caudal vertebrae as well as rib, limb, and skull bones. Found between Skeletons A and D, it was likely also an example of Fraas's *Gigantosaurus*.

Even the lowest level had been investigated in 1910. Where 1909 had produced numerous limb and pelvic bones, 1910 had delivered hundreds of vertebrae, ribs up to 2.5 meters in length, many isolated teeth and skull fragments, and stegosaur defensive spines almost one meter long.

Geographically, the area under scrutiny had been widely enlarged. Bone had been found in abundance in a strip 15 kilometers long by two to three kilometers wide. About 105 sites had yielded material. Kijenjire, two hours north, and Mtapaia, an hour and fifteen minutes east, had received attention. Work had proceeded even further south—a full three-hour march to the eastern slope of the Namyura Plateau at Ubolelo. The two most distant sites were a full hundred kilometers apart. Janensch had picked up a number of stone tools in 1910 while traveling up the Dwanika Stream, adding evidence of ancient human occupation to the list of discoveries coming from the Tendaguru region.

As always, credit for this success belonged to the Africans. Janensch kept a partial list of quarries and overseers, and the 1910 entries that listed individuals included the following:

Quarry XVI	Sefu Abdallah
Quarry XVII	Mohammadi Keranje
Skeleton b	Abdallah Kimbamba
Skeleton c	Issa bin Salim
Skeleton d	Hassan bin Seliman
Skeleton e	Mohammed Ngaranga
Skeleton f	Bakari Liganga
Skeleton l	Abdallah Kimbamba
Skeleton m	Issa bin Salim
Skeleton p	Mohammed Ntandayira
Skeleton s	Salim Tombali
Skeleton u	Selim Kawinga
Skeleton W	Selim Kawinga
Skeleton X	Mohammadi Keranje
Skeleton Y	Mohammed Saidi
Skeleton Z	Saidi Mwejelo
Skeleton Aa	Salim
Skeleton bb	Salesi
Skeleton dd	Isa Salim
Locality St	Sefu Abdallah & I. Hizza
Locality no	Laa Tatu
Locality ki	Salim Tombali
Locality Ng	Mohammed Ngaranga

There was much left over for the third season: unfinished quarries, uncrated jackets, and new possibilities around Mchuya and Makangaga. The German investigators had not concentrated only on sampling Tendaguru's tremendous assemblage of dinosaurs. Janensch had made a particularly thorough entomological collection and shipped this to other museums in Berlin, along with a large variety of zoological specimens. Geological samples and invertebrate fossils had been gathered by both men on frequent excursions out of camp.

Both men had undertaken geological observations and had drawn stratigraphic profiles while on the march to distant quarries. They had documented their trips en route with compass readings, elevation measurements, and photographs. They had also attempted to map fossil-bearing outcrops and quarries by means of triangulation, but had been unsuccessful because of the difficulties of the terrain. There had simply not been enough time to clear a sighting path through stands of trees, so the topographic survey covered only a very restricted area. As always, meteorological records had been kept at Tendaguru.

Back in Berlin, the accountants tallied costs for the season at 54,322 marks, or 40,742 rupees. As in the previous year, workers' wages consumed the largest portion, at 38,676 marks.[42]

The second season, which had commenced in 1910, was closed. It was March 1, 1911, as Hennig's steamship pulled out of Lindi harbor. Once more, the Kaiserhof in Dar es Salaam offered a luxurious home after a 10-month stint in the bush.

Werner Janensch had spent several productive weeks tracking down more sites. He had walked out of Tendaguru on January 2, 1911, and entered the little village of Mchuya two days later, having covered 35 kilometers. As in the environs of Tendaguru, fossils characteristically eroded out of the western foreland of plateaus north of the Mbemkuru. This was true of the Mbalawala Plateau as well, where the rocks resembled the dinosaur-bearing strata of Tendaguru. About a dozen localities seemed quite promising for future excavation. Then he found nothing for three more days, until reaching Bishop Spreiter's site near the village of Makangaga. Another 60 kilometers north of Mchuya, it was surrounded by an overgrown landscape where little bone was revealed. There was a series of stegosaur caudal vertebrae, including tail spikes in their original position. Overall, Mchuya and Makangaga were assessed as worthwhile sites, but the distance to Tendaguru was so great that Janensch recommended the presence of a third European.

Janensch packed up and trekked to Kilwa Kivinje, reaching the port 12 days after leaving Tendaguru. Boheti bin Amrani and the crew that accompanied Janensch were instructed to continue overland to Mikesse on the Central Railway. Janensch would meet them there and together they would march south to investigate Karoo beds around the Uluguru Mountains. Unable to connect with Hennig in Lindi, Janensch reboarded the same ship that day for Dar es Salaam, where he disembarked on January 21, 1911.

-5-

1911

Along the Railway

The months of rough living had taken a toll. In Dar es Salaam, Hennig had three cavities filled by dentist P. Friedrich. At the hospital, Hennig had a sand flea removed from beneath a toenail. He also developed malarial fever, with a temperature of 41.3°C. Back in the Kaiserhof he had just collapsed into bed when a telegram was delivered.

Von Branca announced that another 30,000 marks had been raised, allowing a third field season. Late in 1910, the director had asked royal patron Duke Johann Albrecht of Mecklenburg to convene the Tendaguru Committee in Berlin. The rich results of the Expedition and the undeniable potential for continued success demanded an extension.

Von Branca contacted past donors, who once more supported the venture. Dr. Walther Stechow, a pioneering medical radiologist, transferred another 3,000 marks in 1911. Privy Commercial Councilor von Passavant-Gontard of Frankfurt sent 3,000 marks that year. Paul Bamberg, a former student of von Branca's, owned a factory in Berlin-Friedenau that crafted precision nautical and astronomical instruments. His first donation of 10,000 marks was followed by another of 5,000. The German East Africa Line renewed its offer of reduced shipping fees, with the proviso that duplicate bones be sent to the Hamburg State Geological Institute (Geologisches Staatsinstitut). When the funds remaining from 1910 were added, almost 40,000 marks was available.[1]

The Expedition had received widespread and positive coverage in dozens of newspapers and magazines in Germany. Not only private individuals were in favor of continuing the quest in Africa. At a session of the Prussian Assembly (Haus der Abgeordneten), a deputy of the Progressive People's Party (Fortschrittliche Volkspartei) suggested the government vote for a contribution from the Royal Reserve Funds (Königliche Dispositionsfonds). In April, a member of the assembly's upper house (Herrenhaus) urged the Prussian Ministry of Culture to support the Expedition to ensure that valuable specimens were not lost to erosion.

The Commission for the Geographical Investigation of the Protectorates sent a request to von Branca in January 1911. Dr. Arning, director of the Lindi Prospecting Company, had traveled through Kilwa District before the Maji-Maji Rebellion and had been impressed by a deposit of massive fossilized tree trunks. Arning hoped the Commission would contact von Branca, who might be able to assign a member of the Tendaguru Expedition to investigate.

A protective decree was proclaimed by Governor von Rechenberg in February. It expressly forbade the removal of fossil bones from the colony without written permission. The Expedition was advised to contact the Imperial Colonial Office, which could waive the requirement.

Supplies were ordered well in advance from Germany. Carl Bödicker's regular shipments of provisions included wines like Mouton d'Armailhacq, Chateau Giscours, and Caseler. Tinned foodstuffs included roast veal, mock turtle ragout, sausages and lentils, pickled goose, beef roulade, peas, spinach, and asparagus. Maggi GmbH began shipping free soups and soup concentrates to Africa every few months.

Hennig appeared to have resigned himself to plans from Berlin that specified a program lasting until December. He joined Sattler in Mikesse, to discover that Janensch had already left on February 23. Janensch, accompanied by Boheti and the other Tendaguru men, had moved south of the railway to the Rufiji River. He had received word of fossil bones near the Pangani rapids on the Rufiji. According to the work of Bornhardt in 1896–1897, the strata resembled the South African Karoo Sandstones, which are of Permian to Triassic age. Janensch hoped to more precisely date the beds with the aid of fossils.

Along the way, Janensch picked up fossil plants, including the seed fern *Glossopteris* and the horsetail *Equisetales*. He also found fossilized fish remains, including a partial lower jaw holding six teeth and the imprint of a skull plate. Further southeast and across the river, Boheti bin Amrani packed along a few geological samples.

As they trekked along the Rufiji, Janensch was given a photograph, purportedly of reptile bones. Actually, inorganic erosional structures in the region had been mistaken for bone. Fossil conchostracans, crustaceans with bivalved shells and gill-bearing legs, were found along a creek bed. Though the results may seem sparse, these crustaceans and the fish remains were the first animal fossils ever found in the Karoo of German East Africa. Janensch had collected plant fossils at six localities on his six-week safari.

The group returned the way they had come, to avoid the western side of the Uluguru Mountains, where famine prevailed. Janensch's group reached Mikesse at the beginning of April, about two weeks after Hennig's arrival.

Hennig intended to continue his investigations of Jurassic exposures along the Central Railway. In mid-April, he was dismayed to discover that crates of specimens collected with such effort along the tracks in March 1910 could not be found. The rail hut in which they had been stored had been demolished and the valuable fossils had disappeared. This was a severe setback, as the important Kinkelin quarry could not be found again.

Map E.

Area Investigated by Janensch During Rainy Season 1910/1911

Redrawn from Janensch, 1927, Beitrag zur Kenntnis der Karruschichten im östlichen Deutsch-Ostafrika, Palaeontographica, Suppl. 7

Scale: 1:1,000,000

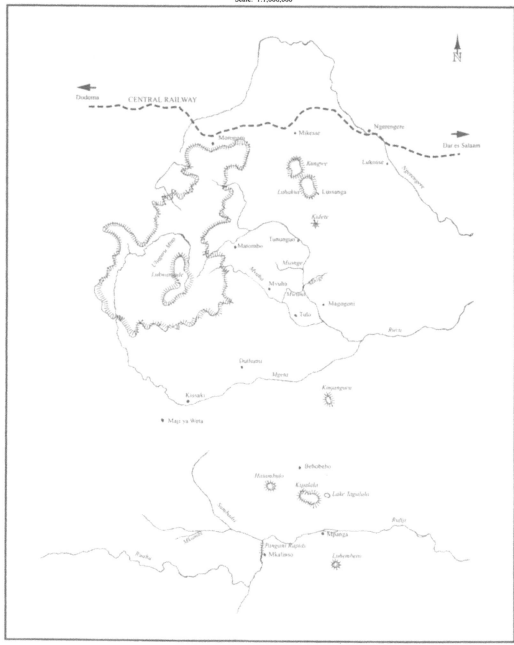

General stores such as Ali Wali & Co. and the Usambara Magazin GmbH also sold the Expedition food and hardware. Golden Birdseye and Rising Hope tobacco were purchased for the consumption of the laborers, who also placed orders for cigarettes, hats, kanga cloth for the ladies, millet, and corn.

Lindi's water was not potable, so boats made a daily trip across the river to a spring at Kitunda. A local pharmacist, J. Schlamp & Co., supplied soda water to the Expedition. The Expedition purchased medicines at this pharmacy for Europeans and Africans, such as boric acid, quinine, bismuth, aluminum acetate, ringworm salts, and thymol. Thymol pills were distributed to the laborers at Tendaguru as a treatment for worms. Dr. Beck visited Tendaguru in late July and left instructions for their use as part of a government campaign to control this debilitating infestation in plantation workers.

Hennig was envious of the geological insights von Staff offered while inspecting the workings at Tendaguru. Von Staff's arrival also had a very personal impact on Hennig:

> Staff is ½ a year younger, had to spend 4 semesters on law, was in Mexico, came to our Institute after I did, occupied himself at least as versatilely as I pride myself, is an experienced alpinist, has in everything a legendary memory and everything at hand for a victorious debate, is already married, docent, has published far more and better material than I have: I once again feel so dumb, so small. And dare I voluntarily allow him to take precedence immediately? Personally, according to my insight into his capability, 1000 times no! Moreover he is in no way completely congenial to me, our temperaments do not harmonize, but there my sympathy is on his side against my reputation. However, he has considerable self-conceit that almost got on my nerves in Lindi.[2]

Von Staff left for Mchuya to confer with Janensch, but contracted malaria.

By June there were already about 290 men on the payroll.[3] Roughly 250 remained at Tendaguru, but soon the extensive Kindope pits were claiming larger and larger crews.[4] The third major quarry, dd, near Ig and St, was located in the middle bone horizon, also a continuation from 1910.

A sharp-eyed local had found a weathered sacrum with ilia near the Muyoya Stream the previous year. Hennig was led north through bamboo thickets and tall grass to the site, and judged it worthy of a quarry. Clearing and digging began on June 2.

A fresh start was made on Skeleton S on June 5. Again, overburden was cut down from the top of the stream bank. Hennig arrived at Skeleton S on June 29, when a skull fragment with teeth appeared in the hard matrix. The quarry was 12 meters long and 10 to 12 meters deep. It yielded giant cervical vertebrae—complex masses of bone with intricate processes—and cervical ribs with tremendously long and delicate shafts. They were removed in bulky jackets, which made for heavy loads. A craftsman was hired from Namwiranye, six hours to the northwest, to build custom-measured crates from local wood at the quarry site. The vertebrae, still encased in protective matrix, were then loaded into the containers.

On the way back from Mtapaia, Hennig was informed of a large femur embedded in a slope near Skeleton I. A preparator with a crew of six was assigned to remove it. Skeleton Aa at Mtapaia yielded an articulated series of sauropod caudal vertebrae. Though the quarry reached a depth of three meters below the surface, extremely hard sandstone discouraged further investigation. Bones at Niongala were well preserved, so a crew would be sent there soon.

Thick grass made prospecting impossible in May and June, so men were sent to Lindi with about three hundred loads that had accumulated at Tendaguru. The bone magazine at Tendaguru remained full, because as many as 10 new jackets were brought in daily.

Hennig estimated that although there were twice as many Europeans on site in 1911, there was three times as much work to do as the year before. The small store was selling food and cloth worth three hundred rupees every week, an indication of the demands of a work force that was rapidly expanding. The renewed stream of field jackets arriving at Tendaguru demanded frequent visits to the coast, to ensure that the storage facilities of the German East Africa Company were not overwhelmed. The expansion of crews toward ever more distant sites meant that many hours of each day were spent traveling from quarry to quarry. This was all related to only one of the Expedition's goals, the collection of dinosaur remains. Little had been done, in Hennig's opinion, to understand the stratigraphy of the local area, let alone the geological mapping of territory to the west.

Furtwängler arrived at Tendaguru on June 24. He was to assist von Staff at Tendaguru. Unknown to Furtwängler, he carried with him the cause of his own premature withdrawal from the Expedition. On his way out to Lindi, bearers carrying his tent, bedding, and mosquito net were not able to follow him into a village as planned. He was forced to sleep in a grass hut, and contracted the severe recurring fever that was caused by spirochetes transmitted through tick bites. The tick *Ornithodorus* was a common scourge in the country.

Hennig and von Staff attempted triangulation with the theodolite around Lipogiro in early July. Furtwängler was too ill to offer any assistance, or even learn the shauri and paymaster routines. Determined to continue, Hennig moved the survey instrument southwest to the Namunda Plateau. At times the bush was so dense, he was compelled to estimate direction from the calls of his assistants somewhere ahead of him. Once on the plateau, the surrounding plain was shrouded in fog that prevented him from sighting on familiar inselbergs.

It was not until July 12 that Furtwängler's fever broke. It had incapacitated him for almost two weeks, and left him so weak that he had to be moved to Lindi for proper care. Von Staff, with characteristic self-confidence, declared himself ready to assume full responsibility for operations at Tendaguru. Hennig marched north toward Makangaga in mid-July. Furtwängler was carried to Lindi in a hammock. Von Staff remained at Tendaguru, relying on Boheti's assistance.

After a brief meeting with Janensch at Mchuya, Hennig bade his partner farewell, in what may have been their last parting in Africa. On July 21, Hennig's safari reached Makangaga. He surveyed the area with disappointment and anticipation:

> To my greatest horror all grass still completely green and for the time being burning inconceivable. . . . Numerous fallen trees and elephant tracks of every kind. In addition, lions and all types of animals are supposed to be abundant here. At least it is true, genuine wilderness in which I am again to establish myself![5]

Makangaga was a low-lying locale, surrounded by rolling steppe where impenetrable thorn brush flourished. The undergrowth was a rampant profusion of lianas, fallen stems, thorns, and entangling roots. All exposed fossil bone was badly eroded. The single trail back to camp was an elephant track, and the animals had destroyed most of the villagers' crops.

The Kilwa branch of the German East Africa Company supplied Hennig while he was at Makangaga. Four thousand marks was transferred to Kilwa from the Expedition account in Lindi's bank to cover the cost of food and drink. Each month, the German East Africa Company ordered a hundred bottles of soda water for Hennig from Lindi.

Hennig's men were now in Kilwa District. The contrast between the web of well-maintained trails in Lindi District and the vast stretches of bush in Kilwa District was striking. Hennig was guided to a dubious bone find: "everything unusable, but I have nothing better. . . . Several of Janensch's bone discoveries could not be traced even by the guide who was there, though they lie close to one another. . . . O grass! o grass!"[6] He ordered the hoes and shovels to be applied on July 25, beginning the first excavation at Makangaga.

Lingering rain hampered prospecting, but once fires managed to take hold, it was possible to probe an array of localities almost as abundant as that around Tendaguru. Most were within sediments that most closely matched the uppermost dinosaur horizon to the south. Bone preservation was poor due to the destructive effects of erosion and plant roots.

On an orientation excursion to Naiwanga to the west, Hennig picked up a few pieces of petrified wood and ammonites. A four-day safari was made to Nalue, to the east, where reefal limestone cropped out, and to the Kiturika Mountains to the south, while all remaining preparators were told to search for fossils and attempt to buy food. More quarries were opened at Kiwala and on the path eastward to Naiwangu in the first two weeks of August.

Hennig prepared maps of the landforms from compass and elevation readings recorded during recent safaris. His feet were very susceptible to blisters, so he was disappointed to receive word that there were no shoemakers in either Lindi or Kilwa. Therefore, the German East Africa Company agent in Kilwa sent his boots to Dar es Salaam for repair. As an avid reader, Hennig greatly missed having books at hand. Yet he had managed to draft as many papers in one month at Makangaga as he had in two years at Tendaguru.

It did not take long for unpleasantness to arise. Hennig's crew was refused all assistance by the locals because they led the German to bone finds, which would only encourage him to linger. Hennig did not command the respect or fear which stemmed from the kiboko whip, nor did he wish to call on the authorities in Kilwa so soon. On one occasion, Hennig offered to trade meat from a hartebeest he had shot, in exchange for cutting a decent track back to camp, but was refused.

Bearers were set to clearing paths and a well was dug to ensure a more hygienic water supply than that available from the swamp. The well water was a milky white, whereas the villagers' water was dark brown.

The supply situation was becoming serious, with little grain remaining in camp. Several men were in Kilwa and Nambunju with strict orders to buy provisions. An uncooperative jumbe sent them back empty-handed, which stiffened Hennig's resolve to sort the matter out as soon as possible. Unwilling to mete out corporal punishment, Hennig rewarded with meat all those locals who supplied millet to the Expedition. About 70 men had been employed in July and August, so the food shortage was serious.[7] In truth, the villagers had little to spare. Crops had failed because of the unreliable wet season. Still worse, elephants and wild pigs ran rampant, devastating the little grain that sprouted.

On August 17, Hennig undertook a three-day excursion northwest to Mitole along the Matandu River. Dinosaur-bearing horizons cropped out but did not merit further effort. Food was in such short supply that locals came to Hennig asking to buy grain. He offered them money so that they could buy food wherever they could find it. They would then bring loads of grain to Makangaga, and would receive a portion for their own use.

Finds in the first quarry were worth saving. Although the preservation left something to be desired, five dorsal and cervical vertebrae of a once complete skeleton appeared clearly. By August 23, twelve articulated cervical vertebrae and a row of ribs lay exposed, which Hennig sketched and photographed.

One morning in late August, Hennig heard lions as he was dressing. He was in the habit of checking the level of the water in the swamp near his tent each morning, and proceeded along a trail. As he stood in elephant tracks nearly a meter deep, a roar froze him. He was alone and unarmed. When he returned safely to camp with no more than badly jangled nerves, he made inquiries. A woman had seen lions near the water hole that very morning, but had not warned the children who were sent to haul water, or anyone else.

On returning from a quarry on August 28, Hennig received surprising news from Janensch. As his notes show, he was not disappointed by it:

> our funds are nearly exhausted and we have to demobilize, myself above all. Am very surprised but also pleased, because I while away too much of my best years here in Africa, and above all here in Makangaga without scenic incentive. . . . I already know what I will have to undertake at home. I must also attempt to strike out a second curator position at our Museum.[8]

The cervical vertebrae in the main quarry were the only hindrance to his immediate departure for Europe. It was pointless to wait any longer for the grass to dry and burn, as it was an unnecessary expense to keep a crew waiting in the field. Once the position of the bones relative to one another was sketched, they were wrapped, removed, and packed.

Twenty-five bone loads were sent to Kilwa in the first week of September.[9] Although there was little work for anyone but preparators and overseers, many loads were accumulating. Bearers were so difficult to recruit that Hennig sought the assistance of the police secretary in Kilwa. He also sent letters to African officials in Mavuji and Nakihu, but had little faith in this approach. It seemed unlikely that anyone would walk four days from Nakihu for a low-paying porter contract.

Porters had still not answered the summons to Makangaga by the middle of September. Hennig sent out messengers, who themselves required a food allowance that the Expedition could ill afford. Other supplies ran down. Wild rubber had long been used as an adhesive for joining bone fragments, but the cloth for protective clay and plaster jackets was not forthcoming, thanks to a mix-up at Kilwa.

The well soon dried up, so he used a water filter, but it became inoperative. Bottled water was the only safe alternative, but it had to be ordered all the way from Lindi and carried out from Kilwa. The next shipment would not arrive for another three weeks so he was forced to quench his thirst with whatever beer or wine he still possessed.

A group of preparators approached him offering to work overtime, as they had no desire to be left behind when Hennig broke camp for the coast. Finally, carriers appeared to transport the 40 loads already stockpiled.[10] Despite the best efforts of the men, the quarry could not be cleaned out by September 17. Hennig felt he could leave, however, and was on the trail on September 18, 1911, after 60 days at Makangaga. A few men stayed to finish packing.

The view from Ssingino Hill overlooking Kilwa Kivinje, or "Kilwa of the Casuarina Trees," was lovely, as the town was nestled in luxurious stands of mango and palm trees. Home to about four thousand people, perhaps 50 of whom were Europeans, it had been a major port from which slaves and ivory had been exported in Arab times.[11] Now it served as an entrepot for cotton, rubber, sesame, copra, and ivory. Since it was the administrative center for Kilwa District, the usual customs, telegraph, and fort buildings were located there. There was even talk of developing a third railway, to run from Kilwa to Lake Nyasa. Hennig found lodging in the facilities of the German East Africa Company. He had spent more than four and a half months in the field, about half of them at Makangaga.

In all, about a hundred loads of dinosaur bone and invertebrate fossils were carried into Kilwa.[12] The main skeleton at Makangaga consisted of 12 articulated cervical vertebrae, complete with cervical ribs from a sauropod. Some poorly preserved dorsal vertebrae were also in situ, lying almost at right angles to the cervicals. The dorsals were of such a low quality that only the condyles of the anteriormost vertebrae were saved. Weathering and root disturbance was partly to blame.

This skeleton rested on its right side—the left rib cage was in very poor condition and the underlying sequence on the right side was intact, articulated, and well preserved. The lower end of a femur and three caudal vertebrae that had drifted in around the cervicals were packed up.

Hennig's brief safaris allowed him to observe and record details of regional geology as well as collecting more specimens of dinosaur bone. He was disappointed, however, that he had not located the contact between Cretaceous and Tertiary beds.

All suitable wooden containers, whatever their condition, were purchased in Kilwa to carry field jackets. Workers were paid, leaving only a few preparators to help ready boxes of fossils for shipment.

A local missionary, Pater Ambrosius, became acquainted with Hennig. Ambrosius had been the first European to visit the caves in the Matumbi Hills to the north of Kilwa and had been employed by Professor Karl Weule of Leipzig University to excavate ancient deposits within the caves in 1906. Weule's guide had been the Swede Knudsen, who was recommended to the Tendaguru Expedition in 1909. Hennig sent two assistants to follow Ambrosius, who agreed to guide them as far as Jusu, where they were instructed to collect fossils. Bornhardt had gathered invertebrates there more than a decade earlier.

On Monday, September 25, a steamer pulled into Kilwa, bringing disturbing news of political events in Europe. The Second Morocco Crisis was intensifying. Germany had ordered the gunboat *Panther* to the port of Agadir after France occupied Fez. As a member of the Entente Cordiale, England supported France's position. According to reports, Germany had called up 400,000 men from the reserves and tension was building.

The last of the 39 crates were filled and passed through customs on September 26, but the 3,368-kilogram load did not leave Kilwa until November 15. Leftover Expedition gear, including hand tools, a lion trap, wire, and the medical kit, was packed into 22 loads and shipped to the German East Africa Company warehouse in Lindi.

Hennig's crew met him in the harbor prior to their departure for Tendaguru. On September 29, the ship bore Hennig north to the capital. He later heard that the Tendaguru men had waited on the beach until the liner vanished in the distance. Would their paths ever cross again?

The self-assured young scientist watched German East Africa slip away on October 1. His encounter with the African continent had spanned more than two and a half years. It had altered his outlook on the world and his awareness of himself forever. In the course of the Expedition, he had been forced to adapt to a foreign climate, an unfamiliar diet, sickness, dangerous animals, homesickness, and the bewildering demands made by over four hundred representatives of a foreign culture and language.

Back at Mchuya, Werner Janensch had not enjoyed much success. Initially, the vegetation was too damp to burn, but eventually an encouraging number of finds appeared over a wide area, including isolated limb bones. Bone was found as far away as two hours to the west and at Kihunda, four hours to the northwest. Elements that were worth collecting, however, were predominantly isolated occurrences in eroded condition.

Prospecting was difficult since the surface was covered by a thick layer of eroded rock overlying topography that was more undulating and less deeply stream-dissected than was customary at Tendaguru. Janensch described only a few fresh exposures as equivalent to the uppermost bone horizon.

To add further aggravation, neither food nor water could be easily procured. As at Makangaga, elephants had destroyed the millet fields and late rains had caused a disastrous failure of the rice crop. Janensch was compelled to direct a steady stream of bearers to the uplands of the surrounding plateaus to purchase rice for his crew. No nearby riverbed offered an adequate supply of water.

Janensch felt it would be more sensible and economical to concentrate on Tendaguru. Roughly 50 loads of bone were packed and carried back to the familiar hill to the south.[13] Janensch walked back to the main camp at the end of July, after less than two months at Mchuya.

Wollf Furtwängler had gamely walked back to Tendaguru after convalescing in Lindi. The effort was futile, as his health was broken, and after a total of two months of hard work at the site, he was afflicted with another fever attack and departed for good on August 11.

As the camp at Tendaguru grew through July and August, serious difficulties were encountered in obtaining provisions. With Janensch's return the camp housed 410 laborers in August.[14] Adding in the women and children, the village at the base of the hill may have numbered eight to nine hundred people. Back in January and February, there had been little precipitation at Tendaguru, the main rains arriving much later. Crops had dried up and had to be replanted. Grain was almost unobtainable in the area, therefore, and had to be purchased at great expense in Lindi. Even coastal supplies were much depleted, and by late August, only beans could be bought. Rice could not be found, and millet was ordered from Mikindani. Once Ramadan started in late September, sugar was unavailable, as it was highly prized during the month of fasting. An appreciable number of bearers were required to transport foodstuffs. The nearest springs were overwhelmed and trips to water sources as much as an hour distant became a regular chore throughout the day. Such were the logistical problems faced by the massive program at Tendaguru.

The stegosaur quarry St at Kindope was one of several bonebeds investigated in 1911. At Locality He another herd of these armored creatures was entombed in the uppermost bone-bearing horizon. Because the bones were so near the surface, their condition was inferior to that of the remains in St. Among the many well-preserved Kindope bones were a complete series of caudal vertebrae, three or four sacra and ilia in excellent condition, extremities, and other vertebrae.

Altogether, St and Ig at Kindope yielded over a thousand bones in 1911.[15] Several articulated groupings of the ornithopod lay in life-like positions. Toward the end of the season, a pleasing series of ornithopod tail vertebrae, as well as limb bones, ribs, and ilia, all belonging to the same individual, were unearthed. Wheelbarrows were successfully used at Kindope for the first time to speed the removal of the tremendous amounts of overburden that large crews were excavating.

Another Kindope site, dd, from the middle bone-bearing horizon, included 80 cervical, dorsal, and caudal vertebrae of a medium-sized sauropod. Another sauropod was buried here too, as well as numerous disarticulated vertebrae from the spinal column of a theropod. In addition to these riches, there were no fewer than three posterior portions of skulls.

Even in 1911, the colossus called Skeleton S absorbed a large crew. Its enormous articulated cervical vertebrae stretched out across a span of 12 meters and ended at the slightly scattered skull. Eventually the list of elements that could be recovered grew to an impressive length:

> 1 femur, 1 tibia, head of the second tibia, 2 fibulae, 2 humeri, 2 ulnae, 1 radius, 5 metatarsals, 5 phalanges, 1 scapula, 2 coracoids, 2 ischia, 2 pubes, 5–6 dorsal vertebrae, 9 cervical vertebrae, numerous vertebral fragments, 20 complete ribs, numerous rib fragments, 1 skull.[16]

By the time workers had pulled the last bone from the site, they had excavated for one and three-quarters years.

Not only immense sauropods populated the graveyard of Tendaguru. In the last weeks of the season, one find in particular generated a great deal of excitement. Not far from the hill were partial disarticulated skeletons of the most rare and fragile of creatures, the flying reptiles or pterosaurs. Unfortunately, it was too late in the year to thoroughly investigate this discovery. Money was running out fast and endless tasks wanted completing.

Fewer quarries had been opened at Tendaguru in 1911 than in 1909 and 1910. This was a deliberate decision to allow all active sites to be cleared of specimens. Large crews were assigned to several major quarries, since it was expected that 1911 would be the last year of the Expedition. Ten pits were situated near the hill: five in the uppermost and five in the middle bone-bearing horizon. Excavations were also carried out at Niongala on the Mbemkuru, three to four hours to the northeast. Sattler had collected here in Fraas's time and instructed locals in the proper methods. Five hours to the east of Tendaguru, at Nambango, an enormous sauropod sacrum was lifted out of the ground. Back in 1900, Bornhardt had picked up a bone here that was later ascribed to a plesiosaur. By this third season, the Expedition was active in over a hundred localities, its quarries spread over an area two to three kilometers wide and 15 kilometers long.[17]

Von Staff traversed the landscape with the analytical eye of a geomorphologist. At Mto Nyangi, the oasis and hippo pool, he found evidence of peat bogs, supplementing Janensch's discoveries. Dr. Paul Vageler, who had asked to join the Expedition in 1909, completed an independent soil survey in the colony in 1911. Concerned about maintaining the priority of the numerous peat discoveries made by the Tendaguru Expedition, Janensch urged von Branca to publish their results promptly.

The dry months of 1911 could not be fully exploited. On August 19, Janensch telegraphed von Branca for more funds. Janensch was shocked to hear from Hans Reck that the money raised so energetically in Germany was exhausted and the work would be forced to halt prematurely. Janensch, Hennig, and von Staff had all interpreted letters sent from Berlin to indicate that they could expand operations since there was hope of even more money. As a direct result,

Janensch had hired almost 500 men. Though they were unprepared for Reck's news, the facts were inescapable. Wages for 480 men, 410 at Tendaguru and Kindope and 70 at Makangaga, as well as the expensive purchases of grain, had rapidly reduced funds. To enable the Expedition to continue, men were dismissed. Their numbers sank to 170 on August 31, when Janensch released 250 men, and then to 120 in October.[18]

Hundreds of specimens still waited in quarries, hundreds of jackets sat at Tendaguru, and hundreds more were stockpiled in Lindi. Quarries, especially Ig at Kindope, still produced good bone, and new prospects beckoned, like that of Nambango. Hennig had not exhausted his allocated funds, so Janensch was able to extend excavation at dd for a time. Ultimately, however, there was no alternative, and on October 14, 1911, Werner Janensch closed the camp that had been his home for three seasons and departed for the coast. Hans von Staff, filled with many new impressions during his six-month stay in the tropics, had preceded Janensch to Lindi by a few days to arrange for the crating of field jackets. At the end of the 1911 season, 2,150 loads had been transported by three thousand bearers.[19]

The tempo of bone caravans picked up markedly in the final weeks. Up to 70 loads were readied at Tendaguru every week in early September. Very heavy loads that had been waiting in storage since 1910 were carried out. A portion of one of the largest femora required a team of eight to carry it. Added to this backlog was the impressive result from 1911. Assessor Dr. Auracher, the new district administrator in Lindi, was more than helpful in assisting with bearers. Schafer, a district agriculturalist on a business safari, generously sent hundreds of carriers whose services he no longer required to Tendaguru.

The volume of bone shipments can be imagined from Janensch's statement that in one week in September 565 men bore 513 loads to Lindi.[20] Furtwängler packed 180 crates in Lindi. This flood of material caused problems in the port. Crates could not be disassembled and returned from Berlin quickly enough. In mid-August, a number of steamers were already full upon their arrival in Lindi, offering little space for Tendaguru freight.

Between August and December 1911, 315 crates weighing 47,450 kilograms reached Hamburg. When another 92 crates of specimens collected in 1910 are added, the total yield rises to almost 62,600 kilograms. Yet, despite the Expedition's frantic efforts, there were around five hundred loads still waiting in Lindi at the end of October. The expenditures in Africa were tabulated by Berlin accountants, and totaled 43,771.58 marks, or 32,828.70 rupees.[21] Workers' wages had consumed 30,534.22 marks.

Judging by the tone of Janensch's letters, it was not a satisfactory end to the Expedition. He asked that empty crates be shipped back from Berlin as soon as possible. Boheti and a few trusted supervisors would be retained on salary until the end of November, to continue searching for skull fragments and remove other valuable material from Ig. They would then pack and ship the remaining jackets at Lindi. Janensch requested holiday time in Italy for the three Europeans. He argued that it would be a great shock to them to arrive in the middle of a north European winter after two and a half years in the tropics, where they had all suffered repeatedly from malaria and dysentery. Von Branca accordingly petitioned ministerial authorities in Berlin to grant a leave in Italy.

A large amount of Expedition equipment was suddenly surplus. The Usambara Magazin GmbH arranged an auction, from which tinned foods, camping gear, and the last load of plaster of Paris realized about 402 rupees from local Europeans. The photographic firm Voigtländer & Sohn asked Janensch to return the camera gear that had been loaned in 1909, but later approved its continued use by the Expedition. The Browning and Colt pistols, a double-barreled shotgun, three Model 98 muzzleloaders, and a rifle were left with a Dar es Salaam firm to sell on commission.

Janensch too, must have felt a wrench as he bade Africa farewell. He and von Staff reached Dar es Salaam on October 20 and left for Naples on November 1, 1911. They had spent two and a half years south of the equator. Janensch had never left his native Germany or slept in a tent prior to his adventure in Africa. As the weeks slipped into months and years, Werner Janensch and Edwin Hennig had witnessed the passing seasons in a harsh but often beautiful landscape. They shared the everyday experiences and also the joyous celebrations and tragedies of thousands of Africans. The lives of these people were affected by the Germans, but it was the scientists who were profoundly and permanently influenced.

Werner Janensch celebrated his 33rd birthday on the return journey. Edwin Hennig was now 29. The two men were leaving a land almost twice the size of their European home of Imperial Germany, but with roughly one-tenth of its population. It is a pity that little survives to indicate how others perceived the myriad changes that the pair had undergone. Hennig did record his feelings toward those who had made his stay so rich and rewarding: "I only wish that one could export the lively spirit of the natives—the colony would richly supply the motherland!"[22]

1. Eberhard Fraas at Tendaguru.
Courtesy of Staatliches Museum für Naturkunde, Stuttgart.

2. BELOW: W. B. Sattler(?) at Tendaguru.
Courtesy of Staatliches Museum für Naturkunde, Stuttgart.

3. Stuttgart Preparator
Max Böck.
*Courtesy of Staatliches
Museum für Naturkunde,
Stuttgart.*

4. BELOW: Tendaguru
display, Stuttgart.
*Courtesy of Staatliches
Museum für Naturkunde,
Stuttgart.*

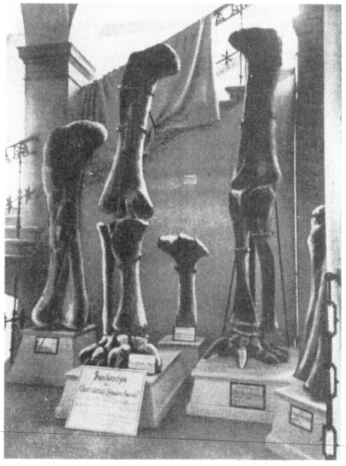

5. Museum für Naturkunde, Berlin.
From Heinrich Loewe, Berlin, Mark Brandenburg und Altmark
(Berlin: Preuss' Institut Graphik, 1919).

6. Wilhelm von Branca.
*Courtesy of Museum für
Naturkunde, Berlin.*

7. ABOVE: Edwin Hennig,
Werner Janensch, Saidi.
*Courtesy of Museum für
Naturkunde, Berlin.*

8. RIGHT: Boheti bin Amrani.
*Courtesy of Museum für
Naturkunde, Berlin.*

9. Werner Janensch, Hans von Staff.
Courtesy of Museum für Naturkunde, Berlin.

10. Above: Edwin Hennig, payday at Tendaguru. *Courtesy of Museum für Naturkunde, Berlin.*

11. Left: Hans Reck. *Courtesy of Museum für Naturkunde, Berlin.*

12. Top right: Tendaguru quarry. *Courtesy of Museum für Naturkunde, Berlin.*

13. Bottom right: Exposing bone. *Courtesy of Museum für Naturkunde, Berlin.*

14. Plastering bones.
Courtesy of Museum für Naturkunde, Berlin.

15. Quarry lg/WJ at Kindope.
Courtesy of Museum für Naturkunde, Berlin.

16. Quarry S. *Courtesy of Museum für Naturkunde, Berlin.*

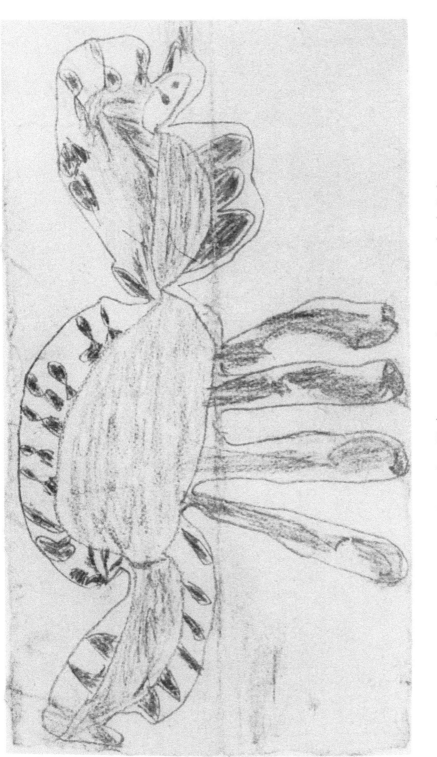

17. Facing page: Quarry St at Kindope. *Courtesy of Museum für Naturkunde, Berlin.*

18. Above: Drawing of dinosaur by African at Tendaguru. *Courtesy of Museum für Naturkunde, Berlin.*

19. Packing bones in "bamboo corsets." *Courtesy of Museum für Naturkunde, Berlin.*

20. Bone caravan en route
to Lindi.
*Courtesy of Museum für
Naturkunde, Berlin.*

21. Crating bones in Lindi.
Courtesy of Museum für Naturkunde, Berlin.

22. Carrying crates to dhows, Lindi.
Courtesy of Museum für Naturkunde, Berlin.

-6-

1911–1912

A Museum Overflows

Hennig made landfall in Naples and spent some time sightseeing in Rome, Florence, and Milan. In Stuttgart on October 31, he met Eberhard Fraas, who must have been pleased to hear firsthand of the richness of Tendaguru.

The longed-for day of homecoming arrived on November 1, 1911. The train came to a halt in the same Berlin station that Hennig had left on March 11, 1909, almost 32 months before. To his great joy, friends, relatives, and colleagues were there to greet him: "How grateful I am, infinitely grateful for this reunion! Much could have been quite different. The wonderful and important epoch has now—an epoch been!"[1]

Germany and the world had changed during his absence. Already in 1910, Berlin's population exceeded 2 million.[2] It was the Belle Epoque in Europe, and technology was accelerating the pace of change. Throughout the continent buses, taxis, and private automobiles were rapidly replacing the horse-drawn street trams and hansom cabs that Janensch and Hennig had used in their daily trips to the Museum prior to their departure to Africa. In April 1909, Peary had reached the North Pole. A month after his triumph, Louis Bleriot had crossed the English Channel in a flying machine. The race for the South Pole would be won by Amundsen in December, and the news must have reached Janensch and Hennig at their homes in Germany. Scandals and tragedies were as much a part of human affairs as ever. Revolution broke out in China and Mexico in 1911, and the Mona Lisa was stolen from the Louvre that year.

The Museum of Natural History had been transformed as well, as Hennig discovered during a meeting with von Branca on November 2. The tireless labor of hundreds of Africans had lifted an estimated 185,000 kilograms of bone from the ancient rocks of German East Africa. About 825 crates had been delivered to Invalidenstrasse 43, Berlin, containing 4,300 individual jackets that had been carried on the shoulders of 5,400 bearers.[3] All available corridors and rooms in the Museum were filled to overflowing. To house the field jackets, the Prussian state funded the construction of a large wooden storage barrack in the open courtyard behind the Museum. Not since the Museum had opened its doors to the public had such a collection streamed in as that from Africa. Several men were employed full-time just to unpack crates and shelve the thousands of bamboo corsets in an orderly fashion.

The task of preparation and assembly was monumental, but a program had been devised even before the first shipment arrived. Some of the enormous bones had been prepared by February 1911, in time for von Branca's final fundraising meeting. A powerful electric drilling machine capable of coring bones had been acquired. A massive 2.15-meter-long sauropod humerus was carefully cleaned piece by piece and each section cored. Sections were then threaded onto a heavy, upright iron rod and cemented into place, and gaps filled with plaster. This project occupied one technician for six weeks.[4] To provide a sense of scale the humerus was placed between a human and an elephant humerus, dwarfing both into insignificance. A reassembled sauropod scapula was mounted between a human and an elephant shoulder blade, with the same effect. Elements from smaller dinosaurs were also prepared and mounted, and together these formed the first display of Tendaguru material to attract curious Berliners.

Systematic preparation began of the myriad jackets, starting with those of the first field season. By October 1911, the contents of the first stegosaur quarry at Kindope, about three hundred bones, were almost completely assembled and restored.[5] Skeleton S received some attention, though progress was much slower due to the size of individual bones. Several men were required to manhandle a single limb element as it was painstakingly rebuilt from numerous weighty fragments. Frequently the highly complex and delicate vertebrae were in poor condition, and long hours of soaking with glue and strengthening with supportive iron rods were necessary.

In April 1912, Dr. August Brauer sent a grateful letter to Janensch and Hennig, listing the specimens they had collected for the Zoological Museum: three amphibians, 102 reptiles, 149 mammals, and 3,032 insects. Among the insects were 1,326 beetles, 902 butterflies and moths, and assorted millipedes, centipedes, spiders, and scorpions. Brauer declared the collection valuable for biogeographic studies, and an enhancement to the East African collection in the Museum.

A committee convened under Duke Johann Albrecht drew up a reckoning of the funds expended in the first three years. Members included Professor Dr. August Brauer, Dr. David von Hansemann, and Dr. Gustav Tornier. The Expedition had spent 183,607.45 marks, of which 127,325.70 had been donated by private individuals.[6]

Von Branca also drew attention to those aspects of the Expedition that were among the most costly. Wages for the indispensable Africans, though cheap by European standards, amounted to almost 90,000 marks, or almost 50 percent of all the funds received. Purchasing wood for crates from Scandinavia and shipping it all the way to East Africa had cost about 19,000 marks, while returning the disassembled empties five or even six times consumed 3,578 marks. Recycling was worth the effort, as the cost of acquiring more lumber for an equivalent number of new crates would have been 6,500 marks.

Freight charges were also significant at 17,000 marks, despite the generous 50 percent reduction extended by the German East Africa Line. Shipping the field jackets used up 13,440 marks of this 17,000.[7] The two hundred tonnes of dinosaur bones were assessed a value of 41,624 marks for customs and insurance purposes. The skulls were separately valued at 20,000 marks apiece.[8]

There was another surprise in store. The Prussian Diet had voted an additional sum of 45,000 marks toward operations at Tendaguru, to be dispensed by a branch of the Ministry of Culture. This was in acknowledgment of the scientific significance of the Expedition as well as of the national pride awakened during the first three years of effort. It was hoped that private donations would be forthcoming, but despite all efforts, only 3,000 marks were raised. When 2,529 marks held over from the 1909–1911 allocation were added, the total reached 50,000.[9] Of the latter sum, 35,000 marks were to be allocated to fieldwork, and the balance held in reserve to cover costs of transportation and preparation.[10]

The sudden turn of events was due to a combination of persuasive advocates in Berlin, the proven success of the Expedition, and the fact that much work still begged to be completed in Africa. Not all crates from 1911 had left Lindi; several important quarries, including the pterosaur and ornithopod sites, were still producing; and it was still desirable to search for additional comparative material.

Hennig had no intention of joining the 1912 season, but von Branca persuaded him to write a popular account of the Expedition, and discussed his future with him over dinner. Uncertainty had clouded the excitement of homecoming:

> A curator position is not granted. . . . Habilitation [qualification as an academic lecturer] above all does not bring me one step forward. . . . I do not really at all have the complete self-confidence for this, my usual mistake. But I do know that I must forge the iron of von Branca for as long as it is hot, therefore I do not say no. Almost 30 years [old] and 1500 M. salary! And no prospect for the foreseeable future![11]

Five months earlier, Janensch had intervened on Hennig's behalf, asking von Branca to establish a curatorial position whose occupant would be responsible for the Tendaguru collection. It would, he argued, be a terrible loss if Hennig, who possessed such invaluable knowledge, left the Museum.

One of Hennig's anxieties was reduced when he attended a Christmas oratorio at the Voice Academy. He met Johanna Trendelenburg, whom he had not seen for three years. They spoke for hours and he had every hope of future meetings.

Then it was Christmas; the happiest one Hennig had experienced in several years. He was reflective as 1911 closed:

> I can look back on the old one [year] with good cheer and gratitude; from a rich cornucopia, it presented me with new impressions, experiences, and pleasures. . . . It gave me my homeland again, all my family healthy and unchanged. . . . Un-

changed I too enter the New Year with all previous strengths, in many things all too unchanged. Of the New Year I now have as the first request a solid life appointment. . . . Admittedly only a true heart and soul union would satisfy me. . . . Habilitation—betrothal—publication of *Jesus*. Well, we will speak again in another year![12]

And what of Werner Janensch? We know little of his yearnings. We can only speculate that he was troubled by less uncertainty. His future as a curator at the Museum of Natural History was secure, and he did not face the same pressure to succeed academically. Tendaguru had presented him with a lifetime of opportunity.

Shipments continued to arrive: another 77 crates, weighing about 17,200 kilograms, left Lindi in January and February. Von Branca recognized that it was imperative to publicize the successful African Expedition to ensure that excavation and preparation efforts could continue. On February 27, 1912, the Society of Friends of Natural Science in Berlin hosted a lecture in a new auditorium of the Agricultural College. Large and small Tendaguru bones were displayed on either side of the speaker's podium. In the program, ladies were requested to remove their hats, no doubt to provide a clear field of view for those sitting behind them.

Von Branca gave careful thought to the list of invitees. The audience included government ministers, members of the Society of Friends of Natural Science, Museum officials, and, according to a contemporary newspaper account, many of Berlin's financial elite. Werner Janensch illustrated his talk on the organization and progress of the Expedition with colored lantern slides. Edwin Hennig followed with a presentation on the origins of the dinosaur deposits. Hans von Staff concluded with a discussion of the geomorphology of the region and the development of present-day landforms over time. The lecture was widely reported in Berlin newspapers.

On the morning of March 14, 1912, Kaiser Wilhelm II, his adjutant general, and the head of the civilian cabinet spent an hour at the Museum of Natural History. Dr. August Brauer, Dr. Gustav Tornier, and Dr. David von Hansemann met the dignitaries. According to Tornier, the kaiser agreed to have the plaster cast of *Diplodocus* remounted in the stomach-dragging, lizard-like pose advocated by both Tornier and the American paleontologist Oliver Hay. This never occurred, however. At a display of prepared bones from Tendaguru, Janensch, Hennig, and von Staff explained the finds of the Expedition to the kaiser. After a tour of the Halberstadt dinosaur remains, led by Dr. Otto Jaekel, the monarch was introduced to Dr. Hans Reck, who had been chosen to lead the 1912 Tendaguru field season.

Hans Gottfried Reck was born into a family of officers in Würzburg, Bavaria, on January 24, 1886. He studied natural history at the universities of Würzburg and Berlin. While he was an undergraduate at Berlin his teacher, Wilhelm von Branca, instilled a passion for volcanoes in the young Bavarian.

Another former student of von Branca, Dr. Walther von Knebel, had vanished during a geological expedition to Iceland in the summer of 1907. Hans

Reck was chosen to lead an investigation into his friend von Knebel's disappearance. A party of four set out in June 1908, including Reck, a woman named Ina von Grumbkow, and two local guides. Ina von Grumbkow had a compelling reason for the journey—von Knebel was her fiancé.

In 11 weeks they covered 1,500 kilometers of territory, spending up to 15 hours a day on horseback. It was established that von Knebel and his artist companion had likely drowned in Lake Askje in July 1907, when their boat overturned. Ina von Grumbkow, or, to give her full name, Viktorine Helene Natalie von Grumbkow, born in Hamburg on September 15, 1872, was an equal partner on this exhausting trek. In 1909, her narrative of this expedition, *Isafold: Reisebilder aus Island,* appeared in print.

Building on the first-hand experience gained in this rugged land, Reck examined the vulcanology of Iceland in his doctoral dissertation, earning his degree from Munich University in November 1910. In the same year, at age 24, he was employed as fourth assistant in von Branca's Museum of Natural History in Berlin. His early papers focused on Iceland's geology, preliminary notices of the Tendaguru Expedition, and descriptions of invertebrates co-authored with Hans von Staff. A colleague described him as a very congenial figure:

> he appeared to me, thanks to his handsome physical appearance, his unusual intellectual agility, and his kindness, frankly as a bright figure. He possessed an extraordinary energy, which enabled him to withstand the greatest exertions for a long time despite a congenital heart condition.[13]

In the ensuing years he also studied at University College, London, and eventually became a private lecturer at the Geological-Paleontological Institute of the Museum of Natural History.

Initial reports stated that Reck would be in charge of a two-year program that would include excavation at Tendaguru and an examination of East African geology. He was to undertake treks as far as possible to the west and south of Tendaguru to determine the furthest extent of the dinosaur-bearing outcrops. At the end of this two-year period, he was to proceed south to Lourenço Marques in Portuguese East Africa (now Maputo in Mozambique) to investigate reports of dinosaur bones. This was another prospect left over from the 1911 Tendaguru field season, like the fossilized tree trunks in Kilwa District. He would be accompanied by the determined and highly capable Ina von Grumbkow, who was now Ina Reck—they had been married in February 1912. Hans Reck was 26 years old and his wife was 39.

On May 1, their steamer reached Dar es Salaam. They went to Mikesse, where Bernhard Sattler offered advice on directing a staff of locals, and provided some rudimentary language instruction during the Recks' five-day stay. On the return trip to the capital, Hans collected about a hundred invertebrate fossils from Jurassic sediments flanking the railway grade.[14]

By May 23, the Recks were in Lindi, where Hans organized a shipment of 120 stockpiled crates containing field jackets from 1911.[15] Boheti bin Amrani had continued excavating around Tendaguru from Janensch's departure in 1911 to the Recks' arrival in 1912, and as a result between seven and eight hundred bearer loads had accumulated in Lindi.[16] It was still raining, for the wet season was uncharacteristically prolonged this year. One advantage would be bountiful crops, easing the pressure on the letter of credit for 20,000 marks that Reck received in Dar es Salaam.

The trail to Tendaguru was almost completely overgrown. By early June, the rains had ceased and the first week was spent refurbishing the large equipment and bone storage hut. Hennig's and Janensch's houses were so infested with insects, snakes, and even a dead leopard that Reck had them burned to the ground. The walls of a bamboo hut were rebuilt on the summit by June 12 and Imperial Germany's red, white, and black flag flew proudly.

A routine was soon established. Jumbe Saidi bin Ali, the camp chief, pounded the signal drum as early as 5:00 A.M. and distributed tools. Boheti bin Amrani was reinstated as head overseer. The only black to be addressed as Bwana by fellow Africans, Boheti assigned daily tasks to the supervisors and selected crews. A scribe took roll call, and within half an hour the teams marched off to the quarries.

The Recks had breakfast at 5:30, and by 6:30 Hans was on the trail to Kindope. He supervised the progress at Kindope until 10:00 in the morning and then returned to camp. For the next two hours, he directed the rebuilding of storage structures and oversaw the production of the "bone corsets." Lunch was taken by the Recks at noon, so that Hans could inspect quarry work by 1:30. When the afternoon sun was at its glaring maximum at 3:00, drumbeats recalled the staff to camp for a welcome meal. Men handed their work tickets to Hans, in a system he had instituted.

Ina and Hans would enjoy the beautiful sunsets until 7:00, when they ate their evening meal. Boheti and others claimed that a mournful cry heard in the trees prior to sunset was the njongoo, or millipede, heralding impending rain. It was typical of Ina to paint her impressions of this peaceful time in words:

> When the sun sinks and paints the distant hills of Namwiranye in violet tones, its last light glides over the plain of the Mbemkuru lowland like a reddish veil. Nearby the contrasts in color fade and only twilight shades remain, that quickly give way to true night. The first bats arrive with quiet clicking; the birds are silent, and in their place, the cricket choir strikes up. The calls of the people resound plainly out of the camp into the clear air, the laughter of the women rings clearly, and soon the rhythmic ngoma song begins.[17]

Hans Reck continued working well into the evening, recording daily observations in field notes, maintaining regular correspondence, and keeping the Expedition accounts current. By the time he was finished, the Southern Cross had risen directly above their hut.

Thanks to a bountiful harvest in 1912, provisions were plentiful. The work force numbered 110 by mid-June and eventually climbed to 150.[18] Excavations were initiated in early June. Quarries left unfinished in 1911 were cleaned of debris, then extended. The focus this season was at Kindope, about three kilometers north of Tendaguru. Three sites from previous years, St, Ig, and dd, absorbed the greatest portion of the labor force throughout 1912. The supply of bone in two of these was exhausted before long. No one knew the final catalogue number of the bones excavated by Boheti during the rainy season of 1911–1912, so Hans Reck assigned new designations to the three quarries to avoid confusion. Accordingly, St became EH for Edwin Hennig, Ig became WJ for Werner Janensch, and dd became HS for Hans von Staff.

Previous work at EH had yielded 844 well-preserved stegosaur elements before Reck appeared on the scene.[19] The final count for EH alone was about a thousand bones. The majority belonged to this armor plate–bearing dinosaur.[20] In the course of excavation the dig had expanded to an imposing size. It became necessary not only to label individual bones but also to subdivide the quarry into 12 sections, lettered from a to m, excluding j. As one section was emptied of specimens it was filled with rubble from another. The approximate final size of this pit: about 50 meters long by 15 to 20 meters wide, or a thousand square meters.[21] Excavation proceeded slowly, since the bone-producing horizon lay four meters below the surface. Hans's total from the gaping EH amounted to two hundred bones, after which he declared the quarry completed.[22]

Preparators at WJ were instructed to detach small blocks of bone-rich matrix. The abundance of bone was so great, at an average of 10 bones per block in some cases, that it was too time-consuming to prepare individual bones.[23] Moreover, the ornithopod remains were delicate and could easily be damaged if removed singly. Adjoining blocks were marked with stripes of ink, so that their positions could be reestablished in the Museum.

By the third week of June, a collection of about three hundred bones in WJ indicated the disarticulated but well-preserved skeletons of two individuals.[24] There was an array of cervical, dorsal, and caudal vertebrae, complete with fragile vertebral processes, as well as a selection of ribs. A few days later, teeth, skull elements, and two forelimbs were found nearby. On June 22, an articulated series of 15 dorsal and cervical vertebrae was uncovered, bent into a curve commonly found among a variety of dinosaur skeletons.[25] This curve is thought to be caused by muscles and ligaments drying and shrinking after an animal's death. In early July, 50 men wielded their tools at the site, 12 of whom were preparators.[26]

Reck attempted to discover how far the bone-bearing horizon at the ornithopod site WJ extended by digging trenches radially outward from the main quarry. The walls were cut back stepwise to prevent them from collapsing on the workers. Once preparators plastered and removed the fossils, rubble was thrown back into the pit. Reck also purchased wheelbarrows in Lindi, had them carried to Tendaguru in sections, and reassembled them there. There were a few mistrials but the men became adept at using these devices. Certain days were devoted solely to shifting rubble with wheelbarrows. Tremendous quantities of earth and stone were transported out of WJ.

The investigations at HS soon yielded interesting teeth as well as numerous large limb bones. After several weeks, it became obvious that the quarry was nearly exhausted of its contents. In early July, trenches were cut outward from the main pit in the four compass directions, but when they too proved sterile, work was halted at the site.

Reck was surprised to discover bones in storage at Tendaguru for which he and Borchert, a preparator, had searched in Berlin, including some from Skeleton S. Altogether, he estimated there were enough loads at Tendaguru for another 50 crates.[27]

Locality MD, the pterosaur bone site of 1911, lay next to a small streambed south of the path to Mtapaia. About 50 well-preserved, disarticulated bones were embedded in soft, fine-grained sandstone. They included a delicate foot element only 1 1/2 millimeters long.[28] Conical teeth and the absence of tail vertebrae, combined with other anatomical features, led Reck to conclude that a group of perhaps six individuals of a short-tailed genus had been buried a few meters from a series of *Barosaurus* sauropod vertebrae.[29] As the excavation continued ever deeper, an articulated series of dorsal vertebrae appeared, complete with processes, which he initially identified as pterosaur. The spinal column measured 50 to 60 centimeters, and each vertebra was about one centimeter in diameter.[30]

Standardized procedures, a trained crew of manageable size, an adequate food supply, and concentration on a few major sites allowed Hans to undertake distant safaris. During these times, Ina was the sole European in an area where the nearest help was three to four days away. She took charge of the shauri when Hans was on safari:

> there is also the entreaty to pay the worker's monthly wage not to him but rather to six others from whom he borrowed. This immediately results in six extra shauris which, due to the number of participants, demands a multitude of witnesses, every one of whom strives to discuss the truth differently, and a public of at least 50 nonparticipating but therefore even more curious observers that gathers.[31]

Hans admitted that he could only manage to act as arbitrator until about 5:00 P.M., and was forced to delegate many of the issues arising from these meetings to Boheti.

Only three or four of the laborers could read or write, so Ina dealt with their business and personal correspondence. Requests for everything from teapots and perfumes to flutes and boots increased around the time a caravan was scheduled to leave for Lindi.

After about a month at Tendaguru, Hans Reck was able to leave Boheti and Ina in charge of excavation while he undertook the first of a series of foot marches. The excursions, which lasted anywhere from a week to a month, formed the second part of his mission in the south—clarifying geological relationships and tracing the extent of dinosaur finds.

On July 9, Hans Reck marched west, meticulously recording the time of each significant geological observation, the character of any exposed rock strata, and temperatures and barometric pressures. He collected geological samples and took photographs along the way. He reached Bernhard Sattler's old campsite near Namwiranye on July 12. Huts had collapsed in the years since Sattler had used them. The abandoned garnet workings consisted of half a dozen quarries, the largest of which was about eight meters deep. Hans returned by the same route, having been absent five days.

Over a thousand ornithopod bones had been unearthed in WJ by July 18. In an area four meters square, Reck counted 150 bones, which at first he thought could be pterosaur remains. Months later he revised his assessment, suspecting instead that they represented small ornithopods. The stegosaur site was far less rich, and appeared emptied of bones. In the hope of finding a skull, efforts were now concentrated on a restricted area that yielded teeth. A second bone transport left for Lindi, and still a hundred loads were stored in camp at Tendaguru.[32]

Hans undertook a second, longer trek about two weeks after his return from Namwiranye. His aim was to better comprehend the geology of the Makonde Formation and add to the understanding of the inselberg structures that were so characteristic of the interior. Ina inspected and recorded findings in the pits at Kindope and localities along the path to Matapua daily. Since the grass had been burned, many more sites were reported by locals, so she made regular trips to quarries near the Namunda Plateau, southwest of Tendaguru. Fossil sites were found close to the hill and within an hour's march in any direction, from Tingutinguti Creek to the pterosaur quarry at Mtapaia.

On July 30, Reck and porters left Tendaguru. Heading west, then south, they followed the route taken by Bornhardt to the Rondo Plateau. The group crossed the dry bed of the Lukuledi River, which opened into the Indian Ocean at Lindi. The Makonde Formation cropped out near Masasi, so a halt was called on August 9. Reck spent two days scrambling to the summit of the Masasi and Nairombo inselbergs. On August 13, they marched into Newala, having left Tendaguru two weeks earlier. This larger settlement hosted a 20-man police force garrisoned in the boma. At this time of year, the Rovuma was reduced to a trickle of water in a broad sandy bed, though lush vegetation grew along its banks. Their main objective completed, the caravan turned north and headed homeward toward Tendaguru. Reck collapsed with fever, so his deputy Alberti was dispatched to collect rock samples and fossils. Reck's condition worsened and he had to be carried. On August 24, he slowly walked back to Tendaguru and a no doubt anxious Ina. The remarkable safari had lasted 26 days.

Fever and exhaustion kept Hans in bed for a few days, and for days thereafter he had himself carried to the excavations. Major earth-moving operations had been completed at the ornithopod quarry at Kindope. Once the overburden was cleared down to the bone level over an area of 12 to 15 square meters, WJ continued to yield excellent results, including partial skulls.

By September 15, more than three thousand individual bones had been recovered from WJ. Over a hundred men were employed at the complex of quarries extending throughout the immediate area. Two distinct bone layers existed, partially overlapping one another, and separated stratigraphically by a few tens of centimeters of deposits. Hennig and Janensch's old Ig quarry had primarily exposed one of the two layers. Reck excavated both, but collected predominantly from what he called the iguanodont herd. Judging from a count of diagnostic limb bones, at least a hundred individuals were represented here. The two concentrations seemed to contain disarticulated remains of similar animals of three different sizes. Mixed into the deposits were theropod teeth and isolated bones of larger dinosaurs. Many other quarries were reported, but there were not enough laborers to investigate them, and further work was deferred.[33]

About two hundred loads had accumulated at Tendaguru, a stockpile that Reck proposed to send to Lindi.[34] He postponed crating them until the Expedition closed down, reasoning that it was more cost-effective to keep his people excavating at WJ than to send them to the coast. He estimated that available finances would allow work to continue until November, but it would be advantageous to dig well into January. To do so would require another five thousand marks, but someone, possibly von Branca, noted in the margin of Reck's seventh report to Berlin that the money was unavailable.

Exciting news reached Berlin in the eighth and ninth field reports, dated September 15 and October 7 respectively. Six lower jaws, complete with teeth, had been found in the ornithopod quarry. To Reck's surprise, a third bone-bearing layer was encountered, and the catalogued specimens from WJ now numbered over four thousand. With many more bones concealed within the blocks, the total could only be estimated. Reck calculated that every thousand bones amounted to about a hundred bearer loads. Even the large dinosaur at WJ provided more skeletal parts, including sacral elements, a vertebra, and rib fragments.[35]

The crew was reduced to about a hundred men by early October, most of whom were assigned to the rich Kindope locality.[36] The most skilled preparators, about 15, were all removing specimens from WJ as quickly as possible. With the reduction in manpower, Reck hoped to be able to continue at Kindope until November, though there was enough material to keep everyone occupied as long as December. Little was being done at the other quarries, but if additional funds became available, Reck would reinstate workers elsewhere.

Enormous pelvic bones, a very large dorsal vertebra, and parts of a pectoral girdle and forelimb were the most notable discoveries from pits closest to camp. The Namunda site, from which a large dinosaur was removed, was now shut down and many of the 25 laborers dismissed. Similarly, 15 of the 25 men at quarries on the path to Matapua were released, after bringing large vertebrae, pelvic elements, and stegosaur bones into camp.[37]

As in the past, operating in this isolated setting demanded ingenuity. When the 50 kilograms of gum arabic were exhausted in the first six weeks of the Expedition, Reck found a substitute in the resin of local trees. Two men were employed full-time to collect the material, and managed to bring back almost a hundred kilograms. Reck had the resin dissolved in hot water and applied to the bones. Luckily, he was still able to find cloth for the jackets, having used over one kilometer of fabric since arriving.[38]

A wonderful find had been made at Kindope, so exciting that Hans telegraphed the news to Berlin while he was in Lindi. He described it as the skull of a carnivorous dinosaur complete with 30 teeth, each six to seven centimeters long. The skull was 35 centimeters in length and was discovered when a worker split open a block of matrix in WJ with an axe.[39] This treasure was carefully wrapped in plaster-coated cloth and packed in its own crate. Inexplicably, no announcement of the discovery was ever published—was the specimen misidentified in the field, or was it lost?

The purpose of Reck's next safari was to examine the stratigraphy of the narrow coastal strip between Lindi and Kilwa as well as the adjoining hinterland. It was hoped that, by collecting fossils and inspecting outcrops, he could clarify the age and sequence of the Tertiary, Jurassic, and Cretaceous strata. During his absence, there would be extensive earth-moving at WJ.

In the course of his march, he produced two extensive north-south geological profiles. He inspected numerous fossil localities, which yielded about 10 loads of specimens. At the marly base of the rock sequence around the Mahokondo Plateau he collected two hundred well-preserved ammonites thought to date to Dogger times, or the Middle Jurassic.[40] This deposit was rich in both numbers and varieties of specimens, yielding perhaps 25 genera and species.[41] All were found in concretions.

On his return to Tendaguru, he met Ina several hours north of the hill and they marched back together. The porters had enjoyed a rich feast of game one evening, and the safari the next morning was an obvious source of pleasure to the Recks:

> With what vehemence, with what glorious wild song did they set off on this morning and move tirelessly further through the glowing Mbemkuru savannah, while the brown-red poker-stiff antelope beefsteaks, dried to ropelike strips that at a distance reminded one of Boer biltong, dangled around our loads and wafted over the entire bearer column in the fiery sun.[42]

Back at Tendaguru, Hans Reck learned that a reasonably complete and articulated ornithopod skeleton had been found. There were jaws with teeth, the posterior portion of the skull, cervical, dorsal, and caudal vertebrae, one forelimb minus the hand, elements of the pectoral girdle, ribs, pelvic bones, and a hind limb minus the foot. The discovery was even more valuable because the bones lay in tight association, making it easier to reconstruct the entire skeleton. Reck reported that the sequence of catalogued elements now reached five thousand.[43]

Somehow, von Branca had managed to convince another wealthy German to support the Expedition. Dr. August Roechling, of a Mannheim iron and steel firm, contributed three thousand marks. Reck was delighted, as this would enable him to continue at WJ until the beginning of January 1913.

Reck rejoiced again two weeks later. In the intervening 16 days, three partial ornithopod skulls had been found, bringing the total to four. The bone count at WJ now stood at 6,000, but the recent massive earth-shifting program had laid bare a large expanse that included 2,000 to 3,000 additional specimens. When roughly 1,000 more exposed bones elsewhere in WJ were added to this number, Reck predicted that the catalogue would record 10,000 ornithopod elements by the end of the year.[44] The 15 skilled preparators had weeks of work ahead of them. They were assisted by five boys who carried away matrix as the bones were removed from the freshly exposed field. So great was the output that a hut was built nearby as a staging area. Five men carried bones to this hut, strengthened them with preservative, and wrapped them, safe from the rain that had begun to fall. They would be transported to the spacious storage hut at Tendaguru, where they would be packed into bamboo corsets for the journey to Lindi.

Excavating teams uncovered the outermost extent of the deposit by expanding to the west, east, and south of the main pit. An additional two thousand marks had been raised in Berlin, and Reck wrote that this would allow him to initiate another campaign to remove overburden to the south, using 50 men for 20 days.[45] Once the bone level was reached, preparators would have a month of work, but it was expected that the entire deposit would be exhausted at that time.

Hans Reck's third multiweek excursion was the last he could complete before the rainy season broke. He hoped to understand how the gneiss and inselberg rocks of the interior to the west of Tendaguru related to the narrow strip of coastal Cretaceous. A second goal was to investigate the report of a fossilized forest, news of which had been passed to Janensch in 1911.

In 1900, Privy Councilor Dr. Walter Busse of the University of Berlin had journeyed through the south of German East Africa at the invitation of the colonial government. Busse was studying agriculturally useful plants and their diseases and insect pests. He engaged 70 porters and marched inland from Kilwa on a safari that covered hundreds of kilometers and lasted almost half a year. Not far from Liwale, northwest of Tendaguru, Busse had come upon a rich accumulation of fossilized logs. A veritable petrified forest lay scattered over a wide area near the Mbarangandu River, but Busse could not bring back any samples. Shortly before the Maji-Maji Rebellion, Dr. Wilhelm Arning also visited the site and published popular articles on the find in 1905. Arning had contacted the Commission for the Geographical Investigation of the Protectorates in 1911, asking that they forward his information to the Museum of Natural History. However, 1911 was the busiest year of the Expedition, and there were no resources to spare for Arning's site. Werner Janensch wrote to Dr. Arning in April 1912 to obtain details, which were forwarded to Hans Reck.

On November 25, Reck marched west of Tendaguru. He walked around and up to the summit of a number of inselbergs: Nahungo, Nakihu, Litohu, and others. Wherever possible he gathered fossils and rocks, all of which supplemented his collections and observations from the Rovuma area.

The men walked into Liwale just over a week after leaving Tendaguru. Like Newala, the town served as a secondary district office and featured a fort with 30 soldiers and a number of well-built houses.[46] It lay on the major road to Lake Nyasa in the interior, the route for the proposed southern railway line. Without accurate maps, Reck consulted with Germans in the town, and was surprised to hear that the well-known fossil locality he was seeking lay another six days to the northwest. The band of men crossed the Matandu River, and as they approached the Mbarangandu River on December 8, Reck found himself in the midst of the fossilized forest. Entire trunks, fractured into sections 50 to 70 centimeters in length, lay scattered about. While some were 3 to 8 meters long, many others stretched for 15 or 18 meters, and ranged from 20 to 50 or even 100 centimeters in diameter.[47] All lay horizontally, occasionally bearing branches and a root system.

A halt was called on December 9. The men had been underway for about 2½ weeks nonstop, yet Reck could not pull himself away from the extraordinary sight of the accumulated wood. The trees weathered out of soft marly sandstone, which encouraged Reck to have his men dig out and expose an entire tree. From the lack of abrasion, he believed that the trunks had not been transported far from the original primeval forest. There were no other fossils by which to date the site.[48] Reck had 18 loads of fossil wood packed. Along with another load or two of geological samples, the packages were sent back to Tendaguru.

So excited was Reck about the fossilized trees that he proposed an ingenious plan to collect several. By his calculations, 10,000 to 15,000 marks would pay for a group of African laborers for two to three months.[49] During the rainy season they would construct rafts and, with two men and one complete tree trunk on each, float down the Mbarangandu and Rufiji Rivers to Salale. At this forestry station, the trees would be transferred onto a German East Africa Line steamer that would carry them to Germany. This would provide the Museum of Natural History with an outstanding display at a reasonable cost, since no excavation was required and labor and food costs were much cheaper than at Tendaguru.

On December 10, Reck turned his column of men back south, reaching Tendaguru by December 23. Hans Reck had spent 29 days on safari, returning in time to celebrate Christmas with Ina.

During her husband's absence on this latest excursion, Ina admitted to the strain of running the Expedition. She dozed off in her hut up on the hill one evening and had an unusual dream. Afterward she wrote wistfully, "Rats and scorpions danced around me joyously, and with the rumbling roar of lions out of the pori [bush], I imagined how nice it would be if my husband were here."[50]

The full extent of the Kindope ornithopod quarry had finally been revealed and the excavating team was distributed among other quarries. The preparators too, had finished in WJ and completed the preservation and wrapping of bones in the storage shelter next to the quarry. There were now over 10,000 entries in the bone catalogue, but it would take another two weeks to assign numbers to the remaining one to two thousand bones.[51] Two more incomplete skeletons had been found, bringing the total to four. Again, the hand and foot elements were missing, but a skull was associated with the skeletons. Equally satisfying were the discoveries of a sixth partial ornithopod skull and two isolated hands, one of which was complete.

With the end of December, another cycle of seasons had flowed by. Monsoons brought increased humidity, but also regular and strong downpours in the interior. Work at the remaining sites on the trail to Niongala and Matapua was hampered when water began to collect in the trenches. Some sauropod material was found there, as well as in the pits near Tendaguru Hill. The last of the

Kindope bones were not brought into camp at Tendaguru until January 3, 1913. It had been a spectacularly successful year—the catalogue recorded 14,482 bones.[52] Since the Expedition had begun, some 15,000 to 20,000 individual skeletal elements had been unearthed. High numbers were attained in 1912 in part because WJ, like other quarries, was already well advanced in the preceding year, and also because the bone deposit was dense and concentrated in a relatively restricted area. In all, 1,400 to 1,500 bearer-loads of bones had been produced during the 1912–1913 field season. Roughly a thousand of these had come from quarry WJ.[53]

Despite increasing rainfall, the pick and shovel men from WJ were set to work on exploratory trenches elsewhere as soon as they were done with Kindope. Channels were dug around all new pits to divert rainwater, an essential step since as much as 75 centimeters of water stood in the season's completed quarries. Hans Reck made a very promising find while prospecting on Lipogiro Hill. The remnants of a medium-sized sauropod rested here. Soon two shoulder blades, ribs, limb bones, a partial sacrum, and 15 articulated tail vertebrae were pulled from the sodden ground.[54]

Even more intriguing material was buried in an underlying sandstone layer—delicate, thin-walled bones that belonged to either birds or pterosaurs. Tibiae, fibulae, femora, teeth, ribs, and vertebrae were carefully wrapped. Among these skeletal elements were 80 to 90 tiny limb bones of animals that Reck could not clearly identify.[55] He suspected they might be reptile or mammal bones, but could come to no conclusion regarding the origin of the deposit.

In the final few days of the season, yet another discovery caught his attention. Enormous sauropod bones were uncovered near Janensch's Quarry X: pelvic bones, cervical and dorsal vertebrae, ribs, scapulae, and metatarsals. A miserable task lay ahead. The damp bones were collected in high humidity caused by regular downpours. Yet it seemed important to have them since they might well be part of the nearby skeleton Janensch had found.

Finally there was no question of continuing. Food for the laborers was becoming scarce, as was the money to pay them. Persistent rainfall frustrated all efforts at collecting. On Sunday, January 19, 1913, the field season was closed.

Hans reckoned he and Ina would walk to Lindi with the last of the many hundreds of loads waiting at Tendaguru by late January 1913. He could not guess how many loads might be crammed into the bone storage hut—perhaps as many as a thousand. It could take until the end of February to crate and ship the material. Then Hans planned to travel to Zanzibar and connect with a steamer that would take him to Majunga, Madagascar, to view the Upper Cretaceous dinosaur deposits reported by the French. When the rainy season ended around late May, he intended to resume his Central African expedition by investigating Jurassic exposures along the Central Railway in German East Africa. The couple had celebrated Christmas and New Year's in Africa. Eight months had been devoted to Tendaguru and environs.

A monumental task lay ahead. About six hundred bearers were required to haul the loads to Lindi, men who had to be hired and organized to march dur-

Map F.
Southern German East Africa
Redrawn from Hennig, 1914, Beiträge zur Geologie und Stratigraphie Deutsch-Ostafrikas, <u>Archiv für Biontologie</u>, 3:3

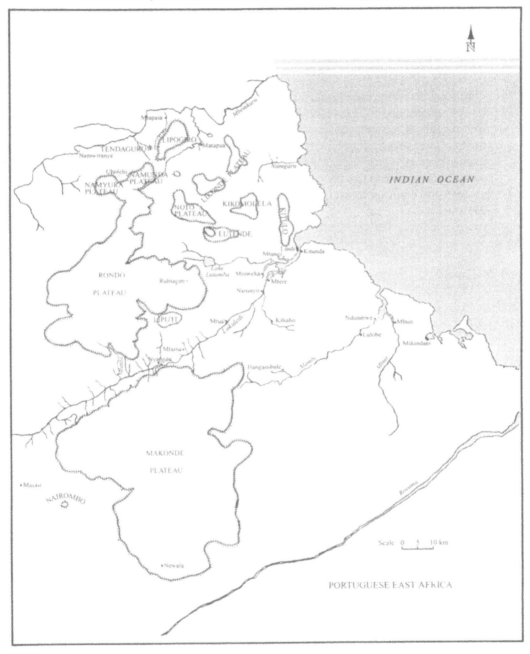

ing the rainy season.[56] Somehow, this was managed, very likely with the assistance of the representatives of the German East Africa Company.

As they turned away and marched to the coast, Ina sadly reminisced, "Tendaguru remained lifeless and silent behind us. The open doors of huts gaped empty and black; up at the house, kilimani, the black, white, and red flag had been lowered and the roof appeared to bend to the ground, gray and sad."[57] Familiar faces were being left behind. Would the Recks ever again hear the daily quarry song—"Kwenda-jojo-kwenda lala. Bwana-kaso-saa-ngoma-lala" (Take it easy, go to sleep. Sir, the dance is over, now we will sleep)?

Ina Reck's warm prose evoked the images that allowed others to share those happy months. Sadly, Hans Reck's delightful sketches would never appear in a published work, nor would Ina's oil paintings. However, they could share pleasant recollections with one another for the rest of their lives, as Hans indicated in his forward to Ina's book: "It is indeed a mutual experience that she portrays, so intimately and inseparably woven, which is only possible for lifelong companions who have endured all situations of life hand in hand, in faithful communion."[58]

-7-

1913–1918

Fresh Discoveries and a Bitter War

Hans Reck auctioned much of the surplus equipment in Lindi on January 26, 1913. The proceeds would pay for crates and transport expenses, since there were only about five hundred rupees left in the Expedition's account. Soon after his arrival on the coast the first steamer carried away five hundred loads of ornithopod material in 80 boxes weighing 17,192 kilograms. Still, the mountain of field jackets was not visibly reduced. So rich was the ornithopod quarry WJ that Reck used every available wooden case in Lindi, only to find he still had five hundred loads that would fill another 80 crates. It was not until mid-February that Hans and Ina steamed away from Lindi. With them traveled another seven hundred loads in 117 containers.[1] For fear of another disastrous fire like that aboard the *Gertrud Woermann* in 1910, insurance worth 20,500 marks was purchased for the shipment accompanying them.

The Recks did not visit Portuguese East Africa or Madagascar, as they had planned. The latter destination was off limits due to plague. Hans and Ina spent two weeks resting in Zanzibar, where they arrived on February 18 or 19, but Hans developed a fever that was so serious he eventually had to be hospitalized in Dar es Salaam.

In the capital Dr. E. Scholz, the government geologist, sought to enlist Reck's assistance with the Second General German East African National Exhibition (Zweite Allgemeine Deutsch-Ostafrikanische Landesaustellung), to be held in Dar es Salaam in late 1914. To celebrate German achievements in the colony, economic and scientific advances, as well as native cultures and natural history, would be showcased. Scholz asked for photographs of the Tendaguru excavations and plaster casts of some bones. He hoped that Reck could support this government-sponsored initiative in the same generous spirit with which the government had backed the Tendaguru Expedition. This request would provide an opportunity for Hans to finish investigating some sites that had been hastily abandoned in January 1913.

Then came what must have been a terribly difficult parting after so many shared experiences. Hans stayed in Africa while Ina returned to Berlin. He sang a song from Tendaguru to hearten her, but Ina was not consoled:

on the next to last evening, we sat on an almost deserted ocean shore and stared as though under a spell at that wicked ship which was supposed to bring me back to Germany. . . . And my heart was so terribly heavy because I had to leave, and it could not be made lighter by the beautiful natural atmosphere and all the comical joking.[2]

Hans was about to embark on what he called the Central African Expedition 1913. The first goal was to journey 1,200 kilometers from the coast to Lake Tanganyika while recording geological profiles and collecting fossils. Government and private authorities proved to be generous patrons. Following discussions with colonial ministers and the director of the railway, Reck was promised a rail car for his exclusive use. It could be stopped and detached at any station and reconnected to a new train as desired. Stationmasters were instructed to loan him their handcars between stations, to avoid time-consuming foot marches. Furthermore, his 25 African assistants were permitted to accompany him free of charge, all freight would be transported without cost, and telegraph and telephone services were placed at his disposal gratis. Holzmann, the firm that was responsible for rail construction, made copies of detailed topographical maps, offered to transport him to the end of steel in supply trains, and provided logistical assistance along the surveyed right of way as far as Lake Tanganyika. Boheti bin Amrani would accompany Reck.

Reck next proposed to record a geological profile from Tanga to Meru, not far from Mt. Kilimanjaro, along the Usambara Railway. From Meru, he would strike out overland to Mwanza on the shore of Lake Victoria. This would provide an opportunity for fieldwork among volcanoes and in the Great Rift Valley. Finally, from Lake Victoria he would move back east on the Uganda Railway, constructing a profile from Port Florence to Mombasa.

While on safari through the Rift Valley, Reck would re-locate a fossil locality reported by Dr. Wilhelm Kattwinkel. Gifted and wealthy, Kattwinkel was a professor of neurology at the University of Munich, specializing in neuropathology and neurophysiology. In Munich he had studied vertebrate paleontology. In 1911, he and his wife Martha had traveled to German East Africa on a self-financed hunting safari and scientific expedition to study sleeping sickness. They had been camped at the southern edge of the Serengeti Plains when they discovered and collected the fossilized bones of prehistoric animals from a gash in the landscape. Local Masai called this feature Olduvai, or Place of the Wild Sisal.

In Munich, renowned paleontologist Ernst Stromer von Reichenbach recognized the significance of this fauna, which was believed to date to Pleistocene times. The director of the Geological-Paleontological Institute of Munich University could not raise funds for an expedition and contacted Dr. Wilhelm von Branca in Berlin. They reached an agreement whereby the cost of excavation and the resulting collection would be split equally between their institutes. The Prussian Academy of Sciences in Berlin and the Society of Friends of Natural Science of Berlin were approached for financial contributions.

Reck contacted the colonial authorities in Dar es Salaam in the spring of 1913 for permission to export any Olduvai fossils he might find. He asked that news of the discovery remain confidential, as he thought a simultaneous arrival of several expeditions would jeopardize a thorough investigation. This last request stemmed from information he had received from Germany, namely that paleontologist Erich Krenkel, now attached to Leipzig University, was also traveling in German East Africa. In discussions with Governor Schnee and his deputies, Reck stressed the desire of the Berlin and Munich museums to have uncontested access to Olduvai.

Reck and his crew left Dar es Salaam station on March 17, 1913, for Lake Tanganyika. They collected the first vertebrate fossil found next to the tracks, the calcaneum of a Pleistocene fox. From the Malagarasi River crossing to Lake Tanganyika would be a hundred-kilometer safari. They found a rich accumulation of fossil snails near the salt-mining operation at Gottorp. At the northern end of Lake Tanganyika, Reck investigated a report of ozocerite, also known as fossil wax.

The homeward journey was completed quickly along the same route. After about 10 weeks, Reck was back in Dar es Salaam on May 25 with two thousand rock samples, three hundred photographs, and 650 pages of detailed notes, profiles, and sketches.[3]

Reck agreed to assist with the upcoming National Exhibition; he would return to Tendaguru for four to six weeks and supply real bones rather than casts, then train Scholz in preparation and mounting techniques in the capital. The German East Africa government would bear the costs of Reck's salary out of its annual budget. There was talk of a National Museum (Landesmuseum) for the colony, to be built with an endowment from the Hans Meyer Foundation (Stiftung). Such an institute would certainly wish to display Tendaguru fossils. Reck felt it was essential to provide assistance from the Exhibition, since the governor had agreed to ensure that the Berlin and Munich museums would enjoy a primary presence at Olduvai. Reck intended to extend his stay at the site at his own expense to complete the excavation of the 1913 quarries that contained delicate bird or mammal bones. At the same time the last of the Tendaguru loads stored at Lindi could be crated and shipped to Berlin.

Reck left Dar es Salaam on June 2, 1913. On a one-day stop in Lindi, he had students at the vocational school assemble crates from boards that had been returned from Berlin. He arrived at Tendaguru on June 10.

In the four months since he had been in the area, the tool shed had collapsed and one end of the bone store had sunk to the ground. Reck immediately burned the old huts of the laborers, along with the building that housed the forge and the shed in which jackets were packaged. Other structures, including his house and the bone store, were repaired and a new packing shed was built.

The first excavation was opened on June 16. Within two days there were 20 men digging. In the next two weeks, several more trenches were extended outward and parallel to one another. The experienced crews produced at least three ditches 28 meters long and 2 to 3 meters deep by June 27.[4] Reck and several men returned to Kindope on the 23rd and soon found good bones. Manjonga was assigned to reopen the purported bird and mammal site.

Manjonga's quarry near Mtapaia, which had first been noticed in 1912 by Boheti, was investigated for eight days. It initially produced three to four bones daily, but when a larger area was cleared of overburden, the preparators were soon finding 10 or even 20 bones a day. The deposit in which the bones were found was a distinct layer about 20 to 40 centimeters thick, overlying hard, sterile sandstone and in turn overlain by sterile marls. The bone concentration was only a few centimeters thick, 20 to 60 centimeters below the surface, and covered a small area, 16 to 20 meters square.[5]

Preservation was good, and although the bone was highly fragmented, breaks were sharp and pieces could be glued together easily. The surface of most of the bones was unweathered. As matrix was removed with a knife, small plant remains and the impressions of skeletal elements were exposed. Altogether the quarry yielded another 70 to 80 delicate and well-preserved bones, bringing the total to 200 or 250.[6] Because they were hollow, Reck suspected that a large number were the remains of pterosaurs, while others might belong to small reptiles or even mammals.

Little display-worthy material was uncovered in the first weeks, which was a concern since providing items for display was Reck's primary obligation to the colonial government. Long trenches were cut around the path to Kindope on July 2. Reck was producing an artificial exposure or geological profile in order to reconstruct the stratigraphic sequence and thickness of the bone-bearing layers. He distributed crews more widely: 13 men were sent to Mtapaia Creek, and an overseer and 10 men to Lipogiro Hill two hours away. A stegosaur quarry, near the Mtapaia camp's water source, was assigned to another group.

The situation improved as a new trench located near old quarries along the Noto trail was revealed to contain 2.5-meter-long ribs, huge vertebrae, two massive pelvic bones, and a large limb bone. Of special interest was a five-centimeter-long spine, possibly from a juvenile stegosaur, a partial sauropod forefoot embedded nearly vertically in the ground, and a giant sacrum. Other old quarries supplied large limb bones suitable for display, one of which was encrusted with fossil bivalves. Invertebrate fossils and fish remains, mainly scales, were saved. Forty-six loads were carried to Lindi on July 12.[7] Reck took about two dozen photographs.

The last day of excavation was July 17. Seven preparators remained at Tendaguru to complete the work, as there were an estimated 40 loads of bones still exposed in quarries. A local jumbe would continue packing, assisted by the boys who hauled water to the quarries. The akida of Noto sent 50 carriers, which allowed Reck to carry almost all the packaged loads to the coast with him.

On July 19, Reck left Tendaguru for Lindi. In town, he supervised the packing of about 80 crates of material left over from the 1912–1913 season, and was back in Dar es Salaam by the end of July.

Reck commenced the next stage of his Central African Expedition as he boarded the train in Tanga on August 2. The journey was initially not as successful as he hoped. There were no sedimentary exposures along the rail line for the first 20 kilometers outside of Tanga. Reck retraced his steps and made two marches of 10 days' duration each.

About seven weeks later, on September 18, Hans Reck was at Moshi, the terminus of the railway, preparing for the trek to Olduvai. A week later, he was camped about two hours from Arusha, near Mt. Kilimanjaro. His caravan of 44 bearers followed his lead through the glowing golden savannah of the Serengeti.[8] Reaching Olduvai Gorge was not straightforward, since it was situated in a little-explored and arid region. Reck considered this among the most difficult treks he had ever undertaken.

The band pitched their tents on the floor of the magnificent Ngorongoro Crater. Reck was forced to rely on a few photographs taken by Kattwinkel and a general map. Lions and rhinoceroses sent the men fleeing for their lives up thorny acacia trees on several occasions, and one evening seven lions circled the camp, roaring only 40 meters away. Reck's party finally located Olduvai Gorge 10 days after leaving Arusha. They did not find fossil bone until October 7.

Fossils were abundant, however, including animals such as antelopes, elephants, hippopotamuses, three-toed horses, and armored fish. Reck stated that over seven hundred bones were collected and packed over the 3½ months his crew spent at Olduvai.[9] Rice and maize for the men was carried from a village four days away.

Most of the crew was composed of veterans from the southern dinosaur fields. Among them were the cook Ali and Reck's servants, Alberti and Issa Makolela. Experienced supervisors and preparators also made the long journey. They included Saidi bin Ali, the Wayao camp leader who enthusiastically beat the morning drum to call men to work at Tendaguru; Mohamadi Tandajira, the Wamwera preparator who adeptly uncovered the most delicate specimens; Saidi bin Manjonga, the slight Wanyamwezi who recovered the pterosaur bones; Issa bin Namanorow, who managed the bone stores, packing, carpentry, and even blacksmithing; Bakari Omari, who found the first elephant remains at Olduvai; Ligwema, the water carrier; and numerous others.

Hans Reck continued on to the next phase of the Expedition, namely the study of the Rift Valley and volcanic structures around Olduvai. Saidi bin Manjonga, one of the preparators from Tendaguru, was left in charge of the excavations. Fifteen of the 44 men were chosen as bearers while the rest continued exploring at the gorge.[10]

Over three hundred geological samples were gathered throughout the trek, supplemented by four hundred pages of notes, numerous sketches, and 50 photographs.[11] Tragedy struck when a bearer wandered away from camp and perished of thirst. The exhausted men returned to Olduvai Gorge by December 17, after seven weeks. Bakari Omari had made a stunning discovery during their absence—a human skeleton. The antiquity of the fossil animals implied a similar antiquity for the human, an amazing revelation in 1913.

The closing weeks of December were spent numbering and labeling specimens, coating them with clay and gum arabic, and recording them in a catalogue. Hans Reck barely stopped for Christmas. The final tally was about 1,750 specimens ready to be carried out in 150 packages.[12] In early January, the bone caravan marched out of the gorge. Loads would be dropped at the fort in Iraku until Reck obtained the money to transport them to Tanga. He would continue to Lake Victoria, where a steamer could bring him to Port Florence and the terminus of the Uganda Railway. With luck, he would be in Mombasa by early February, in time to board a steamer to Germany.

In Berlin, there was no doubt that the German excavations at Tendaguru had been extraordinarily successful. The original participants had been occupied throughout 1912. Edwin Hennig had given a series of lectures on Tendaguru. His popular account of the days at the dinosaur digs also appeared that year in a well-illustrated volume costing four marks. Its title was *Am Tendaguru: Leben und Wirken einer deutschen Forschungsexpedition zur Ausgrabung vorweltlicher Riesensaurier in Deutsch-Ostafrika*, or "At Tendaguru: Life and labors of a German scientific expedition to excavate prehistoric giant dinosaurs in German East Africa." Reviews, both German and American, were positive.

News of the German Tendaguru Expedition spread through the English-speaking world. On May 11, 1912, Sir Arthur Smith Woodward, keeper of geology at the British Museum (Natural History), presented a talk illustrated with lantern slides entitled "The Great Finds of Fossil Bones in German East Africa." The London branch of the German Colonial Society (Deutsche Kolonial Gesellschaft) sponsored the presentation.

Expedition participants were also recognized for their contributions. Von Branca nominated David von Hansemann for the Knight's Cross, First Class, of the Ducal Order of Henry the Lion (Ritterkreuz I Klasse des Herzoglichen Ordens Heinrichs des Löwen). Janensch was nominated for the same medal, second class. The decorations were presented in March 1912. In March, Janensch was also honored with the prestigious title of "professor," again at von Branca's suggestion. Janensch, Hennig, and Paul Bamberg were nominated, once again by von Branca, for the Prussian Order of the Crown, Fourth Class (Preussische Kronenorden, IV Klasse). Kaiser Wilhelm II assented in May.

In July 1912, Edwin Hennig received the Silver Leibniz Medal of the Prussian Academy of Sciences (Silberne Leibniz Medaille der Preussischen Akademie der Wissenschaften). Finally, von Branca also lobbied the Prussian Ministry of Culture in July for recognition of the contributions of Eberhard Fraas and Bernhard Wilhelm Sattler, but it is not known if they ever received these decorations.

A great deal had happened in the German capital during Hans Reck's two-year assignment in Africa. Throughout 1912, a permanent position still eluded Edwin Hennig. The Prussian Finance Ministry had not approved a senior assistant (Oberassistent) position at the Museum of Natural History. Perhaps he really had lost ground during his absence in Africa.

Edwin Hennig married the woman he had so anxiously hoped to encounter upon his return from Tendaguru, Johanna Trendelenburg, on March 15, 1913. He achieved another goal in May 1913, when he became qualified as an academic lecturer in geology. This step had been delayed since the spring of 1909, when he originally intended to complete the requirements but joined the German Tendaguru Expedition instead.

In February 1914, Hans von Staff received the title of Extraordinary Professor at the University of Berlin, and took up new challenges as government geologist in German Southwest Africa (now Namibia). His wife, Dr. Hella von Staff, a dental surgeon, accompanied him.

A variety of scientific papers growing out of the Expedition had been published by 1914. Initially, publications emphasized geology, including stratigraphy and geomorphology, but also described the invertebrate fossils. The first series, *Wissenschaftliche Ergebnisse der Tendaguru Expedition, 1909–1912* (Scientific Results of the Tendaguru Expedition, 1909–1912) appeared in *Archiv für Biontologie,* which was sponsored by the Society of Friends of Natural Science of Berlin. Additional papers were regularly printed in the proceedings of this society.

The foundation for these studies was the tens of thousands of bones, freed from their protective jackets and enveloping marl. With great skill and patience fragments were joined, fragile structures were consolidated and strengthened with glues, and partial elements were restored with plaster. There could be no thought of waiting until all the material was unpacked and prepared, since the task would require years. With Reck's collections added to the total, there were 1,050 cases, or about 235,000 kilograms, of Tendaguru fossils in the Museum. Instead, the results of scientific study would be released as soon as possible and amended if subsequent finds altered earlier conclusions.

Von Branca listed the costs associated with preparation. A single sauropod scapula about two meters long was reassembled from 80 pieces by one preparator over a period of 160 hours. A single humerus, 2.1 meters long and weighing almost two hundred kilograms, required 75 hours of attention while another, which measured 2.13 meters, was only restored to a state of completeness after 12 days.[13]

Cervical vertebrae presented the greatest challenge. Sponge-like in places, they possessed an assortment of blade-thin buttresses, sculptured processes, and arches. A smaller example from Skeleton S took 450 hours of preparation over a two-month period. At a preparator's wage of 75 pfennigs per hour, this single bone swallowed 337.50 marks. Another cervical vertebra situated further down the animal's neck was half as large again, over a meter long and ¾ of a meter high. It possessed twice the surface area and by extrapolation would demand nine hundred hours over four months, perhaps 675 marks.[14] The total expenses generated by just one sauropod can be seen to mount quickly; there were as many as 15 cervical vertebrae, which formed only a part of the spinal column.

The gargantuan dorsal vertebrae from Skeleton S that took several men to maneuver were likely to demand as much time and money as the large anterior cervical. One had already consumed 350 hours of labor in two months, but was missing a large portion, still en route from Africa. Another hundred hours would have to be expended before it was fully prepared.[15] Von Branca was under no illusions about the number of years that would pass before a behemoth such as Skeleton S was available for public viewing. The financial underpinnings of such an enterprise needed attention. Without another 50,000 to 100,000 marks, the ancient beasts of Tendaguru could not be resurrected in the same spectacular manner as those in America.[16]

Hans Reck arrived in Berlin in late February 1914, unprepared for the uproar caused by his discovery of human remains at Olduvai. Newspaper correspondents waited in an anxious queue outside his office at the Museum of Natural History. An avalanche of mail followed, including telegrams from New York, London, and Paris newspapers.

On March 17, 1914, he lectured to a packed hall at a session of the Society of Friends of Natural Science of Berlin. Academics came from as far away as Hamburg and Breslau. Dissension arose immediately. The sediments and animal remains were ancient, dating to the Pleistocene, but the human skull, in contrast, possessed a modern anatomical appearance.

Over the next few weeks, the rest of the skeleton arrived from the field, along with the jackets containing the faunal remains. This could only have aggravated the work of the Museum preparators, who were still occupied with the Tendaguru inventory. At the next Society lecture on May 5, supplemented by the prepared human skeleton, the question of whether the skeleton was a recent burial that had intruded into older sediments was raised again. The find was reported, with numerous photographs, in the *Illustrated London News*.

Discoveries at Olduvai were so sensational, yet inconclusive, that three groups launched expeditions to the gorge in the ensuing months. Professor Kattwinkel of Munich personally financed a program that was to last two years. Reck had done his utmost to convince the millionaire of the need to move quickly. Expenses would be high, given the difficulties of supplying food and water to the isolated location. Kattwinkel was prepared to spend 100,000 to 150,000 marks, and intended to travel to the site again himself.[17] W. B. Sattler, of Tendaguru fame, would provide support in the field. Sattler had left the German East Africa Company, and since August 1912 he had managed the Likwaya Plantation in Lindi District. Scientific direction was the responsibility of Dr. Gustav Schulze of the University of Munich.

Reck's discovery of the human skeleton and the resulting publicity had made it impossible for Governor Schnee to reserve the site exclusively for the Berlin and Munich museums. Professor Dr. Georg Gürich, director of the Mineralogical-Geological Institute of Hamburg (Mineralogisch-Geologisches Institut zu Hamburg) acquired funding from the Hamburg Science Foundation (Hamburgische Wissenschaftliche Stiftung) for an independent investigation. Dr. Wilhelm Arning of Hannover also departed for German East Africa, in part to reevaluate this promising new site.

Just a few weeks after his May 5 lecture in Berlin, Hans Reck was again en route to Africa, on a six- to eight-month assignment. The German East Africa authorities had petitioned the Imperial Colonial Office to second Reck as the interim government geologist for the colony. He would replace E. Scholz, who was on holiday in Germany. Reck's primary assignment was to fulfill the government's request to assemble and mount Tendaguru specimens for the National Exhibition in the capital, now scheduled to open on August 15. Joined by Ina, Hans departed Naples on May 14, 1914.

About 100 to 150 loads of bones had been collected for the exhibition.[18] The sum of two thousand rupees was reserved for preparation and mounting, and

Governor Schnee's deputy indicated that the contract might be extended for the proposed Hans Meyer National Museum.[19] Reck assembled a group of his most skilled African preparators in Dar es Salaam and assigned them the task, but it is not clear whether he returned to Tendaguru in 1914, either for further excavation or to ship stored material. In later publications, Edwin Hennig stated that Erich Krenkel was about to reopen excavations at Tendaguru in 1914.

Reck obtained von Branca's permission to take the collection of thin-walled bones that were found in 1912 and 1913 back to Africa. In Berlin, these had been identified as pterosaur remains. Ina Reck had spent long hours preparing them at Tendaguru, and Hans was anxious to begin describing them while in Dar es Salaam. They weighed about 20 kilograms, as did the copious field notes that Hans brought back to Africa.

Many German firms had shipped equipment to be displayed at the August National Exhibition, and hotels imported extra food and supplies to cope with the influx of travelers from other African colonies. These included influential politicians and retired military personnel. None of the expeditions would ever reach Olduvai. Sattler was escorting the Munich contingent from Tanga, and Gürich was in Arusha negotiating for bearers, when they were overtaken by earth-shattering news. A fateful signal had reached the colonial authorities in Dar es Salaam at 6:15 A.M., August 5, 1914. It informed the governor that England had declared war on Germany.

It is unlikely that the long years of destruction and misery that followed could have been avoided, given the circumstances. The new governor, Dr. Heinrich Schnee, expanded upon the comparatively liberal policies of his predecessor, von Rechenberg. He encouraged Africans to become involved in cash-crop economies and drafted laws to prevent the expropriation of their land. Medical services for Africans had been given priority, including a massive smallpox inoculation campaign that had reportedly reached millions. About a hundred schools were operated by the colonial government in addition to 1,832 established in mission stations throughout the colony. The 115,000 Africans enrolled seem like an insignificant part of the total population of seven million, but the Germans were far in advance of their colonial neighbors to the north and south.[20] Forestry officers placed large stands of timber into reserves, and a massive effort was undertaken to prevent the spread of sleeping sickness by controlling tsetse flies.

Economically, the colony was also the most robust it had been since its inception. Schnee and his counterpart in the East Africa Protectorate were of the same opinion—the extension of the European conflict to Africa would destroy the results of their investments. There were those who feared more dire consequences, namely that training Africans in killing Europeans could only result in a general uprising.

The commander of the colony's armed forces, Lieutenant Colonel Paul Emil von Lettow-Vorbeck, held a different view. He had learned the elements of guerrilla warfare during his service in the Boxer Rebellion in China from 1900 to 1901 and the Herero War in German Southwest Africa from 1904 to 1906. Von Lettow-Vorbeck was keenly aware that there was no hope of victory with the miniscule forces at his disposal. A Royal Navy blockade around both Germany and its largest African possession ensured that he would receive no reinforcements of men or materiel.

If used skillfully, however, the German forces could provoke Britain into continuously reinforcing her colonial army, thereby drawing away from the western front soldiers and supplies that would otherwise threaten Germany. The result of this strategy was a five-year campaign in Africa, involving hundreds of thousands of men, pack animals, trucks, aircraft, warships large and small, armored cars, and even a zeppelin. Elaborate feats of arms were undertaken by both sides throughout the colonial war, which ended in suffering and devastation for all participants.

Dar es Salaam was shelled by a British warship as early as August 8. On September 20, the German light cruiser *Königsberg* sank a British light cruiser in Zanzibar harbor. The German warship, requiring an extensive boiler overhaul, sought shelter in the Rufiji River delta. Another bombardment of Dar es Salaam badly damaged Government House, the Club, the bank, the soda-water plant, and the brewery.

In response to German incursions into the East Africa Protectorate, a mixed force of Indian and imperial British troops attacked Tanga in November. District Administrator Auracher, a supporter of the Tendaguru Expedition, was the government representative at Tanga. Contrary to expectations, the struggle for the city resulted in a decisive and humiliating defeat for the far larger British invasion force.

In Europe, the battles of the Marne, Ypres, and Gallipoli set new standards of horror for the civilized world. On March 6, 1915, Eberhard Fraas died in Stuttgart as a consequence of the illness he had suffered in Africa eight years earlier. He had undergone operations in 1909 and 1913, and rallied both times. The effects of dysentery and malaria weakened him, however, and a brief heart problem spelled the end for him at age 52. His 22-year-old son had fallen in the Argonne Forest near Verdun four days prior to his father's demise. Eberhard Fraas was genuinely mourned by paleontologists internationally, such as Henry Fairfield Osborn: "His nature was most genial and those who had the privilege of journeying with him in the field will most keenly sorrow over his untimely death."[21]

During February and April 1915, South African troops pushed into German Southwest Africa. German soldiers completed a physically punishing retreat across the Namib Desert. On June 8, 1915, Hans von Staff, the brilliant geomorphologist of Tendaguru, died of typhus in hospital at Grootfontein. He had commanded a column that drilled boreholes to supply soldiers with water. His constitution had been weakened by forced marches and was unable to resist the disease. Von Staff's wife, Dr. Hella von Staff, boarded a train for Cape Town after the capitulation and obtained passage to Germany three and a half months later. In July, Werner Janensch enlisted in Germany, joining the reserve division of Airship Battalion 2, or the zeppelin service (Ersatzabteilung, Luftschiffbattalion II).

The British Admiralty was determined to destroy the *Königsberg*. A former South African elephant hunter, Phillip Jacobus Pretorius, trekked inland in disguise and pinpointed the German cruiser's location. Nearly two dozen British warships were stationed outside the delta, and flat-bottomed monitors, towed from Malta to East Africa, shelled the *Königsberg* into ruin with the aid of observation aircraft. The Germans salvaged the ten 4.1-inch guns and mounted them on wheeled carriages.

On October 25, 1915, the supporter of all things of scientific interest in the colony, Bernhard Wilhelm Sattler, died. He had served as a noncommissioned militia officer (Unteroffizier der Landwehr) around Arusha. One account stated that he was returning to Lindi District with native troops that had fought in various battles to the north. At Miteja, north of Kilwa, he is said to have been in a camp when a drunken German askari shot him. The shock of his death was felt throughout the country. Tendaguru's first European discoverer was gone at age 42, buried at Kilwa Kivinje.

Experienced South Africans replaced malaria-ravaged British and Indian troops in mid-1915. Starting in December, an unprecedented supply of Allied artillery, infantry, trucks, armored cars, motorcycles, ambulance vehicles, automobiles, aircraft, mules, oxen, and horses poured out of ships docked at Mombasa. Mechanized war was coming to Africa—all resources diverted from the European war, as von Lettow-Vorbeck had hoped.

South Africa's Jan Smuts, the brilliant and charismatic Boer War commander and politician, was chosen to lead a new offensive in February 1916, at the head of an army of 30,000 seasoned South Africans.[22] Significantly, the Allied askari force of the King's African Rifles was expanded. When Portugal and its African colony joined the Allies in March, German East Africa was surrounded by hostile nations.

Smuts resolved to defeat the Germans by pinning them down in strong frontal assaults with his superior resources, while simultaneously enveloping them from the rear with swift flanking movements. Von Lettow-Vorbeck was willing to concede territory to his opponent, but had no intention of being brought into pitched battle where he would surely be destroyed. Instead, he waged guerrilla assaults and ambushes from positions of his choosing.

In mid-March 1916, a second German vessel broke the British naval blockade and managed to unload its precious cargo of war materiel 30 kilometers south of Lindi. Thousands of porters stockpiled this bounty in a guarded dump on the Noto Plateau.

Four Allied armies forced the gates of German East Africa, starting March 1916. While Smuts moved down the Usambara Railway, South African mounted troops raced south to the Central Railway to destroy the German Northern Corps.

The mounted South Africans performed an amazing feat of endurance as they dashed headlong toward Kondoa Irangi, 120 kilometers north of the rail line. Infantry and supply units were left trailing badly. War came to the lives of Hans and Ina Reck as a consequence of this advance. They made their base at

Ufiome, a small mission station that had become a grain supply depot for German troops. Both Reck and geologist Erich Krenkel of Leipzig were initially sent from Dar es Salaam to the interior to organize these supplies. Hans also worked on various geological problems along the Central Railway and in the Kondoa Irangi area. A few months later, Dr. Gustav Schulze joined them in Ufiome, since his expedition to Olduvai Gorge was now impossible. Schulze then joined the German forces in the colony.

On one of his treks to the southeast of Ufiome on June 28, 1915, Hans discovered another deposit of Pleistocene fossils near Minjonjo. The presence of *Elephas antiquus* Recki made him suspect that their age was similar to that of the Olduvai fossils. He returned to the site again on September 21, and sent two carrier loads of fossils back to Ufiome. He hoped to return and unravel the origin of more of these faunas, so reminiscent of Olduvai.

Ina spent her days painting and exploring the surrounding district with men from Tendaguru and Olduvai. Their idyll in this scenic region of mountains and lakes was destroyed forever by the second week of April 1916. The South African cavalry swept down from the north, riding hell-for-leather. The German evacuation was sudden and caused another painful separation for the Recks, as Ina fled alone:

> One evening we were ordered to pack up our nearly 40 loads within a few hours and then we marched away in the middle of the night in the doubtful light of a kitchen lantern. Off we went into the rain-blackened darkness unlit by any moon —the well-mounted Englishmen a few hours behind us, and a totally uncertain future ahead of us.[23]

Hans Reck joined the German forces as a volunteer and probably fought under Lieutenant Colonel Holtz in the defense of Ufiome on April 12. Outnumbered four to one, the Germans were driven off the high ground they occupied, and fought a running battle south into the Masai Steppe in the pouring rain. By the time Hans saw Ina again for a few minutes, two days after her hurried flight, he was already a veteran of two battles. With the exception of two brief meetings, they were not to see one another again for four years.

One of the most astonishing Allied advances of the war in East Africa collapsed completely due to the northern rainy season. The skies opened, the landscape flooded, and the exhausted horses and troopers languished and sickened, so badly had they outrun their supply line, which was by now also bogged down. Horses died by the thousands as logistics broke down.

A combined British-Belgian advance from the west commenced in April 1916. Hans Reck, volunteering in the 23rd Feldkorps, commanded a squad (Patrouillenkorps) of 2 Europeans, 10 African soldiers, and several carriers. Their mission as of mid-June 1916 was to organize local Wagogo people to provide intelligence and offer resistance to the Masai before the eastern length of the railway inevitably fell.

Reck was concerned about the fate of his notes, sketches, and photographs, the result of two years of geological investigation in the colony. Equally important was the collection of pterosaur bones from Tendaguru. As the only pterosaur remains known from the continent they were priceless, and could not be allowed to fall into the hands of the Allies. Burying them was not feasible, nor was leaving them in a village—if discovered, they would become the property of the enemy. His solution was risky, involving an acquaintance by the name of F. G. Rikli. Rikli, a Swiss engineer working for the firm that had constructed the Central Railway, agreed to carry them back to Switzerland, should he be allowed to leave the colony. Hans Reck turned over ownership of fossils, notes, and personal valuables to Rikli on June 24, 1916.

The late northern wet season that halted the South Africans ended Smuts's advance as well, as it was one of the heaviest rains on record. The Central Railway was eventually straddled and Mikesse and Morogoro were taken on August 26. Hans Reck's tiny detachment was ordered to withdraw into the Uluguru Mountains in late August.

Smuts was drawn ever deeper. Allied columns attempted yet another envelopment as they moved through the wild summits of the Uluguru Mountains. Numerous inconclusive engagements were fought through the area Janensch had explored in March 1911, on his quest for Karoo fossils along the Pangani River.

German demolition of the railway and bridges, coupled with sporadic rains, severely hampered Allied motorized transport. Supply lines stretching to Mombasa, 480 kilometers to the north, were near collapse. Allied soldiers faced starvation, and disease ravaged the ranks: malaria and blackwater fever were endemic, and typhoid, typhus, tick fever, and amoebic and bacillary dysentery decimated the soldiers. The tsetse fly destroyed tens of thousands of transport animals. The soldiers eventually reached their breaking point as well, and 12,000 to 15,000 South African and British troops underwent a slow and uncomfortable evacuation to South Africa at the end of 1916.[24] Von Lettow-Vorbeck pulled back to the Rufiji River.

To assist their invasion efforts, the Allies resolved to take control of coastal ports out of German hands. Tanga was occupied on July 7, 1916. Throughout August, Dar es Salaam was pounded by battleship gunfire. Around five thousand shells rained down in 26 separate bombardments, smashing the railway yards and other targets.[25] After a cautious landing by two thousand Allied troops, the capital was occupied on September 4. Kilwa was the next to fall, on September 7. Mikindani was occupied on the 13th and Lindi on the 17th.

A Portuguese incursion across the Rovuma in November 1916 was met by the Tunduru Division (Abteilung Tunduru). When the German force stormed the fort at Mkama its leader was seriously injured, and died in Tunduru on December 1. He was Wendt, the former Lindi district administrator, who was so deeply admired by Hennig and Janensch.

In Germany, Werner Janensch had completed basic training, being promoted from private in January to noncommissioned officer in April 1916. He was posted to a weather station at Varna, Bulgaria, in September, where he served until transferred to Constanza, Rumania, in November. Edwin Hennig's war experiences were mixed with joyful news. The title of "professor" was bestowed upon him and he celebrated the birth of his first daughter.

The Allies finally, grudgingly, admitted the African soldier's value over the European, because of his higher resistance to disease. The King's African Rifles were greatly expanded. The native Carrier Corps was substantially enlarged, by draconian methods, to replace the animal and mechanical transport Smuts had lost. By early 1917, it totaled 135,000, up from the previous 7,500.[26]

Smuts began a big push on January 1, 1917, but was called to London. In the wake of the rainy season, a new South African general began an offensive from Kilwa and Lindi. Lindi District had suffered a serious famine in 1913–1914 and again in 1915–1916, when crops failed because of insufficient rains. Thousands of locals had starved in the latter crisis since the war made it impossible for the governor to transport grain from the rest of the colony. With extensive replanting and regular rain, the situation had improved considerably by 1916–1917.

An Allied buildup was taking place in Lindi, so long a sleepy coastal settlement. Ammunition, food dumps, troops, tents, barracks, even an airfield appeared as if out of nowhere. In the harbor sat vessels of all sizes. The captain of the wrecked *Königsberg* shelled the town so effectively with one of his own guns that in June a British force moved up the Lukuledi River to push him out of the area. By mid-October, the Germans had been driven eighty kilometers up the river. Toward the end of June 1917, the British were pushing on to Nakihu, a familiar settlement in the happier days of the Tendaguru Expedition.

A shrewd Afrikaner who had already performed great services for the Allies by locating the *Königsberg*, P. J. Pretorius, was recruited by the British. From mid- to late 1917, he skillfully fomented rebellion among natives in Lindi District, especially on the Makonde and Mwera Plateaus. He supplied guns to this former Maji-Maji hotbed and overran local farms, taking the German managers prisoner.

Von Lettow-Vorbeck drew closer and bitterly contested the region in a series of bloody and confused confrontations. Plantations along the Lukuledi, where Hennig and Janensch had been welcomed in their search for fossils and peat bogs, were now the sites of carnage. When Allied forces at Kilwa and Lindi linked up, and the Nyasaland-Rhodesia Field Force blocked escape to the west, it was assumed that von Lettow-Vorbeck would lay down his arms and surrender. Instead, he marched further south, invading Portuguese East Africa and drawing Allied resources with him through territory more trackless than any he had previously been through.

Anyone deemed incapable of these punishing treks would be left behind, including the sick and lame. Almost two thousand soldiers and porters stayed in German East Africa to be captured by the Allies. On November 25, 1917, 200 to 300 Germans, including Governor Schnee, 1,700 to 2,000 askaris, and 3,000 porters crossed the Rovuma en route to Portuguese East Africa.[27]

Among those remaining behind was Hans Reck, who was taken prisoner toward the end of 1917. At the beginning of the year, he had led a transport column from the Makonde Plateau to Liwale, the village from which he had launched his investigation of fossilized tree trunks in 1912. In February, he had been relieved of duty due to illness, and his squad had been dissolved about a month later.

After his capture Reck was taken to Dar es Salaam and held in the large POW tent camp on the outskirts of the city. The Pleistocene fossils that he had collected in 1915 were lost during the course of the war, along with many of his field notes. Christmas was spent in these cramped and bitter conditions. His only consolation was a two-hour visit with Ina. She lived in the capital unharmed, though subject to numerous restrictions. Altogether about three hundred women and children and about 60 noncombatant men were relocated to houses in one district, around which were posted notices forbidding entry to Allied soldiers.[28] This quasi-captivity lasted about eight months for Ina, though she spent longer in the city.

The capital had undergone many changes, which not only the Germans found disagreeable. Cars, trucks, and motorcycles, virtually unknown before the war, raised a pall of dust as they roared through the streets, now that the city was the main Allied command center. German residents accused the Allies of looting German property, and similar charges were also leveled by British officers against their own soldiers. Black-marketeering became widespread and confidence tricksters abounded, allegedly reselling army supplies. It was later claimed that Tendaguru fossils brought to the capital for the August 1914 National Exhibition were taken away to South Africa by the occupying soldiers.

Hans Reck spent three days in Dar es Salaam before being shipped to Egypt. At El Ma'adi, south of Cairo, he whiled away two years in a fly-infested prison camp. Reck would have been astonished to hear of Germany's audacious final attempt to supply the Defense Force. A 226-meter-long airship, the L-59, was prepared in Germany and sent to Africa. On November 23, 1917, while floating two hundred kilometers west of Khartoum, the zeppelin crew received a radio signal claiming that von Lettow-Vorbeck had surrendered. Not realizing this was false, they reluctantly turned back to Europe. The dirigible had spent almost four days aloft, having covered 6,757 kilometers.[29]

The United States entered the war, and the October Revolution in Russia took that country out of the conflict. Germany announced unrestricted submarine warfare. Werner Janensch had been transferred once more, to the weather station at Braila in Romania, in March. On November 19, 1917, he was appointed a war geologist (Kriegsgeologe). Edwin Hennig was also a war geologist and was operating in the Carpathian Mountains when some encouraging news reached him. An academic position had opened in Tübingen with the departure of Josef Felix Pompeckj. Pompeckj had replaced von Branca at the Museum of Natural History when the venerable vulcanologist retired.

In addition to the misery of dysentery and malaria, the German Defense Force suffered from waves of smallpox, pneumonia, and meningitis. On September 28, a ragged force recrossed the Rovuma. They had left their colony nine months earlier and traversed perhaps 2,500 kilometers. While recovering in the Songea area, a number of the war-weary native soldiers deserted. The German Defense Force invaded the third Allied colony as it entered Northern Rhodesia in early November, unaware of events in Europe.

On November 13, 1918, captured Allied dispatches indicated that Germany was defeated. The Armistice had been signed two days earlier. Revolution broke out in Berlin on November 9, the kaiser abdicated the same day, and a republic was proclaimed. All this was inconceivable to an army that had prevailed through unimaginable privations for almost four and a half years. On November 25, 1918, von Lettow-Vorbeck laid down arms at Abercorn, Northern Rhodesia. The forces under his command consisted of about 155 Europeans, 1,168 askaris, and about 3,100 porters with their wives and children. While the German officers were interned in Dar es Salaam the Spanish influenza killed 10 percent of the European officers and many more askaris.[30]

From a military viewpoint, von Lettow-Vorbeck's campaign was a brilliant feat of arms. He had recruited into his modest Defense Force only 3,600 whites and 14,600 askaris over the course of the war. With this force, he had withstood an estimated 210,000 to 240,000 men, of which 80,000 were white.[31] The cost of this loyalty and bravery was appalling. Death, destruction, and disruption, followed by political changes, transformed the country. The lives of as many as a million East African natives were directly affected by their harsh treatment in carrier or labor battalions.[32] They were forced to participate in a war about which they knew little and cared less, and their society was disrupted as they were transported to foreign territories. Forced requisitions of cattle and grain, coupled with occasional drought, caused famine. Disease was rampant. Constant movement spread smallpox and Spanish influenza, among other scourges. Between 1918 and 1920, the latter was responsible for 50,000 to 80,000 deaths.[33] The tsetse fly greatly expanded its former range and took a toll of animal and human lives.

Socially, the balance of power shifted in many native areas with the disappearance of German rule. For more than four years commerce and trade had ceased, adding economic stagnation to the physical destruction.

Politically, there were far-reaching changes. In article 119 of the treaty signed at Versailles on June 28, 1919, Germany renounced all rights to her overseas territories in favor of the Allied powers. The European nations that were occupying portions of German East Africa, namely Great Britain, Belgium, and Portugal, proposed to administer the country as a mandated territory under the Covenant of the League of Nations. Belgium retained Ruanda-Urundi, Portugal the Kionga Triangle, and Great Britain the remainder. Thus, everything that Germany had built in Africa was lost to her. Any plans for future developments or scientific studies were shattered.

Ina Reck returned to Berlin by boat, along with other refugees. Hans had a longer wait. On October 25, 1919, he and two thousand other POWs were freed. Janensch was demobilized on November 22, carrying home his Iron Cross, Second Class (Eisernes Kreuz, II Klasse). All were in for a shock. The Germany to which the Recks, Werner Janensch, and Edwin Hennig returned had been forever altered from the country they had known. Wilhelmine Berlin was a thing of the past. Its population had escaped the utter devastation inflicted upon France and Belgium, but the strain of war had nevertheless toppled a regime.

Rationing had begun in 1915. Consumer goods had disappeared or risen dramatically in price as a black market developed. Soup kitchens had appeared in the capital city in the "Turnip Winter" of 1916, as the Royal Navy blockade took hold. The food blockade was maintained until June 1919, resulting in malnutrition among the young and old.

Political factions maneuvered for control of the new republic, causing disorder that eventually deteriorated into virtual civil war. About two million troops returned to Germany within a month of the Armistice, many to be stationed in Berlin. Christmas 1918 saw army artillery firing on mutinous sailors in the heart of the city. Starting in January 1919, the right-wing Free Corps (Freikorps) proceeded to crush the left-wing Spartakists in a series of bloody encounters.

Workers' strikes, involving up to a quarter-million people, were brutally repressed. Martial law was declared in Berlin and was not lifted until December 1919. Unable to maintain authority under these conditions, the National Assembly relocated to Weimar in February 1919.

Unemployment increased to alarming proportions and a stagnant economy began to suffer from creeping inflation as early as 1919. One-quarter of the nation's unemployed lived in Berlin. Staggering national debts had been incurred in the assumption that the cost of war would be paid by reparations extracted from the vanquished. Now Germany was the vanquished nation and the victorious Allies were demanding billions in gold. Widely viewed by Germans as a deliberate attempt at humiliation, the Treaty of Versailles not only stripped Germany of its colonies but also redrew national boundaries. World opinion had shifted from admiration of German achievements to outrage and hatred for the horrors she had inflicted. She was an international pariah.

Such was the scene in Berlin that greeted the veterans of Tendaguru as they were demobilized: a shabby, demoralized city with a hungry, war-weary population yearning for order and relief from shortages.

-8-

1919–1924

The British Museum in Tanganyika Territory

Half a world away from Berlin, events were unfolding that would ultimately lead to a resumption of excavations at Tendaguru. German paleontological successes in the region had not gone unnoticed by scientifically inclined Europeans in East Africa. Closest on the ground were members of the Allied occupation forces, who recommended action to their governments even before the Armistice was signed. Von Lettow-Vorbeck was still campaigning when the district political officer stationed in Lindi, Major Granville St. John Orde-Browne, sent a lengthy note to the curator of the Palaeontological Section of the British Museum (Natural History). Orde-Browne had been an assistant district commissioner in the East Africa Protectorate in 1909, and was a Fellow of the Royal Geographical Society and the Zoological Society.

Dated July 28, 1918, Orde-Browne's letter reported that Major Pretorius of the Intelligence Department knew of fossils at Niongala, an area previously worked by the Germans. Phillip Jacobus Pretorius had been instrumental in locating the *Königsberg* in the Rufiji River delta and had heard of the German excavations at Tendaguru from locals while he was a military scout with the South African forces. He likely passed through the site itself during his last period of action in the war, before he was shipped home due to poor health. He had organized Africans to overrun German plantations in the Lindi area from May to December 1917. Subsequently, he had engaged locals to reopen the quarries or collect stored or abandoned bones, and informed the commander in chief of the occupying forces of the significance of the sites.

Orde-Browne visited two localities on the orders of his superior officer. He brought back a report, photographs, and fossils. The first excavation dating from German times that he discussed measured 6 by 5 meters and reached a depth of 3 to 4 meters in pits with two stepped levels. The second quarry was less extensive. Two grass huts sheltered bones that were labeled in Arabic script. One hut had burned down, ruining the contents. Orde-Browne listed vertebrae, teeth, and partial femora among the contents of the storage buildings. His enthusiastic report was forwarded to the assistant adjutant general in Dar es Salaam and the British Museum:

work should be continued on a small scale, so as to save any of the more exposed bones from the effect of the rains. . . . This will mean an expenditure of about £10 monthly, and I have offered to bear this myself. . . . I would, however, urge the desirability of some expert visiting the spot. . . . Meanwhile, I am doing all that I can to safeguard the specimens, and any work done will be carried out most carefully.[1]

Charles William Hobley also wrote to Arthur Smith Woodward at the British Museum (Natural History) in the summer of 1918 regarding fossil bone around Tendaguru. Hobley was trained as an engineer and geologist, and after lengthy civil service in the East Africa Protectorate, was acting as chief political officer to the British Expeditionary Force operating in German East Africa. In this position he may have received a copy of Orde-Browne's report, but also would have known him personally from the days of their civil service in the East Africa Protectorate.

Hobley had already collected fossils in East Africa for the British Museum. A Miocene proboscidean, *Deinotherium hobleyi*, had been named in his honor. His interest in natural history and anthropology, his profound knowledge of East Africa, and his willingness to offer assistance to any scientific effort recall the same qualities of Bernhard Wilhelm Sattler. Once more, a fortuitous sequence of circumstances ensured that Tendaguru would not be neglected.

The British Museum (Natural History) was the logical choice due to its preeminence among museums worldwide. Natural history museums had developed similarly in Berlin and London, though progress in Britain predated that in Germany. Individuals of wealth and substance in both countries had established private collections of "curiosities" as early as the 1600s. As sciences such as zoology, botany, and geology matured, order was brought to these accumulations of unusual objects.

In England, Sir Hans Sloane wished his collections to remain in London, and in 1753, the British Museum was created through an act of Parliament. Throughout the 1830s and 1840s, the specimens were moved into newly built wings of Montagu House in Bloomsbury. In time, they were housed in a building constructed for the purpose, at the present site of the British Museum in Bloomsbury. By the mid-1800s, these collections were growing as Britain's worldwide colonial empire expanded. As superintendent of the Department of Natural History, Sir Richard Owen suggested to British Museum trustees in 1859 that a separate building for the department was essential. After considerable debate, land was purchased in South Kensington in 1863. Queen Victoria firmly supported the plans of her late husband Albert, the prince consort, to develop the area with cultural and scientific repositories.

Construction had commenced in 1873, though problems surfaced, delaying completion by several years. The Romanesque-style facility was completed in 1880, and officially opened to the public on April 18, 1881. The final cost came to £412,000, or £602,000 when all internal fittings were installed.[2] The attractive new museum possessed a 205-meter-long facade running east and west, with a recessed central entrance. Two towers, each 58 meters in height, flanked the main entrance. At either end of the three-story facade wings was another tower. A 52-meter-long, 22-meter-high main hall ran north at right angles to the east-west wings. Much use was made of natural lighting for the galleries, supplemented first by gas, then by electrical lighting. Terracotta ornamentation featured representations of animals and plants.

The transfer of millions of natural history specimens from Bloomsbury began in 1880, and took six years. The east facade wing housed geology and paleontology displays on the ground floor. Paleontology and mineralogy occupied workrooms in the basement, where the general library was located. Four subject departments each had a keeper, public display galleries, an area for the research collections, workshops, study rooms, laboratories, and a departmental library. Keepers reported to the director, who was appointed by the three principal trustees.

Hobley urged Arthur Smith Woodward, a distinguished paleontologist and keeper of geology at the BM(NH), to exploit the German discovery that was now in British hands. Woodward was aware of the Tendaguru finds, having given a lecture about them in 1912. He approached the director of the British Museum (Natural History) on August 12, 1918, with the draft of a plan:

> a collection should be made of the important fossil bones of Dinosaurian reptiles which are known to occur there in great abundance. The only collections hitherto made in the district are now in Germany, and it would be of great scientific interest to have a similar collection in the British Museum. The Trustees B.M. would highly appreciate the favour if the War Office would . . . employ some military native labour in such work, and to entrust the supervision of the collecting to Mr. A. Loveridge (in civil life curator of the Nairobi Museum).[3]

Arthur Loveridge was a competent field naturalist, at ease in remote camps with African staff, and capable of organizing collecting programs under difficult conditions. Had he accepted an offer to excavate at Tendaguru, the outcome of the British Museum effort might have been very different.

The task of launching an expedition was initiated by Woodward. The approval of several administrative levels within the British Museum was necessary. Agencies responsible for colonial affairs had to be contacted, as did the military government of the occupied territory in East Africa. The war had disrupted Museum affairs, for of the 65 BM(NH) staff members who had served, 13 had died in the line of duty.[4]

Funding had to be secured, equipment had to be obtained, and an excavation program needed field workers. The Berlin Museum of Natural History might send out its curators, but it was not practical for someone of the age and stature of Arthur Smith Woodward to undertake fieldwork in Africa. No one at the Museum possessed the robust constitution and experience essential to collecting enormous and fragile fossils in a remote African setting. Fortunately, Woodward had dealt with a potential candidate for years: a man named Cutler.

William Edmund Cutler was born in London, England, on July 23, 1878. He had spent three years of his late teens in Breslau, Silesia (now Wroclaw, Poland), and had become fluent in German. He later immigrated to Canada, possibly during a campaign by the Canadian government to attract settlers. Three million immigrants, a third of whom came from Britain and British colonies, arrived between 1896 and 1914.

Cutler and a fellow immigrant had formed a partnership to raise cattle on the prairie grasslands of the North-West Territories, destined to become the province of Alberta. In 1908, at age 30, Cutler filed claim for a 160-acre (64-hectare) homestead along Three Hills Creek northwest of Drumheller. For several years, he lived the free and rough life of the range, but the West was changing, and a combination of circumstances ended his ranching days.

Cutler worked in the Drumheller coal mines and picked up seasonal employment by joining ranching and harvesting crews. By 1910, the badlands along the Red Deer River had attracted paleontologists from the American Museum of Natural History (AMNH), and dinosaur fossils began to stream out of the country. In response, the Canadian government hired the fossil-collecting Sternberg family to preserve some of the spectacular Cretaceous legacy for Canadians. Perhaps through meeting with or hearing of the museum teams who worked near Drumheller, Cutler too resolved to make his living by unearthing and selling prehistoric remains. It is also conceivable that he was in contact with the BM(NH). Arthur Smith Woodward had dealt with the elder Sternberg and was well aware of the tremendous potential of the Alberta badlands.

In 1912, Cutler discovered a horned dinosaur with skin impressions. Barnum Brown's AMNH team subsequently removed and described it as the type specimen of *Monoclonius cutleri*. The following year, Cutler approached the Calgary Natural History Society, which contracted him to obtain specimens for a new museum in the city. In April 1913, he was out along Little Sandhill Creek, alone, when he discovered a small hadrosaur. It was very difficult to hire horses to transport heavy plaster jackets but a local man, Albert F. Johnson, helped Cutler move his crated specimens and his camp. Johnson eventually joined Barnum Brown's team and stayed on with the American Museum of Natural History to participate in the Central Asiatic Expeditions to Mongolia.

By mid-July, Cutler had removed about four thousand kilograms of fossils along the Steveville badlands of the Red Deer River.[5] He was fortunate in joining Barnum Brown's team from August 1 until September 23, and absorbed a great deal of practical knowledge of anatomy and museum collecting techniques. He was lucky to be working at all, since the prairie provinces had slid into an economic recession.

Under the sponsorship of the Calgary Syndicate of Prehistoric Research in 1914, he discovered an ankylosaur skeleton, complete with dermal scutes. It, too, was found in the Steveville badlands. Arthur Smith Woodward had offered £50 in support of the work. Cutler broke his shoulder or collarbone while excavating the skeleton and was incapacitated for six weeks.

Later that summer he recovered the right lower jaw of a Cretaceous marsupial mammal near Little Sandhill Creek. Woodward described it as the type of *Cimolestes cutleri* (it was later redescribed as *Eodelphis cutleri*). Eventually the massive ankylosaur was pulled from the ground in a 1,800-kilogram block.[6] It was sold to the British Museum (Natural History), where it was prepared and mounted many years later, and described as *Scolosaurus cutleri* (later redescribed as *Euoplocephalus cutleri*).

Though the work was backbreaking and not especially remunerative, the scientific significance of the objects he collected was important to him. The First World War brought great changes. On June 28, 1915, he enlisted in the 50th (Calgary) Battalion, Canadian Expeditionary Force, as a private. On his attestation papers he listed his trade as "scientist" and was duly registered at 165 centimeters in height, with brown eyes and black hair, 36 years and 11 months of age, with Church of England listed as his faith.[7] His position was a humble one: "for many months [he] washed dishes in the sergeant's mess. According to his comrades, their educated scullery servant was one of the finest fellows they ever knew."[8]

Following basic training he shipped out to London and underwent additional training in the large Canadian camps in the south of England. In mid-March 1916, he was transferred to the 10th Canadian Infantry Battalion and sent to the battlefield in France and Belgium.

In June 1916, the Fighting Tenth, which would gain widespread respect as the Calgary Highlanders, was positioned near the Ypres Salient in Belgian Flanders. Cutler suffered a gunshot wound to the left arm on June 4, possibly during a counterattack on Mont Sorrel, or perhaps from the ever-present sniper fire. The wound must have been serious, as he was evacuated. In September, his draft back to France eventually placed him back in the Fighting Tenth. This assignment was canceled, as he was admitted into brigade hospital with bronchitis. Shortly thereafter, he was seconded to the Canadian Records Office in London, where he spent the remainder of the war.

Cutler was elected a Fellow of the Geological Society of London on December 6, on the recommendation of Arthur Smith Woodward. While in London, Cutler took every opportunity to visit the British Museum (Natural History) and Woodward, its keeper of geology. Threatened with closure first as an economy measure, and later by space demands for the large wartime administrative staff, the Museum never shut down permanently. It enjoyed great popularity among service personnel from all corners of the British Empire.

In February 1919, Woodward asked Cutler to lead the British Museum East Africa Expedition to Tendaguru. On March 20, Woodward recommended that Museum trustees engage Cutler for a period of two years. On April 5, Cutler was demobilized in Calgary, having been overseas for over three years. He had earned the Good Conduct Badge, the British War Medal, and the Victory Medal.

Cutler turned 41 in 1919, and soon made his way back to the Red Deer River fossil fields. Though without sponsorship, he discovered another partial horned dinosaur and began to collect it alone. The work dragged on through the winter of 1919–1920. Camp conditions were terrible, but Cutler was captivated by the desolate landscape around him:

> I well remember one evening in January [1920], whilst in Eo-ceratops camp, the ground and buttes being buried in snow on lesser slopes, and the moon being full,

when in spite of a temperature of 20 degrees below [−29°C] . . . I was only drawn to the fact that I was freezing by the bark of a coyote.[9]

He failed to return to Steveville, a town of 90 people at a ferry crossing on the Red Deer River, to purchase supplies. The townsfolk found him immobilized by serious illness in his tent and brought him back into the settlement to recover. The dinosaur eventually filled fourteen crates and was stored in Calgary, but found no immediate buyer.

By 1921, Canada was in the grip of a serious postwar economic depression and Cutler temporarily gave up trying to make a living by selling fossils. An employment card lists him as an assistant forest ranger in the Crowsnest Pass of southwest Alberta. He also joined the United Mine Workers of America, and may have picked up winter employment in the collieries there as well. Regular strikes rendered this career unreliable, however. There was another strike in 1922, so he returned to Calgary.

Arthur Smith Woodward may have suggested collecting in Manitoba, since he knew of Ordovician fossils outside of Winnipeg. Cutler purchased a small wooden rowboat in Calgary and set off down the first of three rivers that would take him to Winnipeg. He was admitted as a special student for one course at the University of Manitoba. The invertebrates that he found at the Stony Mountain Quarry are still stored at the university's Geological Sciences Department. In late 1922, he was employed in the rail yards of the city. However, there had been considerable activity behind the scenes in London, and his luck was about to improve.

In August 1918, the BM(NH) had pursued the matter of the East African dinosaur site with the Colonial Office in London. In mid-December 1918, this agency requested that the administrator of Tanganyika Territory arrange for the protection of the stored fossils until an expert could be sent out from London. Woodward made his proposal to William Cutler in February 1919. The British Treasury approved an application for £2,000 in June.[10] A year later, the Tanganyikan authorities, having heard nothing, raised the matter again: "When may British Museum expert be expected. Governor General of South Africa is enquiring whether possible secure specimens for South African Museum."[11]

Woodward too, had received no news in the intervening year. In March 1920, C. W. Hobley provided rough cost estimates for ship passage, camp equipment, local labor, rations for workers, excavation tools, packing materials, local transportation, servants, and medicine. Expenses would be substantial, considering that the Germans had spent the equivalent of £11,000 or roughly $50,000 at Tendaguru, exclusive of preparation costs. Hobley suggested other excavation sites in both Tanganyika Territory and the East Africa Protectorate, as alternative locations to keep the Expedition active when the rains fell in the south. Quotes for supplies were requested from various firms.

Woodward met with the trustees of the Percy Sladen Memorial Trust, but they declined any financial support until they received assurances that the South African government would not be providing funds. Fagan, the museum secretary, suggested that Dr. David Meredith Seares Watson, a zoologist at London University, could participate in the venture. Fagan asked if someone in England could do the job, to save the cost of Cutler's ship passage from Canada. Woodward, however, firmly backed William Cutler:

> Mr. Cutler . . . is a most robust man, full of energy and enthusiasm, accustomed by long experience to work in hot sunshine, and specially trained in the handling of Dinosaurian bones, of which he has excellent knowledge. He has also had much experience of moving and packing heavy weights.[12]

There was no question that Tendaguru had attracted attention. The Air Ministry in London had surveyed a route from Cairo to the Cape in 1920. The *Times* of London sponsored an attempt to fly this route. Dr. Peter Chalmers Mitchell, secretary of the Zoological Society of London and special correspondent for the newspaper, was a passenger aboard the Vickers-Vimy aircraft.

The plane departed London in late January 1920, and after several forced landings due to engine problems, crashed on takeoff from Tabora, Tanganyika, in late February. Chalmers Mitchell traveled to Dar es Salaam, where he reported that enormous bones, including a 3.5-meter-long humerus, had been found near Lindi, and that they were being protected for British science. It was emphasized that a skilled collector was urgently required, but the project stalled. Throughout 1921 and most of 1922, no correspondence in the British Museum East Africa Expedition file indicates any progress.

Postwar conditions in Britain were not conducive to expenditures on ventures such as vertebrate paleontology in Africa. While 1919 was a year of economic growth, as millions of returning veterans were absorbed by war-expanded industries, production dropped abruptly in 1920. Working hours and wages were reduced. Foreign competition made it difficult for some traditional industries to maintain their prewar market share. Almost two thousand strikes took place in the U.K. that year.[13] By mid-1921, the country slid into a severe economic depression during which 2.4 million people lost their jobs.[14] Servicing and reducing debts incurred during the course of the war now consumed about 40 percent of the government's annual budget.[15] Despite the postwar dreariness which affected the country, there was some economic improvement by 1922.

The thread of Tendaguru continued. The protocol of proceeding through the Colonial Office delayed progress. In late December 1922, Sir Sidney Frederic Harmer, the 60-year-old director of the British Museum (Natural History), again wrote to the secretary of state for the colonies. Harmer had been named director of the Museum in 1919, and the transition in directorship caused additional delays. Harmer was known for deliberate attention to detail.

Harmer's letter to the administrator in Tanganyika, Sir Horace Byatt, included an apology for the gap of almost three years, from February 1920 to January 1923. It also posed a number of logistical questions. Governor Byatt replied that

> certain small excavations were made in that neighbourhood during the course of the campaign in this Territory; but in view of transport difficulties it was decided to cease operations and place the remains already discovered in safety and under cover.

. . . These fossils, which are approximately 300 in number and weigh about half a ton, are at present stored in the Administrative Office at Lindi. The site is now completely overgrown: and therefore your representative would have to deal not only with the remains already secured, but would have to undertake fresh excavation.

The governor assured Harmer that tools would be loaned by the Tanganyikan authorities, that Museum supplies would be exempt from customs duties, that local steamer passage would be available at reduced fares, and that labor costs were sixpence per man per day.[16]

In January 1923, Woodward recommended that the South African Museum in Cape Town be approached to join in the work and share in the costs. Three months later, its director declined to participate due to "grave financial stringency" in the Union of South Africa. Intriguingly, he also mentioned that the museum possessed "a fair quantity of fragments of large limbs etc. etc. part of a pelvis etc. of possibly among others Brachiosaurus brancai, etc. . . . . boast(?) of a 214 cm humerus."[17] The Transvaal Museum in Pretoria also pleaded inadequate resources.

Harmer raised the possibility of American participation but was opposed by Woodward. The largest private American museum, the AMNH, was currently sponsoring the ambitious Central Asiatic Expeditions to Mongolia.

In July, Harmer came to a formal arrangement with Cutler, more than four years after Woodward and Cutler had first discussed Tendaguru. Cutler would be expected to leave Canada around mid-February to start excavating before the beginning of May 1924, in order to maximize the time available in the dry season. He was offered a salary of £25 per month (£300 per year), plus traveling expenses and a £10 per month "subsistence allowance."[18] The collections were to remain the property of the British Museum and no publications or press releases were authorized unless approved by the Museum. Cutler's first task would be to select and ship specimens from those already in storage and then commence further excavation for dinosaur bones for a period of 12 months.

Wages and rations for excavators, bearers, an interpreter, and a headman, about 50 men in total, were estimated at £50 per month. When Cutler's salary and allowances were added, the monthly operating expenses totaled about £100. Equipment, medicines, and lifting tackle was estimated at £220, and traveling expenses for Cutler from Winnipeg to London to Dar es Salaam and back added another £490.[19]

Fluent in German, Cutler read Edwin Hennig's popular account, *Am Tendaguru,* and was impressed by the scale of that operation. Cutler intended to employ the same methods at Tendaguru that he had used in the badlands of Alberta. Conditions were considerably different, however. He would have to modify his approach in the same way Eberhard Fraas had, when faced with the contrast in geological exposures and collecting techniques between the American West and Tendaguru, or Hans Reck, who moved from Tendaguru to the very different Olduvai Gorge.

Cutler and Harmer were concerned about transporting specimens from the quarries to the coast. Orde-Browne confirmed that having them carried by bearers was still the sole method available. Hobley concurred for the most part, but also thought that a Ford van could be used on a portion of the route.

In early November 1923, Harmer confirmed to William Cutler that the Expedition was approved. He included a sobering qualifier: "I am afraid that with the funds at our disposal we shall have to carry out the work on a much smaller scale than that you indicate as having been done by the Germans."[20] Funding for the Expedition came exclusively from the Museum's Purchase Grant, money the Treasury allocated for the purchase of specimens.

Despite the lack of external partners, the Museum continued with its plans. In November and December, Assistant Secretary Smith contacted several shipping companies to request reduced transit costs for Expedition members and gear, and discounted freight charges for the field jackets from Africa. Agents for the British India Steam Navigation Company offered a 10 percent reduction on passage and Expedition equipment, and a 50 percent reduction on return shipment of specimens from Mombasa to London.[21]

The membership of the Expedition was still unsettled at the end of 1923. It was hoped that a paleontologist or zoologist could be found to accompany Cutler. Marquess Curzon of Kedleston sought a position with the Expedition for a young relative, and raised questions regarding Cutler's character. Curzon was the former viceroy of India, a Conservative member of the House of Lords, president of the Royal Geographical Society, chancellor of Oxford University, and foreign secretary from 1919 to 1924. He was narrowly defeated for the position of prime minister in 1923. Harmer passed the question to Woodward with alacrity. Woodward remained steadfast in his support of Cutler:

> If companions are sent, Cutler should have absolute discretion to dismiss loafers + hindrances. He would have no more respect for an Oxford graduate than for a western cowboy if the man were loafing and incompetent. . . . I know nothing of his leadership—I only know he would not suffer fools.[22]

Lord Curzon attached his relative to another expedition, and the matter was dropped.

Shortly thereafter, another individual stepped forward to express interest in joining. Captain William Hichens had served in the Intelligence and Political Service of the East Africa Protectorate and was for a time assistant political officer of Lindi District. He mentioned that he had personally carried out excavations in the Tendaguru area.[23] Perhaps Hichens had been sent to Niongala by Orde-Browne. But Hichens disappeared from the scene, the only surviving comment indicating that he suffered some sort of speech impediment.

Museum officials assembled quotes and placed orders for gear. Those suppliers that had been approached in 1920 were again asked to provide current prices at the end of January 1924. The Hardy Patent Pick Company of Sheffield sent listings for crowbars, axes, billhooks, shovels, hoes, picks, and mattocks. Herbert Morris Ltd. offered heavy lifting tackle. The Export Department of the Army & Navy Co-Operative Society quoted on provisions. Burroughs Wellcome priced out the medical supplies. Richman, Symes & Co., Colonial Agents,

Merchants and Shippers, responded to requests for a myriad of camp equipment as well as surveying apparatus. A. W. Green and Jonathan Fallowfields each provided information on suitable cameras. Harrod's was asked to supply burlap. Sutton, Carden & Co. offered to supply industrial methylated spirit to preserve zoological specimens. James Gregg & Co. sent estimates for manila hemp slings.

On February 2, Cutler departed New York, bound for Liverpool. He was expected to arrive in London with sufficient time to check the equipment and remedy any deficiencies before leaving for Africa. The embarkation to Dar es Salaam was set for February 21, 1924.

As late as January 27, 1924, C. W. Hobley suggested another possible Expedition candidate, a young man born in Kenya who was well acquainted with African languages and customs. Two days later Assistant Secretary Smith asked the man in question, L. S. B. Leakey, whether he was interested in the Expedition.

Leakey was interviewed at the Museum and, on February 3, was accepted as an assistant to William Cutler. Leakey's salary was set at £7 per month from March until about December 1924.[24] He was fortunate to be offered that much, since all previous discussions of assistants for Cutler stipulated that only unpaid volunteers could be accepted. So, like Hennig (who had been precipitately assigned to the German Tendaguru Expedition), Leakey was selected at the last moment.

Louis Seymour Bazett Leakey was born at Kabete Mission, fourteen kilometers from Nairobi, Kenya, on August 7, 1903. His parents were members of the Church Missionary Society. Louis grew up playing with local Kikuyu children. On his parents' second leave to England, in 1910, Louis was enrolled in a formal school for the first time. By May 1911, the family was back at Kabete. The Leakey children were taught Latin, mathematics, and the Bible. The mission was always home to a group of dogs, monkeys, bush babies, mice, and other wild animals.

At the outbreak of war, the Leakeys were forced to cancel another leave to England, and spent a six-year period in Africa that truly formed Louis's character. He went for hikes with his siblings and their tutor, and was encouraged in his interest in natural history by his father, who was similarly inclined.

Louis turned 11 just after the war began, and became acquainted with another natural history enthusiast who had recently moved to Nairobi, Arthur Loveridge. This was the same Loveridge who in 1918 had been recommended to supervise excavations at Tendaguru. An expert zoologist, he frequently visited the Leakeys, making a tremendous impression on young Louis. The boy was determined to become an ornithologist. From the many Kikuyu boys of his age group, he learned to trap wild birds and animals to either prepare their skins or sell them to zoos. Loveridge then added patient lessons on classifying and preparing specimens for museum collections. After receiving a book on prehistory as a gift, Louis vowed to make archaeology his life's work.

In the summer of 1919, Louis's African idyll came to an end as the family sailed home to England. He was enrolled in a boys' public school in Dorset.

Louis was 16½ years old and miserable. He simply did not fit in to the conventional British school system after his freedom and independence in Africa. His dream was to study anthropology at Cambridge, and after struggling with many unfamiliar subjects and financial difficulties, he was awarded a scholarship. October 1922 found him at St. John's College, where he began his first year as an undergraduate.

Leakey was trying to qualify as a Cambridge Rugby Blue in October 1923 when he suffered severe blows to the head during a game. He was advised to postpone his studies for a year, a catastrophic break in his plans. Louis did not have the money to take a year off, and approached a family friend, Charles William Hobley. Hobley suggested the British Museum East Africa Expedition to Tendaguru, and Leakey was hired.

A common postwar problem was about to throw the breathless arrangements off track. News arrived that the ship would sail a week or even more before its scheduled departure date to avoid a threatened dock strike. As a result, Leakey and Cutler would be inoculated for typhoid aboard the vessel.

Leakey arrived at the Museum on February 12, and Cutler on the 13th. This left Cutler a week to inspect and finish organizing the supplies. He immediately ordered a thousand pounds (450 kilograms) of plaster of Paris, divided into metal tins. There was no hope of assembling all the required materials in time for the sailing date. A set of German scientific results published in *Archiv für Biontologie* was ordered mid-February. The special picks that Cutler wanted were not available in London. The Museum telegraphed Dr. William Diller Matthew, curator at the American Museum of Natural History, to have them sent from New York. Matthew replied that the picks, originally designed for O. C. Marsh, were no longer available, but sent one as a sample.

On February 14, 1924, came the news that the vessel was leaving in two days. Since Cutler's preparations were incomplete and the Museum staff had been unable to procure all the necessary gear, the booking was canceled. A flurry of letters and telephone calls were required to explain new arrangements. Cutler and Leakey were then booked on a Messageries Maritimes French mail steamer. Carrying their personal baggage with them, they would take the train to Marseilles and board the ship on February 28.

Some equipment had already been loaded on the first vessel, so Governor Byatt was asked to arrange for its unloading in Dar es Salaam. Excavation tools that had been sent to London from Sheffield had missed the boat and now had to be retrieved from storage at the Royal Albert Docks. A considerable portion of the gear had been neither ordered nor delivered. Cutler was assured that heavy gear would be forwarded with the first steamer, and should reach him within two weeks of his arrival in Africa.

Advances of £100 and £25 were forwarded to Cutler and Leakey respectively. To cover further expenses in the field £500 was deposited with the Standard Bank of South Africa in London, which would wire it to Dar es Salaam.[25] Finally, the first-class rail and ship tickets were handed to Cutler and Leakey. They met with Harmer on February 26, and were wished the best of luck.

It had been a hectic few weeks and the dockworkers' strike was unfortunate, but high hopes traveled with the Expedition members as their train pulled out of London on February 26, 1924. The British Museum East Africa Expedition was officially launched.

-9-

1924–1925

Cutler, Leakey, and a Difficult Start

A pair of strong personalities were suddenly thrown together on the long ocean voyage to Africa. Cutler, the experienced collector, was accustomed to operating independently, often under demanding physical conditions. A self-made man eager to contribute to scientific knowledge, he was 45 years old. Louis Leakey, equally independent, unconventional, and enthusiastic, was described as very self-confident and intense. He was 20 years old.

Both men had had unusual upbringings, having spent years outside of England. They wished to be respected for what they had accomplished—Cutler for his hard-won expertise in dinosaur collecting and Leakey for his intimate familiarity with Africa and Africans. Pride, a 25-year age difference, and the disjointed start to the Expedition created profound challenges.

Conditions in Tanganyika Territory had slowly improved. Horace Archer Byatt had become the first governor in 1920, the same year that "Tanganyika Territory" was adopted as the area's official name. Britain's weak economy and ill-defined responsibilities to its overseas mandates made it difficult for Byatt to reconstruct this war-ravaged country. A public works plan was developed to rebuild bridges, rail facilities, ports, and public buildings. In Dar es Salaam, the rail yard engines that supplied electric power were overhauled. Rolling stock and locomotives were reconditioned or replaced, the floating dock was salvaged, and ocean liners like the *Feldmarschall* were repaired and pressed into service.

Administrative order and legal process were reestablished with the help of government officers from neighboring colonies and the armed forces. A police force was created, and customs and immigration departments were soon in place. In 1922, the East African shilling replaced the German East African rupee as legal tender.

Under the terms of the Treaty of Versailles, all German settlers and missionaries were expelled by 1922. Their property was auctioned off and the proceeds applied toward the war reparations levied on Germany. Adding to Byatt's difficulties was a postwar slump that was keenly felt in East Africa, where most countries depended on high prices for agricultural exports. These conditions prevailed until late 1923, and Byatt's tenure is remembered as a period of economic stagnation. The Territory was widely regarded as a backwater at the time the British Museum East Africa Expedition arrived.

In the capital, Cutler encountered problems clearing Expedition gear through customs. Armed with letters of introduction to Tanganyikan government officials, Leakey convinced an officer to review the regulations. The governor was empowered to grant exemptions, and the Expedition gear was allowed to enter without hindrance. No one appeared to be aware of the arrangements that Harmer had discussed with the governor in January 1923.

While Cutler waited in Dar es Salaam for missing supplies to arrive from London, he directed Leakey to build a camp at Tendaguru. The first crates to arrive carried a dozen provision boxes. Each contained Indian tea, sugar, matches, Danish butter, French sardines, herrings, sausages, roast beef, jam, potted meat, salt, pepper, mustard, marrowfat peas, headcheese, green peas, Lazenby's soup cubes, corn starch, semolina, baking powder, Lea & Perrins sauce, tinned fruit, Oxo bouillon cubes, cheese biscuits, tongue, cocoa, and candles.[1]

Another consignment, of 22 crates, included excavating tools like shovels, billhooks, axes, hoes, and picks. To maintain the implements, a portable forge with tongs and an anvil was shipped. For collecting and preserving fossil bones and modern plants and animals there were shellac, burlap, methylated spirit, and Jeyes fluid; there were also shotguns, shells, and a botanical collecting outfit.

Camping gear included a 3.3 x 2.7-meter ridge tent with separate canvas bathroom, verandah, and groundsheet, folding canvas cots, mosquito nets, mattresses and pillows, bed sheets, a table and folding chairs, a bath and washstand, canvas buckets, and tarpaulins. Oil lanterns would supply light. Cooking gear was packed in a wooden Venesta box and comprised aluminum saucepans, a kettle, and a frying pan, as well as water containers. A wicker "tropical tiffin basket" held enameled plates, cups, and saucers, a teapot, cutlery, glass jars, a can opener, and a corkscrew. To prevent scurvy, lime juice was supplied. Miscellaneous Expedition equipment included binoculars, a thermometer, and a magnifying glass.[2]

Cutler carried with him a helmet, wool bush shirts, blankets, knives, and a Colt .38 pistol, as well as a butterfly net and a tin of cyanide crystals for specimen collecting. The ubiquitous felt spine pad, a colonial-era device considered essential for preventing heat stroke, was supplied in triplicate.[3] The Expedition's progress was to be recorded with a Sanderson hand-held plate camera with an Aldis f6 lens and a tripod. A complete darkroom setup would allow Cutler to develop and print the films in the field.[4]

Another dozen crates contained rope slings, double pulley-blocks, manila rope, and plaster of Paris.[5] The abundance of equipment was a contrast with Cutler's days in Alberta, when he worked without assistance and pleaded with sponsors for supplies. The gear was transshipped to smaller coastal steamers, like the S.S. *Dumra,* a 2,090-tonne cargo and passenger vessel that plied the Mombasa-to-Mikindani route.

When not dealing with the shipments, Cutler walked along the beach and picked up corals, gastropods, and bivalves. The invertebrates filled four crates, which were stored with the African Mercantile Company.

Leakey, who had been expecting Cutler for two weeks already, had been productive. On his arrival in Lindi in late March, Leakey had outfitted himself with a camp cot, a mosquito net, lanterns, food, and cooking pots. He was billeted in the German Club, which was in poor repair after many years of disuse. Reading of Boheti bin Amrani in the volumes of the *Archiv für Biontologie,* Leakey made inquiries in the African district of town. Boheti arrived shortly. "He was not in very good health and said that he did not think he could come up with me and show me the site, but he gave me a helpful description of how to find it."[6] The jumbe of the region around Tendaguru, Ismaeli, did not know the location of Tendaguru, as he had not resided there before the war, but agreed to help search for the site and select men to build a camp. Within two or three days, 15 porters and a cook were hired.[7]

The safari left Lindi on April 17, 1924, and Leakey stood atop Tendaguru Hill on Easter Sunday, four days later. He sent carriers to scour the area, and soon had a selection of bottles and rusted tins, confirming the site of the German camp. A spot was cleared for the tent, and a drum was procured from a nearby village to signal that laborers were needed. Leakey offered to hire men at 14 shillings per month plus rations.[8]

> It was with a strange mixed feeling of pressure, triumph, expectation and loneliness that I retired to my tent that night. I was more than fifty miles away from the nearest white man, in the heart of a district known to be inhabited by lion, leopard, elephant, and many other kinds of game. I had only a few natives with me, and although I was but twenty, I was undertaking a responsible job for the British Museum.[9]

In the course of the next two and a half weeks, Leakey organized the camp. He ordered a dwelling measuring 11 by $3^{1}/_{2}$ meters to be built partway up Tendaguru Hill.[10] The structure was divided into three rooms of equal area, with a bedroom at either end for Leakey and Cutler respectively and a living room between them. A large expanse at the base of the hill was cleared for the workers' huts. A water source was located after a few afternoons of searching.

Leakey made two more round trips to Lindi, picking up stone implements on his journeys. He received orders from Cutler to clear a 2.4-meter-wide roadway from Lindi to Tendaguru, an impractical assignment. Instead, he requested the assistance of Wyatt, the provincial commissioner, in recruiting 50 porters to carry the Expedition gear that had begun to arrive. Leakey was granted authority to hire locals at six shillings per day to clear trails.[11]

William Cutler steamed into Lindi nine weeks after arriving in Africa and soon met the master overseer, Boheti. "I observed that he is a light-coloured tall thin native, with a clever lean face, of Arab type. Contrary to my expectations, he speaks neither English nor German but he certainly gives the impression of a very intelligent native + one who knows the region."[12] For almost two weeks, Cutler remained in town, picking up fossils and speaking to officials about a road to Tendaguru. When Cutler had been interviewed in New York before leaving for London, he had spoken of laying down tracks for a miniature rail line.

Cutler soon identified a stumbling block that would trouble the Expedition, namely the level of commitment the Museum and the mandate's authorities had to the venture:

> The administration will certainly not be able or willing to spend money on making us a <u>road,</u> but . . . they would feel that it were more justifiable, did they know our stay in the territory would extend over . . . say two years, to justify such improvements. This region is <u>not</u> of much importance to them to warrant anything more than native tracks, but they are fully alive to the outside interest + notice that will be focussed upon the region, if the expedition has sufficient time + backing to get results.[13]

Another issue would result in a strained relationship between the Museum and its field party. Cutler dismissed the bone remnants that lay in a Lindi storehouse as not worth shipping to London. They likely lacked provenance data, rendering them all but useless scientifically. Cutler was adamant that he would neither ship nor collect broken fragments of bones, as he thought the Germans had. Yet it was this material upon which the Museum was relying to justify the initial expenditure of the Purchase Grant.

Leakey returned to Lindi and on June 11, 1924, a string of 46 Africans, including three women, a baby, and a boy, followed William Cutler and Louis Leakey out of the port.[14] Cutler had spent three weeks around town. Leakey had reached the area nine weeks previously and was on his fourth trek to Tendaguru.

Cutler's curiosity about natural history and unwillingness to heed his young assistant's warning about some seed pods led to a nasty experience. Leakey made light of the incident:

> Mr. Cutler . . . had apparently picked a number of the pods and tied them up in his handkerchief. . . . we were all very tired and hot and sat down under the shade of a mango tree to rest, and Mr. Cutler proceeded to take the pods out of his handkerchief and then wipe his face and neck with it. In a moment the little hairs were all over him and he was dancing about in agony, yelling like a madman and cursing like a trooper.[15]

The caravan stopped at the foot of Tendaguru Hill on June 13. Although he was laconic in his notes about reaching the famous site, almost three months after setting foot on African soil, Cutler was impressed with the camp his young assistant had organized. In addition to the living quarters, complete with slatted bamboo blinds, there was a large kitchen and storehouse, a darkroom for developing film, a grass outhouse, and disposal pits for refuse.

Engrossed by his surroundings, Cutler recorded plants and animals in one notebook and geological observations in another. Butterfly collecting became a passion. Despite his interest in natural history, Cutler displayed indifference to animal suffering when he killed a hornbill in a grisly fashion. He offered no explanation, but Leakey, who had aspired to become an ornithologist, was surely repulsed by such callousness.

The neighborhood around Tendaguru Hill was scoured immediately. Recent heavy rains made it impossible to set fires to eliminate the grass, which grew up to three meters tall. Including personal servants, there were sixteen Africans on the payroll.[16]

The pragmatic Cutler was convinced that it would be impossible to move heavy specimens through such rough country by using human bearers alone. He wanted to apply North American techniques of fieldwork. The Sternbergs and Barnum Brown had exposed bones of an articulated or semiarticulated skeleton, separated the skeleton into manageable sections, and encased the blocks within burlap and plaster jackets. These blocks were dragged to more accessible ground on a wooden sled by horses or manpower. The field jackets were crated and the crates loaded onto wagons with tripod and block and tackle. Teams of horses hauled the crates to railroad sidings.

Janensch, Hennig, and Reck had adapted to African conditions. They had divided up even articulated portions of dinosaurs into much smaller units, marked adjoining fragments of bone, meticulously labeled the jackets, and drafted accurate quarry maps. Very heavy loads such as complete sauropod cervicals had been packed out over rough terrain using manpower alone, alternating teams of bearers.

Leakey developed dysentery while treating an outbreak among the African crew. It was compounded by an attack of malaria. He lay on his cot with a 40°C fever, and was vomiting and passing blood. Cutler sent a runner to Lindi with a request for Dr. Blackwood to provide medicine and advice. A frightening incident occurred while Leakey was incapacitated:

> about midnight I suddenly heard footsteps in the living-room that divided my bedroom from Mr. Cutler's. Thinking it was my chief I tried to call out to him to bring me a glass of water, as the fever made me terribly thirsty. . . . I then heard deep regular breathing and realised that it was not Mr. Cutler at all but some wild animal. . . . The next minute a leopard leapt on a little pet baboon of mine that was sleeping by my bedside and jumped away with it through the open window.[17]

Satisfied that he had pinpointed several promising sites, Cutler put eight men to work in Ditch I on June 22, ten days after his arrival at Tendaguru. The quarry was a few hundred meters to the right of the narrow path that led to Tingutinguti Creek. After six days, the men had cut a trench 5 meters long, 2.5 meters wide, and 2 to 3 meters deep.[18] Not one bone had been recovered from the hard white sandstone.

The crew was shifted to Ditch II, along the road to Matapua, about three kilometers or half an hour east of Tendaguru. Large bones lay on the surface near a German quarry. Despite the efforts of 10 men, progress was disappointing and the site was abandoned on July 12. Dissatisfied, Cutler reopened the trench three days later. Quarries were often shut only temporarily, and when this one was restarted it rewarded Cutler's perseverance. In 13 days, an average of eight men dug an area measuring over 13 meters long by 6.5 meters wide by just over 2 meters deep.[19] Another limb bone and ungual came to light. The total haul: a metatarsal, a vertebral centrum, an ungual, a femur, a humerus, a metapodial, two bones which resembled the radius and ulna, and sauropod and theropod teeth.[20] Cutler experimented with dynamite on July 25 to loosen overburden. Ditch II, restarted on August 5, was again abandoned on September 27.

Cutler had his suspicions about the quality of the work in Ditch II, noting that the best bones were incomplete, with ends displaying clean fractures. Yet he was proud of the extent to which his crews had worked the site: "Judging by their ditches here, with few exceptions, the Germans did not begin to pursue their excavations with thoroughness as our ditch two, of richness, borders one of their long very shallow trenches."[21]

Ditch III was opened on June 29. This quarry, northwest of Tendaguru on the way to the village of Ruanika (possibly the Dwanika of German times), was also near another old German excavation. A hefty limb bone almost one meter long was found on the surface. Some time later, an ilium was uncovered. When Ditch III was abandoned on July 25, it had been open for seventeen days under a crew of five men. They had removed overburden from an area 8.5 meters long by 8.4 meters wide by 1 to 3 meters deep.[22]

A headman and six others opened Ditch IV on July 14, at the site of weathered bones along the Nguruwe trail, 2½ kilometers south of camp. A 1.34-meter-long humerus and a 1.7-meter-long femur were found. A large sacrum was uncovered and plastered. The main block was undercut and overturned on September 18, to expose a large pubis. Four men spelled by another four carried this jacket, the heaviest yet, into camp.

The northern end of this quarry was expanded, revealing a 1.8-meter scapula and ilium near the elements of a forelimb: radius, ulna, and humerus. Metatarsals as well as ribs and possibly an ischium were revealed later. Twenty bones were exposed by September, and Cutler thought they indicated the presence of two animals. When Ditch IV was extended outward on three sides, a layer of overburden between .75 and 1.5 meters deep was shifted.

Work was begun on Ditch V, 1.2 meters from Ditch IV, on July 19. Like all the others to date, it was near an older German quarry. Eight men set out with their hoes and shovels. An enormous heap of waste dirt was produced when this quarry was connected to Ditch IV. When a series of cervical vertebrae appeared in Ditch V, excavation proceeded toward this pile.

Ditch VI was shut down on August 2. Other than surface finds of one dorsal and four caudal vertebrae and a metapodial, the 5.4-meter-long, 2.7-meter-wide, 2.4-meter-deep quarry was barren. An average of six men had toiled seven days.[23]

Ditch VII was commenced on July 31st, about 10 minutes southeast of Tendaguru, between the paths to Lindi and Matapua. The new quarry was an extension of a German excavation. Five men were assigned, and the ditch was abandoned on August 7 after one week. The pit was 5.4 meters long by 2.5 meters wide by 2.5 meters deep, but the haul was disappointing: a small caudal vertebra and a couple of bone fragments.[24]

Ditch VIII was located behind Ditch V and again in the vicinity of German diggings. Eight men worked for three days, but when absolutely nothing was found Cutler pulled them off on August 9. Its area, like that of so many others, was substantial: 17 meters long by 3.3 meters wide by .6 meters deep.[25]

Ditch IX, immediately behind Ditch II and at right angles to Tendaguru Creek, was begun on August 8 with a crew of five. Cutler closed it on August 19, after seven days. Only a barren hole 8.5 meters long by 3.6 meters wide by 1.3 meters deep had been produced.[26]

Ditch X was opened between the paths to Lindi and Nguruwe. There was a German quarry on the southern side of the trails. A worker named Marco was placed in charge of the crew of eight men. The pit was of impressive dimensions—17.5 meters long by 2.4 meters wide and from .75 to 2.5 meters deep—when it was closed after eight days on August 19. Only an ungual phalanx, a carnivore phalanx, and a few fragments were found.[27]

Ditch XI was only 32 meters from Ditch X, to the right of the junction of the Lindi and Nguruwe roads. It was investigated for 35 days, by about 10 men. A man named Juma directed 20 to 23 men here for a few days in late October before XI was closed for good. The site gave up nothing beyond fragments and bits of ribs.[28]

Ditch XII, 46 meters from Ditches VIII, IV, and V, was begun on September 8 on the promise of eight caudal vertebrae found on the surface. Eleven men were employed, but after 16 days of removing overburden, the quarry was abandoned as barren.

All these excavations followed the same pattern. Billhooks were employed to clear surface vegetation, and crews with picks and mattocks moved in to excavate. Trenches were cut in a diagonal or zigzag pattern, and the earth shoveled into metal pans or kerais, to be carried off and dumped. Wheelbarrows were purchased but were never the success they had been for Hans Reck, and were abandoned. Once the bone level was reached, a larger area was excavated to the same depth. Those who were judged to be the most careful workers now took over to expose and pedestal the finds using awls, knives, and brushes.

At this stage Cutler decided which specimens to collect by plastering and which to "parcel." Parceling fell to Leakey. He wrapped small bones first in soft paper, then brown paper, and finally labeled the package with the quarry number. As the Expedition's leader, Cutler insisted on plastering alone, once the specimens had been preserved in shellac and covered with tissue paper.

Larger bones were plastered in sections, with a small gap left between the edges of adjoining jackets. Once pedestaled and overturned, the partially encased bone was broken at the gaps in the jacket, and the individual sections completely plastered. When hardened, the jackets and parcels were carried to camp on litters and placed on shelves in the bone storage hut. Despite Cutler's fears, even heavy jackets could be dragged out of quarries to Tendaguru by means of a wooden sled.

Cutler complained of being limited to working five quarries at once, whereas the Germans had opened as many as ten simultaneously. He argued that the

Germans had been able to more quickly reject a unit that was not producing, and move onto other areas for greater productivity. Cutler attributed their superior ability to explore Tendaguru to exceptional support from their colonial officials.

Leakey had hired nearly a hundred laborers, whose long-distance migration across the country had resumed after the disruption of the war. Men were arriving from as far away as Songea, a 30-day march, only to be turned away. The camp ran short of foodstuffs until a group brought produce from Matapua, five hours to the east, and Mandawa, 29 kilometers to the north. Cutler was greatly relieved: "saved the situation as the men were severely grumbling at short rations + having to go hunting food for sale, after a day's work."[29]

Runners were dispatched to settlements up to 64 kilometers distant to offer money for food, but this effort was only temporarily successful. As deliveries of millet, cassava, and maize diminished, the Expedition offered cloth and other items in exchange. The British Expedition seems to have been facing the same problem the Germans had: economic conditions made it more profitable for African farmers to sell their crops to Indian businesses on the coast.

The water source at the rock pools was shrinking visibly, and the most distant source was "polluted by leaves + boys washing clothes."[30] Leakey described the precarious situation:

> Of the three pools, one was set aside for washing purposes, and . . . the men who wanted to wash had to walk ten miles to this pool. . . . A second pool was set aside to supply water for cooking and drinking. Each day a gang of fifteen men walked over to the pool and came back in the evening, each carrying five gallons of water in a can on his head. Of this 75 gallons a day, a part had to be set aside for the plaster of paris work, and the remainder was divided out among ourselves and our workmen. The third and smallest pool was left as a reserve supply for emergencies.[31]

Though no one would perish for want of water, the scale on which the excavation could proceed was directly related to the availability of water.

The population of Tendaguru in mid-July included Cutler and Leakey, their two personal servants, a cook and assistant, a headman, a bird collector, three boys, three water carriers, 25 excavators, four bone experts, and the game scout Salimu, for a total of 42.[32]

July 23 was William Edmund Cutler's 46th birthday. Marco returned breathlessly from Lindi with 16 letters, including one from E.M.L., a woman to whom Cutler referred frequently in his field journal. It had been a satisfying day: "Gave the men some bananas + the boys some fish, being my birthday, it being customary in the East to <u>give</u> upon one's birthday. Leakey presents me with a box of chocolates."[33] This was a thoughtful gesture, as the gift must have been purchased on the coast from Leakey's meager allowance.

A daily routine was established, in which Leakey rose at 5:00 A.M. and beat a drum to signal the start of work. Half an hour later, he took attendance, distributed tools from the storage hut, and assigned crew leaders their tasks. While the laborers marched off, Leakey offered medical treatment to the sick. He then left for the quarries, returning to camp at 8:00 A.M. to breakfast with Cutler. The two men would trek to a quarry and shellac and plaster bones.

Work lasted from 5:30 in the morning until 2:00 in the afternoon. Cutler demanded that his crews apply themselves as vigorously to their tasks as he did, and anything less than maximum effort was met with a reprimand. An alarm clock in camp notified the cook when it was time to pound the signal drum that announced the end of the day.

Cutler and Leakey would separately inspect more quarries, and return to camp around 4:00 for a meal. While Cutler preserved and reassembled small bones in camp, Leakey supervised the bird collector's work and labeled and catalogued specimens. Before the sun set, daily rations for the laborers were distributed to crew leaders.

Fluent in Swahili, Leakey likely supervised the monthly payment of wages. Disagreements amongst the work force were discussed and probably also settled by Leakey, as Cutler did not understand Swahili.

While Cutler collected corals and butterflies on Sundays, Leakey would rise at 4:00 A.M. and hike north about 19 kilometers to the Mbemkuru to stalk game that was attracted to pools of water.

Cutler intended to avoid delay in the coming season, when the grass would be again all-obscuring, by adopting a German stratagem. Prospecting would be carried out prior to the planned December departure and promising sites would be marked with signs.

Coincident with Boheti's arrival at Tendaguru around mid-August, Cutler drafted a letter to the Museum, only his second substantial report in two months. Cutler addressed his report to Harmer, the museum's director, and opened with a complimentary assessment of Louis Leakey: "Mr. Leakey has and does assist me greatly by his knowledge of the Swahili dialect. . . . He has proved very energetic and assiduous in obtaining for you, birds and mammals. . . . If he should stay he would be more useful to me than a newcomer."[34] He also sent a list of medicines needed to treat the workers' ailments.

Both the museum director and the assistant secretary had raised the issue of bone shipments to justify expenditures. Cutler countered that the lack of transport and crating materials was the cause of the delays. Baltic timber for crates was going to cost 280 to 300 shillings for 50 cubic feet (1.4 cubic meters), and since 1,800 kilograms of lumber had only just been ordered from Zanzibar, he could not say when a bone consignment could be sent.[35]

There was encouraging news in mid-August, when local jumbes mentioned that the Lindi government had ordered them to begin road construction. The assistant secretary of the BM(NH) assured the governor in Dar es Salaam that the Museum intended spending at least two seasons in the country, and hoped that the administration could render some assistance by grading a road suitable for motorized transport.

With four quarries in progress at the beginning of September, six more promising sites in hand, and four types of dinosaur remains identified to date, Cutler felt that the Expedition was moving along as well as could be expected.[36]

Cutler and Leakey were stimulating the local economy by buying ebony "walking sticks" carved by Africans. The going price was an economical sixpence each, but this trade did circulate additional currency in the region, as the two Europeans had amassed over two hundred by the end of September.

An assistant political officer named Currie arrived in camp on September 15 to investigate the Expedition's chronic provisioning difficulties, as local headmen had complained angrily at being compelled to deliver food.

By the last day of September, many of the workers were anxious to return to their distant homes and plant their crops. About 80 people received their wages that day, and almost half departed for Songea the following morning. The leave-taking was amicable, and most men expressed the desire to be rehired the following field season. The departure of so many laborers simultaneously had a serious effect. There were only about 30 men left in camp, and some massive specimens were ready to be lifted from the quarries.

Leakey received word from Cambridge around mid-October that an extension to his leave of absence would not be granted. He informed the Museum and sent a runner to the coast to determine sailing dates for his passage home. The messenger brought news that Louis would have to board the *Dumra* on November 2. This left him scant time to close out his obligations at Tendaguru, and also left Cutler short-handed.

On October 29, Cutler and Leakey walked away from camp, the first time Cutler had left Tendaguru in about 20 weeks. The site had been under excavation for about 18 weeks. Leakey was deeply concerned about the departure date of his ship. He would forfeit another £60 toward Cambridge tuition if he did not get back in time for his courses, and could not afford the loss. He devised several fundraising schemes, including selling his ebony walking sticks and giving public lectures in Britain.

In Lindi, Leakey urgently recruited porters for what was to be an epic journey. The *Dumra* was out of service with boiler trouble and was not expected for several weeks. Leakey's liner was leaving Dar es Salaam for Europe on November 16, so he had no option but to walk the 430 kilometers to the capital. Neither Leakey nor Cutler wrote of their parting. After an exhausting 12-day push, driven by concern over unreliable transport, Leakey reached Dar es Salaam. On December 17, he was back in London, after an absence of almost 10 months.

This energetic young man, who had proved to be of tremendous value to his older colleague, if not always appreciated by him, passed from the story of Tendaguru. But in years to come, he was to build on the knowledge he gained there, and meet a man with whom he shared common experiences in the quest for African dinosaurs.

Cutler sought out the political officer and magistrate to discuss the progress of the road and the possibility of hiring a truck to transport crated specimens. Lumber had finally arrived, so Cutler tied the boards into bundles that could be carried by two men.

After a break of 11 days, Cutler was back at Tendaguru. The bone storage hut was much enlarged and now featured two tiers of shelving. Fresh grass roofs crowned most of the buildings. He was worried about losing specimens to the impending rain, and immediately began plastering nonstop and removing dozens of parcels from the ditches.

On November 13, 12 men broke ground at Ditch XIII, situated on the bank of a creek behind Ditch V. Four days later, Juma was given 19 men to expand the quarry. Only two days later, however. it was closed. Cutler made no mention of any finds in the horseshoe-shaped trenches after six days of digging. The nearby creek, a branch of the Nitongola, now had trial excavations on either side of it, extending down to bedrock.

Ditch XIV, under Juma, was opened on November 20. This site was on the left bank of what Cutler called Tendaguru Creek, 90 meters from Ditch II. As always, there was an old German quarry close by, this one 2.4 meters deep with good-quality bone eroding out of its floor. Ditch XIV was abandoned on November 28, after nine days of digging. The yield was not noteworthy.

Another group of men was assigned to Ditch 2X on December 8. It was an extension of an old German quarry sitting parallel to Tendaguru Creek, 15 meters from the end of Ditch II. Having discovered a calcaneum and scapula in its walls, Cutler ordered two sides cut back. Boheti led the team a few days later, finishing on December 26.

On December 10, Cutler viewed Ditch IV for the first time since October 28. One bone resembled a skull element, and there was also an ilium and a few vertebral centra. Once the rain tapered off around January 10, a crew strained to lift the weightiest jacket, which contained unidentified elements, possibly skull parts. A team of two cut a path around a hill while six men struggled with the load, spelled off by another six.

Ditch XV was commenced on December 27. Like virtually all the British excavations, this pit was situated between two German quarries, this time on the right of the trail toward Lindi. It was closed January 6, after 11 days of digging,

A new quarry, 15X, was begun near Ditch XV. It was shut on January 13, after about six days of investigation.

Ditch 16 was opened by several men on January 14, possibly because numerous exposed fragments had been found down a slope near a small creek. It was soon closed, but yet another was begun near the jumbe's house on January 28. There were five test pits in a small area, to be continued during the next season.

Rain fell sporadically—sometimes lightly, sometimes torrentially. A letter from another political officer, Anderson, dismissed Cutler's intentions of building rain shelters for bearers along the route to Lindi.

Cutler held an uncompromising attitude toward people with whom he dealt. He expected the same level of dedication and effort from anyone who worked with him as he gave to his own work. He placed a great deal of pressure on himself to succeed, and had little European support to draw upon. Working ceaselessly and in isolation, he was losing his objectivity. The last weeks of the season at Tendaguru were marred by numerous misunderstandings and confrontations with his workers.

On payday at the end of November, there were disagreements because Cutler cut wages for some reason. On December 6, the hottest day of the year at 40°C, he dismissed the cook after a series of arguments. He did, however, send him off with a letter attesting to his skill. Prone to suspicion about even his most reliable co-workers, Cutler made an inexplicable remark about Boheti: "also has been acting strangely of late, A sly cunning + possibly dangerously intriguing Mohamedan."[37]

Linguistic and cultural differences added to Cutler's problems. A number of episodes in which he complained of lack of cooperation or obedience from his work force almost certainly stemmed from his inability to make himself understood in Swahili. He, too, speculated that his instructions were misunderstood due to his poor command of local languages.

Christmas Eve brought him little peace. In his notes he mentioned, in German, that "boyish pranks" had been played, and then warned his laborers that there would be no Christmas presents unless a "stolen" piece of burlap was returned. Christmas Day was no different than any other—he plastered, and was irritated by careless work that damaged a bone. His diary entries were certainly not the homesick yearnings of Edwin Hennig, but then Cutler was older and seemingly unencumbered by close family ties.

> Christmas Day. Spent last in Hospital, W'peg, Canada, with badly scalded foot from accident whilst working at C.N. Railway E yards. Spent this day, in my most interesting, but still, routine work.[38]

Cutler's mood had not improved by December 26:

> Rats are at work in storehouse + Termites in my room. I wonder how the magnificent L.S.B. Leakey is faring? He is sure to be blowing his trumpet hard.[39]

New Year's Day brought only a brief attempt at reconciliation with his laborers:

> Gave all men packet of cigarettes after big row because so little work on new ditch.[40]

A large packet of letters, including one from the Museum in London, arrived, normally a heartening event. However, Secretary Smith's message contained clippings from British newspapers that, despite giving Cutler high praise, irritated him:

> press cuttings stating that my finds represent the largest yet found, (I do not believe them), Brachiosaurus of USA + the German's Dicraeosaurus, were equal to mine + skeletons at that. They stated that I was the best collector extant + now state that I shall have the skeleton, both erroneous and burdening me with onus of a damaging negation.[41]

In a letter to the *Times*, Harmer had reiterated how little the Museum could achieve with the available funds. He outlined the generous expenditures of the Germans and the tremendous support Americans had given the Central Asiatic Expeditions to Mongolia, and made a forceful appeal for public support. He would have cause to complain in the *Times* repeatedly in the coming years about the dearth of private donors.

A freelance journalist for the *New York Times* had contacted the director, but despite the Museum's attempt to arouse public support, he received a lukewarm response from Assistant Secretary Smith:

> suggest that you postpone the article until a consignment of specimens has reached the Museum and we are in a better position to appreciate the results of the Expedition. . . . They [an article and photos in the *Times*] would not be of much interest to persons in other countries.[42]

Perhaps Museum officials were sensitive about having British efforts compared with the unparalleled fundraising for American expeditions in Mongolia.

Harmer returned to a strategy that had been used in the past, namely an appeal to the Percy Sladen Fund of the Linnean Society in London. There is no record of the Fund's reply. Newspaper articles did have some effect—the Duke of Bedford and William Christy each contributed £100. The first donations totaled just over £250, of which Sir Sidney Harmer himself donated £25.[43] Several of his relatives also gave modest sums.

Calculating expenditures as of December 1924 at £1,797, Harmer realized the Expedition could not be sustained on what remained of the original grant. He requested that the Museum trustees approve an application to the Treasury for an additional £2,000 from the Reserve Fund of the Museum Purchase Grant. He acknowledged that the administration in Tanganyika Territory had invested £200 in road improvements between Lindi and Tendaguru.[44]

The director also tried to help Cutler get his specimens to the coast. In late December 1924, he wrote to the undersecretary of state at the Colonial Office, "I am informed that there is a suitable car at Lindi and that it might be more easy to obtain the use of it for Mr. Cutler if the authorities were to be assured that the Colonial Office is willing to support the Expedition."[45] C. W. Hobley recommended that the Museum contact Leo Weinthal, who had published a four-volume description of the Cape-to-Cairo route and was therefore familiar with transport in Africa.

Upon his return to London, Leakey drew attention to several points that had concerned Harmer. Foremost was the question of money. Cutler had rejected the specimens stored in Lindi, and decided to use all Museum funds to support his own fresh excavation program. Since all further expenditure was dependent upon the shipment of bones, a crisis was approaching. Cutler had raised this point months ago: "if my getting more [funds] awaits your receipt of specimens, then I fear that I shall go bankrupt."[46]

Cutler had been working especially hard since Leakey's departure, though he was making things more difficult for himself than necessary. "I am out, since Leakey left, often before six + at work in the bone field by 8 am until 2–3 pm. blazing as it is this is a good spell + I don't rest, though I let the men do so."[47] In his doggedly independent world, few men could be trusted to do a job as well as he could. He held that plastering specimens was a skill that could only be attained after considerable time. As a result he rarely allowed Boheti, who had prepared, supervised, and undoubtedly plastered in the German fashion for five field seasons, to plaster bones.

Cutler was doing everything in his power to maintain the momentum of the field season. Huts were set up at the active quarries so that half-plastered blocks could be completed when downpours normally interrupted this activity. He went on to provide the assistant secretary with the rationale behind his approach: "Had I stopped saving exposed bones, still numbering some thirty to fifty, in order to make boxes + ship, safari down to coast and all, I should certainly have lost a deal of material which now will hardly happen."[48]

Nevertheless, one could ask why he continued to have his men open fresh trial quarries so late into the season rather than training them to plaster or pack, or having them carry small items to the coast or even build crates. The trial pits would of course give him an early start the next season, but so would a systematic prospecting sweep by his workers. He may have felt the need to retain a labor pool for carrying specimens to the coast, but he was eventually supplied with fresh porters by the administration.

One is tempted to conclude that he simply could not bring himself to trust his men with anything more than basic duties that required little supervision, namely moving earth. And to be fair, he may have developed some of this attitude through his association with Barnum Brown's AMNH crews, or because of a bad experience with assistants in the past, rather than because of professional jealousy. Even so, the number of people employed to shift overburden meant that more area could be opened and, consequently, more bones revealed than one person was capable of dealing with, especially a perfectionist like Cutler.

Despite his disagreements with his workers, he again requested medicines: "I badly need <u>carbolic</u>, <u>lint</u>, <u>bandage</u>, <u>iodoform</u>, <u>Epsom salts</u>, <u>pot. permang.</u> to treat native sores + ailments, at least enough for 100 people. This is an absolute need here where they rely upon us for such things. A little <u>laudanum</u> also + <u>Spanish Fly</u> for blister." Cutler closed with a remark to which Harmer felt compelled to reply: "In spite of Leakey's idea of his indispensability I shall be able to carry on well without any assistance of Europeans from home."[49]

This unyielding independence left Museum officials uneasy: "I cannot help thinking that a really competent man, with a head on his shoulders and of good constitution, might prove invaluable in taking routine work off your shoulders, and in being ready to stand by you in case of trouble with the natives."[50] Harmer considered sending out another scientific assistant, someone with training in osteology.

Of serious concern was Cutler's health. As he was the sole representative of the Museum, any threat to his well-being would have major consequences. Louis Leakey had expressed disapproval of Cutler's cavalier attitude to health. Harmer was diplomatic but direct:

> I cannot help feeling from what he [Leakey] said that you are taking liberties with a tropical climate. . . . Your solitary position would give rise to much anxiety if you were to have a serious illness. The organization of your camp might be broken up and the whole work brought to a standstill.[51]

Harmer clarified another point of contention, namely the Museum's position regarding contact with the media. It was the source of confusion and some hard feelings on Cutler's part. Being so far removed from the public eye, he could easily feel that an unfair share of publicity had been garnered by the ever-enthusiastic Louis Leakey. An article in the *Times,* to which Leakey had contributed, appears to have catalyzed a long-simmering situation. Harmer forcefully clarified that Leakey had only written the piece at the Museum's insistence, to publicize the Expedition in the hope for donations: "You need not fear that any attempt will be made to deprive you of credit to which you are entitled. . . . Mr. Leakey is thoroughly loyal and fully recognizes that you are the Leader of our Expedition."[52]

On the political scene there were important developments. A new governor, Sir Donald Cameron, had been appointed for Tanganyika Territory in September 1924. Harmer had met with Cameron in January 1925, and the governor had expressed support for the Expedition.

Cutler's diary entry of January 5, 1925, had a wistful, almost poignant tone to it: "A year ago yesterday at day-break 8 am. −35° Fahr. Standing on the ice of the Saskatchewan river said goodbye to them."[53] It is likely that "them" referred to the Lingard family, to whom Cutler wrote many letters. They lived near the hamlet of Wingello, about 50 kilometers southeast of Saskatoon, Canada, and Cutler may have met them on a fossil-collecting foray.

The pressure to complete the field season before weather destroyed specimens robbed Cutler of impartiality, and he lost trust in his workers: "I suspect that my headman Juma is cheating me on time, coming home early etc. If I could get another I quickly would."[54]

Difficulties seemed to assail him from every side: the paths were running with water, the administration in Lindi insisted that all government tools be brought back for inventory, and, finally, Cutler was feeling unwell: "I for first time, today unable to go to field. Vertigo and nausea."[55] In compliance with the administration's order, shovels, picks, pangas, crowbars, and metal pans were carried to Lindi.

Cutler began building crates in late January, stenciling them with the letters A to I. Most measured 51 centimeters square by 35 centimeters deep. It was a trial attempt, and perhaps he was unconsciously avoiding the inevitable: "I tried to freshen up my onetime knowledge, always feeble, of making boxes, + accomplished two after trouble. . . . I never was any good at carpentry + the Amer. Museum hands, when I was with them on Red Deer River, used to ridicule my efforts."[56]

The Tendaguru camp was readied for the end of the field season. The tent was emptied and many ebony walking sticks were packaged. Cutler packed his extensive insect collection along with the invertebrate fossils he had gathered. A crowd of 94 bearers arrived at Tendaguru on January 29. Crates were nailed shut and labeled, but Cutler still wanted to bind them tightly with wire. Juma protested that there was insufficient food to last the five-man caretaker crew the estimated three months that would pass before excavation resumed.

Finally, all was to Cutler's satisfaction, and he turned his back on Tendaguru on January 31, 1925. He had reason to be pleased with the results of the first year in the field. About 19 quarries had been opened, from one to four per month, and virtually all the contents preserved. Insects had been gathered, and fossil invertebrates sampled. Leakey had sent back a representation of the bird, small mammal, and reptile life of the district, as well as ancient human artifacts. Tendaguru had been under investigation for almost eight months during the first season of the British Museum East Africa Expedition. Even though only nine crates of field jackets were on their way to London, many more specimens were safely stockpiled for future shipment.

Since 94 porters were not needed to transport nine crates from Tendaguru, and since he was able to later construct and fill a total of 25 crates at Lindi, Cutler must also have had individual field jackets carried to the coast on this safari. On February 15, the *Alchiba* dropped anchor in Lindi harbor, and at long last William Cutler's first consignment to the British Museum (Natural History) was loaded: 23 crates of fossils, 11 bundles of carved ebony walking sticks, and two other crates, weighing 1,958 kilograms.[57] An arrangement was made with Richman Symes in London to market the walking sticks on Cutler's behalf.

Cutler was tormented by what he diagnosed as neuralgia, as well as dental and sinus problems. Furthermore, he suffered from high fever, which Dr. Blackwood soon diagnosed as malaria. Cutler was not anxious to take medicine, due to its severe side effects: "I take a new medicine, also of quinine + it has a similar effect upon me, an unbearable itching + burning."[58] Dr. Blackwood warned that there was little he could do to treat the disease without quinine. The days dragged on in abject misery, chills alternating with fever, as Cutler waited for another steamer to Dar es Salaam, where he would spend the rainy season.

Although Cutler had insisted that he was capable of managing the Expedition single-handedly, he surprised the Museum by unilaterally engaging Howard Leslie Lachlan as an assistant at £15 per month. Lachlan, who had experience in the tropics, including Sumatra, had been employed at the Kikwetu Sisal Estate as a bookkeeper until a motorcycle accident injured his shin and heel.

The Museum's director was taken aback by the turn of events. Misunderstandings were bound to occur given the isolation in which Cutler operated. Slow communication rendered any consultation with London a ponderous task. Cutler also interpreted his original agreement with the Museum as allowing him a wider latitude in the field than his employers were willing to grant.

Boheti brought Cutler some fruit on the day of his departure, March 14. Cutler had spent 39 days in the port. The tribulations of Tendaguru were falling from his shoulders. William Edmund Cutler stepped ashore at Dar es Salaam on March 18, 1925. His first challenging season in Africa was behind him.

-10-

1925

Berlin Builds Dinosaurs

And what had occurred in Germany between 1919 and 1925? Social, political, and economic conditions had remained chaotic. The poor and middle classes had suffered grievously. The effects of food rationing and unemployment were felt most keenly by the poor and elderly. Food riots were not unknown, and children especially were malnourished. Strikes were commonplace. Participants in a 1919 general strike in Berlin had been brutally attacked by the ex-Army Free Corps, who used armored vehicles, heavy artillery, and even bombers. The carnage had resulted in more than 1,200 deaths.[1] The capital had been paralyzed as buses, streetcars, shops, factories, schools, and government offices stopped operating, and supplies of electric power, gas, and water were shut off. A right-wing coup was attempted in March 1920, but failed to unseat the republic.

Yet somehow, throughout all these difficulties, the vast rows of field jackets from Tendaguru were methodically removed from their shelves and prepared. Before the war, von Branca had lobbied influential patrons for funding to prepare and mount the hundreds of tonnes of bones. Geologist Wilhelm Bornhardt encouraged senior mining official Carl Gruhl to contribute more than 15,000 marks in 1914 and 1915. Pompeckj, von Branca's successor, approached the Emergency Association for German Science (Notgemeinschaft der Deutschen Wissenschaft), which granted 30,000 marks for the preparation of the African finds in April 1922. If these sums are added to the 1912 grant from the Prussian Academy of Sciences, they approach the 50,000 to 100,000 marks that von Branca had deemed necessary for the task.

The saga of Reck's priceless pterosaur bones had a delayed but fortuitous conclusion. He had last seen them in April 1916, as he went to war. The Swiss engineer F. G. Rikli, to whom Reck had entrusted the specimens, reached Tabora. Rikli and his family marched to Lake Victoria in October 1916, several weeks after Belgian forces captured Tabora. The cost of their trek rose dramatically due to a shortage of bearers, many of whom had been conscripted into the Carrier Corps. In Mombasa, the Riklis were delayed by British censors, who spent a great deal of time examining Reck's notes and maps. As a result, the

Swiss family missed one of the rare steamers still plying the ocean to Europe and were forced to wait in a Mombasa hotel for the next ship. Their odyssey continued to Durban and Cape Town, South Africa, and then to Europe, but the interrogation was repeated in France. They were back in Bern, Switzerland, four and a half months after leaving Tabora. In mid-April 1917, Rikli presented an invoice to Pompeckj for 1,248 Swiss francs, asking to have his expenses reimbursed in exchange for the fossils.

Upon his return to Berlin in 1919, Reck resumed work at the Museum of Natural History and explained the matter to Pompeckj in December. An agreement between Reck and the Museum settled the account in early February 1920. The paleontologist offered to pay Rikli immediately out of pocket; possibly because of postwar inflation, the sum now stood at 1,651 Swiss francs or 26,523 marks. Presumably the Museum repaid Reck, and the pterosaur remains were returned safely to Berlin. They had journeyed from Tendaguru to Berlin and back to German East Africa, then circumnavigated Africa to Europe.

William Diller Matthew, curator of vertebrate paleontology at the American Museum of Natural History in New York, witnessed conditions in Berlin in the immediate postwar years. In the fall of 1920, he spent three months visiting European museums. Matthew reached the Museum of Natural History on September 26, 1920, and was disappointed at the state of the building, compared to what it had been on his last visit, 20 years earlier:

> at present very dingy and dirty owing to lack of care. A large number of limb bones of the Tendaguru collection have simply been laid out on the floor behind the cases and roped off.
> . . . This whole town looks miserably dingy and dirty; very much down-at-heel. . . . The people look very much run down, a large part of the population, I think, is much underfed. The bread is of wretched quality, heavy, soggy, and full of grit. Prices of everything in the stores allowing for the present value of the mark (about $1^3/_5$ cents), are a third to a half lower than in New York; everything, however, is of the very cheapest and most inferior quality.[2]

An expansion program had been begun at the Museum as growing collections from scientific expeditions strained the capacity of the building. The arrival of 225 tonnes of fossils from Africa exacerbated matters. In 1907 and 1912, it was suggested that the Palaeontological Museum and other departments relocate to the district of Dahlem. These plans never came to fruition, and in late September 1914, construction had begun on extensions to the original building.

The plans included fourth-story additions to three northern wings. Two courtyards were to be given a glass roof. The galleries that stored the research specimens were to be divided into two levels by a second floor, and the main atrium was to be reserved for dinosaur skeletons. The atrium currently housed whale skeletons, but the fruits of rich dinosaur quarries such as Halberstadt and Tendaguru now demanded more space. At the end of 1917, the War Ministry put a halt to the project, long before it was completed. Still, an additional 14,000 square meters were now available.[3] The Museum had never been threatened with closure during the war, and wounded soldiers were toured through the displays. But postwar political and economic difficulties had left the building in poor condition when Matthew visited.

Matthew could not help but acknowledge the enormity of the commitment to ready the finds for research and display:

> The Tendaguru collection is likewise an immense task in preparation, and when I saw it in Berlin . . . it was far from being completed, after more than ten years' work.[4]

Clearly, he was deeply impressed with the African material:

> It is a magnificent collection, far larger and more varied than I had any expectation of finding, and in preservation equal to the best of our Morrison material.
>   . . . the skull of Brachiosaurus (the biggest sauropod skull ever found and, I am inclined to add, the best).[5]

Given postwar conditions, any progress would have to be considered heroic. By the mid-1920s, an impressive body of literature was available to both the public and the scientific community. Numerous reports in newspapers had introduced the aims and progress of the Expedition. Soon accounts appeared in popular magazines that presented the scientific and technical advances of the era. Germany was especially well served in this regard, for in 1914, 4,200 newspapers and 6,500 different magazines were published in the country. Berliners had an overwhelming choice of 93 newspapers per week in the 1920s, more than any other city in Europe.[6] Scientific journals soon had numerous submissions from Janensch, Hennig, Reck, and many other scientists.

The Expedition had greatly enlarged the understanding of the late Jurassic vertebrate and invertebrate fauna of East Africa, and the geology of the southern and central parts of the former colony. Sediments could be more accurately dated, and stratigraphic successions and paleogeographic reconstructions were refined.

As scientific results were published, previous work was revised. Major manuscripts by Krenkel and Dacqué had described invertebrate fauna that forced a modification of Fraas's age estimates for Tendaguru. In 1914, Hennig, too, revised Fraas's sequence of one marine Lower Cretaceous horizon and one terrestrial Upper Cretaceous horizon.

Hennig proposed a 120- to 150-meter-thick stratigraphic sequence that formed an uninterrupted succession from Middle Jurassic to Lower Cretaceous. Three continental beds composed of freshwater to brackish water marls contained dinosaur remains. They were separated by marine beds and transition layers. The sequence began with the Lower Saurian Bed (Untere Saurierschicht), which was considered Middle Jurassic, Dogger Epoch. This was overlain by the shallow marine *Nerinea* Bed (Nerineenschicht), named for the predominant gastropod. A transitional bed containing ammonites was deposited next. The Middle Saurian Bed (Mittlere Saurierschicht) was dated to the Upper Jurassic, Malm Epoch. Another marine sandstone section, the *Trigonia smeei* Bed (*Trigonia smeei* Schicht) followed in the profile, and was named after the abundant bivalve, which served as an index fossil. A second transitional bed was then overlain by the Upper Saurian Bed (Obere Saurierschicht). Hennig believed the Jurassic-Cretaceous boundary was found in this stratum. The marine *Trigonia schwarzi* Bed (*Trigonia schwarzi* Schicht) dated to the Lower Cretaceous, Neocom Epoch. North of Tendaguru, the Newala Sandstones or Makonde Beds cropped out. They were Upper Cretaceous, but unfossiliferous. Janensch established a new geological series, the Tendaguru Beds, which included the sequence from the Lower Saurian Bed to the *Trigonia schwarzi* Bed.

Hennig, Janensch, and von Staff also published their conclusions on the geology of other regions of the country. Hennig's monographs on the coastal Tertiary outcrops appeared in 1913, and his findings along the Central Railway were published in 1912, and in greater detail in 1924. His discovery of Lower Cretaceous Urgon deposits while en route to Makangaga was in print in 1913, and the fauna from the Urgon was described in 1916. This was the same year he dated the folded shales in the Pindiro Valley as Middle Jurassic, Dogger Epoch.

Von Staff proposed that tectonic mechanisms had shaped the landforms in southern and coastal areas. In 1912 and 1914, he posited cycles of epeirogeny, or geological uplift and subsidence, that had been accompanied by marine transgressions and regressions. Fluvial erosion further altered the relief.

Dozens of new genera and species of fossil invertebrates were established by paleontologists like Dacqué, Krenkel, Dietrich, Lange, Oppenheim, Hennig, and Zwierzycki. The invertebrate fauna, which was crucial to dating the sediments and understanding the environment of deposition, included cephalopods, gastropods, bivalves, brachiopods, annelids, corals, bryozoans, echinoderms, and foraminiferans. Fossil arthropods and paleobotanical remains would not be discussed for several more years.

A diverse vertebrate fauna, previously unknown in the region, was also described. *Lepidotus,* a holostean fish, was found at Tendaguru, a first for the colony. Teeth of the shark *Orthacodus* were collected. Near Lindi, teeth of sharks like *Corax, Lamna,* and *Scapanorhynchus* were found in Upper Cretaceous sediments. Hennig also picked up Tertiary-age teeth of the white shark *Carcharodon* on the coast. Crocodile teeth were found, but not identified precisely. Similarly, pterosaur bones and fragments of a small reptile awaited further study. A highly significant mammal jaw had been unearthed, but was still unnamed.

The overwhelming majority of Tendaguru fossils were dinosaurs. A new faunal assemblage, the first to be comprehensively collected from the continent, had been discovered. The scientific community was surprised by the size of some groups and the abundance and completeness of their remains.

Several forms of enormous herbivorous sauropods had been identified, probably the first since *Algoasaurus* from South Africa. They were considered sluggish swamp-dwellers. Their small heads and seemingly simple dentition were interpreted as adaptations to a diet of aquatic plants. The controversy over their posture had faded when W. J. Holland of the Carnegie Museum demonstrated the impossibility of Hay's and Tornier's skeletal articulations in 1910. As early as 1916, von Branca had speculated that the sauropods' tremendous size might have been due to a superior ability to extract nutrition from vegetation, or a slow metabolism.

In 1914, Janensch had established two new species of a very large sauropod known from the United States, *Brachiosaurus brancai* and *Brachiosaurus fraasi*. Compared with the North American genus, the African varieties were much more complete, allowing Janensch to suggest a giraffe-like posture that allowed the animal to feed from greater heights than other sauropods. Two species of a medium-sized genus were recognized, *Dicraeosaurus hansemanni* and *Dicraeosaurus sattleri*, again by Janensch in 1914.

Fraas's 1908 paper establishing *Gigantosaurus* was questioned by zoologist Richard Sternfeld in 1911. Sternfeld asserted that the name was preoccupied and proposed the genus *Tornieria* in honor of zoologist Gustav Tornier. In 1922, Janensch reassigned *Gigantosaurus africanus* to *Barosaurus robustus*. *Barosaurus* had also originally been discovered in North America.

Stegosaurs, previously known in Africa only from the problematical South African *Anthodon* (= *Paranthodon*), were represented by whole herds of a relatively small form at Tendaguru. In 1915, Hennig published a preliminary description of a new genus, *Kentrosaurus aethiopicus*. Baron Franz Nopcsa charged that the name was preoccupied by the North American Cretaceous horned dinosaur, *Centrosaurus*. The following year he suggested *Doryphorosaurus* as a replacement, but Hennig renamed the animal *Kentrurosaurus*. The defensive armor was less derived than that of the North American stegosaur, consisting of smaller plates and more spines. A curious feature common to stegosaurs, a greatly enlarged cavity in the sacral vertebrae, was described in the Tendaguru genus.

Common in other parts of the world, especially North America, bipedal herbivorous ornithopods formed another abundant component at Tendaguru. Hans Virchow first used the name *Dysalotosaurus lettow-vorbecki* in 1919 for this small, swift dinosaur. In 1920, 1921, and 1922, Felix Pompeckj formalized the name and described the species's anatomical features. The skull structure implied an orthal jaw movement, in which vegetation was sliced by teeth that were replaced in waves. Related forms, hypsilophodonts, were sometimes reconstructed as capable of climbing trees.

Carnivorous dinosaurs were represented at the site by several types of varying sizes, though they were among the least abundant forms. *Elaphrosaurus bambergi*, a lightly built bipedal animal, was named in 1920. It was classified as a coelurid by Janensch. This taxonomic position would be revised several times in the ensuing decades. Much larger genera were present, though their remains were scarce. In 1925, Janensch published on theropod teeth, tentatively erecting *?Allosaurus tendagurensis*, *?Ceratosaurus roechlingi*, *?Labrosaurus stechowi*, and *?Megalosaurus ingens*. He believed there were several more theropods at Tendaguru.

A similarity between the Tendaguru fauna and that of the North American Morrison Formation led Janensch, like Fraas, to suppose that a land connection had existed between the continents during the Upper Jurassic.

The depositional environment at Tendaguru had been reconstructed in preliminary notices by Hennig in 1912, and by Janensch in 1912 and 1914. Both men agreed that the rich accumulations of fossil remains had been deposited near lagoons or sheltered coastal bays. Shorelines had been exposed for considerable distances by ebbing tides, and sandbars, coral reefs, or barrier islands isolated the lagoons from the deeper ocean. Modern analogues were seen at the Great Barrier Reef of Australia and along the coasts of Florida, Texas, and Brazil. These shallow coastal or neritic facies represented quiet-water or low-energy conditions. The transitional beds had been deposited at times of higher energy, when storms or tectonic or isostatic processes resulted in much stronger wave action. This had caused the ocean to override the lagoonal barriers and sweep in coarser materials. Mollusc colonies had formed during these episodes.

Quarries had been analyzed to understand the taphonomic processes that had affected the skeletons. Janensch proposed that the dinosaurs had foraged on mudflats, where herds and individuals had been mired and drowned by incoming tides. Mass accumulations of ornithopod and stegosaur remains in the Kindope quarries pointed to the catastrophic death of entire herds. Ornithopod long bones, such as tibiae, fibulae, and femora, had been oriented northwest to southeast, parallel to one another. After the animals died, admittedly from unclear causes, their bones had been disarticulated and sorted by waves before final burial. The restricted area in which they were found and the unabraded condition of their thin-walled bones indicated that they had not been transported far after death.

Stegosaur remains had been recovered from many more localities than the single ornithopod site. They were also usually completely disarticulated and restricted to relatively small areas. Lighter skull and foot elements had been transported further. Janensch thought these animals had been trapped in soft sediments and drowned by returning tides.

Sauropods were not represented in the same numbers as ornithopods, but several quarries indicated that these large animals might also have associated in groups. Whether the discovery of 18 to 20 femora in Quarries IX and XVI indicated monospecific herds could not be established. The cause of sauropod deaths seemed less obvious. In several cases an articulated forefoot or hind foot had been found in a vertical position. The same was true of the humerus and tibia of Skeleton S. These examples may have been older or diseased animals that had been trapped in very soft sediments.

After death, wave action had disarticulated skeletons and sorted the bones. Scavengers had also scattered the skeletons. Rivers in flood had dropped sediments into the lagoons and swept in bones of animals that had died further inland. Cadavers bloated with the gases of decomposition had floated further, a possible explanation for the occurrence of bones in deeper water sediments. Bivalves like *Mytilus* and *Cyrena* were associated with the dinosaurs, indicating brackish water. In a few examples, dinosaur bones had been deposited with marine invertebrates like belemnites, crinoids, and bivalves.

Janensch discussed his discovery of peat bogs in 1914. The physical characteristics of the bogs, almost unknown in the tropics at that time, were enumerated. Chemical analysis indicated a higher concentration of silicic acid than was found in bogs in temperate climes.

Lithic artifacts, collected by Janensch along Dwanika Creek in 1910, had been examined by Emil Werth in 1916. He declared the scrapers, bifaces, and unretouched flakes Paleolithic in style. They were the first such objects identified in the country, an identification that had been delayed by the immense task of dealing with the Tendaguru dinosaur remains.

Hans Reck returned from African internment to resume his duties as an assistant at the Museum. He and Ina had moved into a house in Hermsdorf, in north Berlin, which sat on a height with a view of distant forests and lakes.

Matthew met Reck and Dietrich, but missed Janensch because he was getting married. The 42-year-old Janensch chose 29-year-old Pauline Sophie Renate Helene Henneberg as his wife. His salary had risen from 5,400 marks in 1919 to 25,200 marks in 1920, with various supplements.[7] No one could guess what an impossibly insignificant sum this would represent in just a few short years.

Who were the men behind the scenes, the preparators who had overcome wartime and postwar hardships to process fragments of prehistoric life for the past decade? The most senior of the preparators was Gustav Borchert. He was born on March 22, 1863, in Mohrin, northeast of Berlin. In 1893, he joined the Museum of Natural History, which had opened its doors to the public only four years earlier, as an attendant (Diener). He was about 30 years old at the time and had the good fortune to prepare the famous Berlin *Archaeopteryx*. This specimen had been discovered in 1877, and was purchased in 1880 by Werner von Siemens, the industrial magnate whose sons later contributed to the Tendaguru Expedition. In 1908, Borchert was promoted to the position of preparator. He was married and was about 47 when the Tendaguru crates began arriving in 1910.

Ewald Siegert was born on December 23, 1883. He joined the Museum as an institutional aide (Institutsgehilfe) in 1912, near the close of the Tendaguru Expedition. A cabinetmaker or carpenter by trade, he did not serve in World War I due to a heart condition. He was about 27, married with a daughter, when his skills were harnessed by the African project.

Johannes or Hans Schober was born on July 3, 1886. He was a mechanic or locksmith by trade, and his specialty was blacksmithing. After demobilization from the army he continued his preparation duties at the museum. He is remembered for his "spicy humour and witticisms."[8] Schober was a young man of about 24 when Tendaguru specimens required his attention in 1910.

These men would all soon face the most devastating conditions that had yet prevailed in peacetime. Assassinations continued in Berlin. From June 1922, the nation's economy was beset by a rapidly growing inflation that reached crippling levels. Political uncertainty, the battles of rival factions, strikes, the attempted coup, unemployment, and anger over the punitive terms of the Treaty of Versailles, combined with other factors, caused a loss of confidence in German currency.

In January 1923, French and Belgian forces occupied the industrial Ruhr in response to Germany's nonpayment of reparations. The mark plummeted even more swiftly. Immediately before the war, it had taken 4.2 marks to buy one U.S. dollar. On June 14, 1923, it took 100,000. The government commandeered the printing presses of media giant Ullstein to print yet more bills, which only fueled inflation. At the end of September, the figure stood at four billion marks to the dollar. The madness peaked on November 21, 1923, when 4.2 trillion marks were required in exchange for one U.S. dollar.[9]

The effects of this swift devaluation were devastating. Middle-class artists, civil servants, and academics were in desperate straits. Savings were wiped out and anyone on a fixed income lost all hope of secure retirement. The Berlin municipal government reported that less than 10 percent of families in the city could maintain an acceptable standard of living.[10] Food and all other essential commodities suffered the same astonishing price rise. Workers received absurdly high wages as often as five times a week, and when possible, one family member would rush out to purchase whatever could be obtained in the shops, even if just to trade the item for another. Barter became an accepted economic exchange. The crime rate grew as people struggled to survive. Anyone with access to foreign currency could live like royalty on a handful of dollars or pounds.

Against this backdrop of social and economic upheaval, a multidisciplinary team of researchers deciphered the overwhelming tonnage from Tendaguru. Borchert and Siegert, who had continued preparing fossils throughout the war, and Schober subsequently, had unpacked and freed hundreds of skeletal elements from field jackets and marl, assembled sections, and reconstructed missing ones. Most of the stegosaur bones had been prepared by a woman, L. Dietrich, née Trendelenberg, Edwin Hennig's sister-in-law.

This work had progressed relatively quickly compared to that on the huge and intricate sauropod vertebrae, or the mighty limb bones that took several men to manipulate. As the number of prepared stegosaur specimens grew, von Branca, then the Museum's director, determined that this dinosaur would be the first to be mounted for public display. He assigned the task of scientific description and mounting to Edwin Hennig.

Hennig had completed several papers on the stegosaur in 1915. Following military service, he accepted Pompeckj's post in Tübingen in 1917. Though he was no longer in Berlin to work with the collections, his major monograph describing the anatomy of *Kentrurosaurus* was published in 1925. Pompeckj, the new Museum director, arranged for financial support from the Prussian Academy of Sciences to produce the illustrations for this publication.

Hennig keenly regretted that a planned trip to examine stegosaur material in museums abroad had been canceled due to conditions following World War I. He returned to Berlin in March and April of 1921, to review the progress made in preparation and to revise his manuscript.

Pompeckj transferred the mounting project to Werner Janensch after the war. Only two other free-standing stegosaur skeletons had been mounted in the world, both in the United States. The first had been completed at Yale in 1910, under the supervision of Richard Swann Lull, and the second at the National Museum of Natural Sciences in 1917, under the direction of Charles Whitney Gilmore.

Sufficient material had been readied in Berlin to enable elements to be selected for a composite skeleton, a task complicated by the lack of articulated skeletons. Stegosaurs were common enough at Tendaguru; Hennig estimated that 40 to 50 individuals were represented at the 28 quarries and two isolated finds that produced evidence of the animal. The largest and most prolific quarry, St at Kindope, yielded the majority of the 1,200 stegosaur bones in the collection. But it was a bonebed deposit, a mix of animals of every size, age, and sex. Bones had also been found in another bone-bearing horizon, adding a difference in geological time to the puzzle. Janensch spent a great deal of time deliberating, but only infrequently consulted with Hennig, due to the difficulties of the postwar era. Their interpretations of the animal did not differ substantially.

To obtain an idea of fundamental features, such as number of vertebrae and foot structure, Janensch referred to monographs by Gilmore on the American genus, *Stegosaurus*. Then there was the question of interpreting how joints articulated and what range of motion they allowed.

Once bones had been selected from animals of a similar size, missing elements were modeled in plaster. To add to Janensch's difficulties, the placement of the defensive armor was not obvious, since only a few articulated series of bones were found in situ. He admitted that while he had made careful decisions, there was a degree of conjecture in his proposed assembly.

Some of the bones chosen for the mount were from the type specimen described by Hennig, and originated from Quarry St. From this site in the middle bone-bearing horizon came dorsal and caudal vertebrae and the sacrum, ilia, right pubis, left ulna, and left femur. The left radius came from site r, also in the middle horizon. Forefoot elements and the defensive spines and plates were chosen from sites in the upper horizon. Janensch felt justified in using this material because he and Hennig agreed that the same species was represented in both deposits. The mount of *Stegosaurus stenops* served as a guide, though Janensch admitted that the African dinosaur may have been mounted with one extra vertebra due to differences in the number of true sacral vertebrae (three in *Stegosaurus,* according to C. W. Gilmore's description of the stegosaur in the U.S. National Museum, and four in *Kentrurosaurus*).

Cervical vertebrae were chosen from Quarry St and Skeleton H in the upper horizon. The axis and three cervicals were restored, as were the neural arch of one dorsal vertebra and the centra of two more cervical vertebrae. A series of 23 articulated caudal vertebrae was selected to complete the tail. The first six caudals came from a different animal. Six fused vertebrae were chosen to represent the posterior end of the tail. Janensch used a vertebral formula of 25 presacrals, two dorso-sacrals, one caudo-sacral, and 42 caudals. Only five left ribs and six right ribs were considered complete enough and of the appropriate size to be mounted.[11] All the rest were modeled in plaster. No cervical ribs were modeled or mounted because Janensch felt they were too imperfectly illustrated in the literature to be copied. The chevrons were almost exclusively reconstructed in plaster, as they too were rarely recovered.

The forefeet were assembled from elements uncovered in Quarry X, except for the ungual on digit one. A decision was made to use the bones found in the quarries for the phalangeal formula of 2 2 2 1 0.[12] Only those bones that were already known but missing from one foot or the other were modeled. Janensch admitted that this interpretation could be challenged, especially for the greatly reduced phalangeal count for the middle three digits. Hind feet were based on a complete example found in Locality Ki, but modeled in plaster to match the smaller size of the rest of the bones used in the composite skeleton. The question of ungual phalanges was equally problematic on the hind feet.

The anterior portion of the skull and the dentaries were modeled in plaster and based on *Stegosaurus*. The posterior part, with braincase and condyle, was derived from Quarry St.

Janensch cautioned that the placement of armor, which consisted of plates and spines, was only an interpretation, open to revision with future discoveries. There was no articulation of these bones in the field, with the sole exception of a pair of the terminal tail spikes. Though the majority of the examples of armor came from St, they were also present at a number of other sites. Three types of spikes and plates were recognized by Hennig, and their presence in two geological horizons added to his conviction that there was only one species of stegosaur at Tendaguru. Janensch proposed nine pairs of presacral plates and spikes, positioned as opposed pairs in a double row along the spinal column. The three anteriormost pairs of plates over the neck were entirely conjectural and modeled in plaster. The final six pairs of armor plates were assembled from quarry sites. Only those spikes or plates that were too fragile, or which did not approximate the size of their mates, were reconstructed in plaster, based on the shape of their opposite number. Five more pairs of spikes were positioned along the tail. Another type of spine, with a round flattened base, was rare, and of a shape that Janensch felt would be best positioned over the ilium rather than along the vertebral column.

Janensch pondered the question of whether the three types of spines were evidence of sexual dimorphism, since one form was wholly absent from Quarry St. He concluded that it was more likely a case of selective preservation due to differences in the depositional environment. He based his reasoning on the almost total absence of foot elements and plates at Kindope, which he believed had been washed out of the site.

Hans Schober worked closely with Janensch on the mounting process. His forging and metal-smithing skills were especially useful. Space was cleared in a workroom in the basement of the Museum, and a wooden framework was constructed. The bones were suspended from it with rope and wires to maintain the skeleton in a provisional pose. Vertebral centra were cored so they could be strung in the correct sequence onto a metal armature for support. Measurements were taken and duplicated in the bends and curves of the external metal framework that held the limbs, ribs, and other elements in place.

Devising a method that could meet conflicting demands was an art. The structure had to be strong enough to rigidly maintain even the heaviest bones, such as the sacrum and limbs, yet light and unobtrusive enough so as not to draw attention away from or obscure parts of the skeleton. Fragile and small bones required the same sturdy yet invisible support. A final condition imposed upon Schober was that the bones be removable for future study. This meant that individual elements or small groups of elements had to be independently connected to the main framework so that minimal effort was required for their removal.

Once this was accomplished, a podium was built in the large atrium. The free-standing skeleton was disassembled in the workshop, moved to the main floor in the freight elevator, and reassembled on its metal framework. The job was completed in early 1924. The majority of the mounting must have taken place at the height of economic and social upheaval in Berlin, a testimony to the dedication of Museum staff. Visitors were invited to view the new display for a nominal 25 pfennigs.[13] The achievement was considered a concrete step on the long road to rebuilding national pride.

Later that year, former Museum director von Branca celebrated his 80th birthday and was honored in publications describing the first assembled Tendaguru dinosaur. The first stegosaur remains had been uncovered at Kindope 14 years earlier. Reck and others pointed out that, considering the wealth of material and the disruptions of the war and the postwar period, the lapse of time was no cause for shame. Even the Americans, in their undisturbed efforts, had taken eight years to excavate, prepare, and mount a *Stegosaurus*.

What did the Germans think of the recent British efforts to reopen Tendaguru? There does not appear to have been any communication between the BM(NH) and the Museum of Natural History in connection with a new Tendaguru expedition. Considering the millions killed or maimed during the war, this comes as no surprise. The Germans were aware of the British expedition from reports in English and African newspapers and English-language scientific journals. Initial reactions from Janensch and Reck were muted, and expressed caution:

> its success . . . must above all depend on whether it will succeed in obtaining in requisite quantity skilled, willing, and dedicated helpers among the natives. . . . [14]
> . . . Tendaguru does not easily and willingly unlock its treasures. Only in hot toil does it allow its secrets to be wrested away. Accompanied by the loud chorus of the newspapers of the world this expedition disappeared into the bush of the Lindi region. Since then there has been silence about it.[15]

The loss of the colony was a grievous insult to German pride. Hennig and Reck, too, were swept up in bitter postwar rhetoric, but also repeatedly complained of having lost specimens that had been collected prior to 1914:

> The culture-destroying effect of the colonial predatory war of the British has unfortunately also revealed itself in this case. Further excavation efforts on the part of German collectors were thwarted by the outbreak of hostilities. Highly important discoveries, which were supposed to be displayed at the exhibit in Dar es Salaam in August 1914, were carried off by the enemy and are obviously hopelessly lost to science.[16]

Anger and disillusionment notwithstanding, Museum staff carried on through all adversity. The first step in the overall plan of Tendaguru had been to gather data through a comprehensive field program. An enormous wealth of specimens, notes, samples, profiles, and photographs had been safely brought to Berlin and now required analysis and interpretation. Preparation and description of the skeletal remains proceeded hand in hand.

-11-

1925

A Death in Africa

The popularization of the British Museum East Africa Expedition took an unexpected turn. In early February 1925, First National Pictures offered to donate £250 to the Tanganyika Expedition Fund in exchange for 15 newspaper articles to be written by the director and scientists at the BM(NH). The articles were to appear together with a film the company was about to release, and were to heighten public interest in the world of the past. The film was a dramatization of Sir Arthur Conan Doyle's 1912 novel *The Lost World*. Sculptors Willis O'Brien and Marcel Delgado had created miniature clay models that were painstakingly animated in the silent black and white picture by stop-motion photography. Harmer was expected to address the audience on opening night in London, and Bather, the Museum's keeper of geology, was expected to produce a radio program on prehistoric life. Museum trustees approved this course of action.

By February 18, the British Treasury had approved the latest application for increased expenditure on the Expedition. The Museum was now free to draw a total of £4,000 from the Reserve Fund, part of the grant for purchasing specimens.

After he was featured in the *Times* and the *Illustrated London News,* Louis Leakey was approached to give a lecture about Tendaguru at Cambridge. He planned to show lantern slides and convinced First National Pictures to lend him a clip from their as yet unreleased film, in the interests of additional publicity. Half the proceeds from the lecture would be donated to the Museum's Expedition Fund. On the evening of February 25, the nervous student surveyed the audience that had gathered in the Guild Hall in Cambridge, a venue with seating for a thousand. As his talk began, however, Leakey lost his nervousness and described his participation at Tendaguru enthusiastically.

Harmer was dubious about a suggestion that Cutler should author a popular book about Tendaguru: "I do not feel confident that he would be competent to produce a book of the kind you want. He is an admirable collector . . . but I do not think that his literary qualifications are likely to be very great."[1]

Cutler must have been concerned about his health, since he visited a hospital in Dar es Salaam for blood tests. The medical officer informed him that he was suffering from gallstones that he had been carrying for many years.

Between the 21st and 24th of March, Dr. H. M. Fisher, a dentist, extracted six of Cutler's teeth, including molars. This may have been a continuation of problems that had plagued Cutler since 1913, when he had paid $75 for dental work in Alberta, a fortune at the time.

Acknowledgment for the specimens recovered by the Expedition was a sensitive point for Cutler:

> Mr. Leakey was instrumental in getting camp built-up and in organising the safaris of loads of supplies, but the bones have not been his work. . . . please understand that I would not belittle his activities, he was extremely energetic and capable in many ways. However in a bone camp there are capabilities only to be gained by long experience. . . . I managed to bring into Lindi one of the largest safaris seen there viz 125 porters + that in the mid rains + with loads of 120 kilos.

It was the next statement that truly summed up the essence of Cutler's character:

> I desire no publicity only that the scientific world shall be informed of the results of my efforts.[2]

He also disputed Leakey's claims that he had neglected his health.

On Friday, April 3, the new governor, Sir Donald Cameron, arrived in Dar es Salaam. He was decisive and efficient, and his tenure promised change. The governor lost no time communicating with Cutler, as he had been briefed on the Museum's aims while in London. Four days later, Cutler drafted a proposal, estimating that the one-ton Morris truck that the government had offered to place at his disposal to carry loads from Tendaguru to Lindi would be required for six to eight weeks. The road had been completed to within 16 kilometers of the quarries, and Cutler thought porters might haul the crates this distance, in short stages.

Price quotes were requested from the Equator Sawmills in Nairobi, because Cutler felt that at 225 shillings per tonne, he had paid a premium price for the Baltic or Swedish lumber he had ordered from the Rufiji Delta Trading Company in Dar es Salaam.

Cutler advised Harmer that more equipment was required. Shoemaker's awls, of the pattern used by the AMNH, had proved to be excellent excavation tools and Cutler requested three dozen. A beam scale would allow jackets, loads, and foodstuffs to be weighed. Since time had been lost last season in preparing maize and millet grains for food, he requested a hand-operated grinder with a large-capacity hopper.

Unaware of the dismay he had caused by hiring Lachlan, Cutler explained that the £50 bond he had posted for Lachlan was required by a regulation that relieved the administration of the need to support "distressed British subjects."

Cutler dined with a well-known longtime resident, Clement Gillman. In 1905, the young Gillman had joined the German firm Holzmann, builders of the Central Railway. Survey work provided the engineer with the perfect opportunity to pursue his passions for geography, natural history, clouds, and geology. Gillman's Peak, a feature on Mount Kilimanjaro, was named in his honor.

Commissioner Wyatt had indicated that the quarters occupied by Cutler and Leakey in Lindi might be requisitioned for railway officials. Concerned, Cutler stressed the importance of reserving the larger area on the main floor for the Expedition's supplies and field jackets. Privacy and security were afforded here for the final packing and labeling of the crates.

On May 3, Cutler was again in Lindi, after spending seven weeks in Dar es Salaam. Commissioner Wyatt informed him that he would not get the Morris truck to haul field jackets. No vehicles had yet arrived in Lindi, and when they did, Cutler would have to reopen the matter. Cutler expressed his disappointment with the attitude of local officials privately to Harmer: "the Administration is little interested in our quest or in the Expedition's success. . . . they appear more considerate here of the native than of the struggling European." He hastened to add that he was grateful for the lodging, tools, and porters that the government representatives had offered, but that he was not alone in his assessment: "the sentiment of most Europeans here, [is] that the Administration does not want them as I have heard expressed by Senior Commissioner Wyatt, here, to me."[3]

One hundred and fifty carriers had assembled in Lindi, swelling the population of the little coastal town. Loads were readied, including about 1,800 kilograms of lumber.[4] This was a safari the likes of which Lindi may not have seen since Cutler's large troop had marched back from the interior in February, and the German Expeditions before that. Cutler was back at Tendaguru on May 29, after an absence of almost four months.

An impressive consignment of materials had been transported: 67 loads of boards, eight loads of provisions, three loads of salt, woodworking tools, burlap, the heavy forge, 10 mattocks and picks, kerosene, two tents, two tarpaulins, two canvas "bed chairs," crockery, and a host of miscellaneous items for another field season.[5]

Boheti began uncovering bones in Ditch 2X on June 1: two vertebrae, a possible ischium, and a pubis within five days, and metatarsals by mid-month. Three men extended Ditch 2X outward, exposing metacarpals and a radius. Another seven in this quarry were excavating into a sheer three-meter-high wall, while Cutler plastered daily at the site.

Foot elements such as unguals and proximal phalanges appeared, leading Cutler to believe he was dealing with a bipedal dinosaur. By June 30, 17 men were applying their picks and shovels to the overburden in another large extension.[6] One of the several additions to 2X was designated 2X+. It ran at an angle from a creek branch back to the original quarry. Shortly thereafter, a femur appeared in the fresh cutting. A "megalosaur" tooth of great size and in good condition came to light on July 9. Lachlan and Cutler drew a plan and profile of Ditch 2X, the only record made of the position of the finds in the BM(NH) quarries to date.

In mid-June, a crew of 35 was put to work at the grass-obscured Ditch IV of 1924. Others were starting on an extension that had been outlined by Cutler. Sauropod teeth emerged as the extended quarry was cut down to the bone horizon. The specimens were oriented to the northwest. At the end of June, a crew of 31 was engaged at the site.[7] There were further interesting discoveries on July 16. The fresh cutting that had been initiated into the hill produced a coracoid, a humerus, a radius, and three ribs. Ditch III had produced an ilium, and Ditch V was reopened.

Lachlan struggled to build crates, but his right shin was infected and so troublesome that he fainted. Although he had assembled 15 crates by the end of the month, there were enough jackets to fill 80. Cutler was unhappy with the speed at which this vital task was progressing and asked for Anderson's assistance:

> I have nobody here who can make boxes for my specimens except myself, who am otherwise too occupied. Lachlan can do very little being an utter novice. . . . Do you know of a good man who will do this work? . . . I would pay him 50¢ per box or more and food.[8]

The first disagreeable incident of the new season was not long in coming:

> I discharged the Swahili cook Selimani Msau after having to throw him out of my Banda + he almost struck at me.[9]

Selimani was originally Lachlan's conscientious servant, but disliked Cutler. A runner was sent to Lindi to explain the incident to the authorities in case a complaint was lodged.

With camp activities in a routine, Cutler and Boheti conferred about future safaris to Kindope and Niongala, and further afield to Namwiranye and Kilwa. As of July 3, 67 men were employed at Tendaguru.[10] Food was being delivered regularly, mostly millet and maize. A relative of Boheti applied to open a store at Tendaguru but was refused by Cutler. Boheti had earlier asked for a one-month leave of absence to attend to his millet crop near Lindi. This would leave Cutler in a bad situation, as Boheti was to lead the bone caravan later that month, and Cutler probably refused this request as well.

Twenty-five crates had been filled with field jackets, labeled, addressed, securely bound with wire, and banded with "hoop iron" by mid-July.[11] Their contents were almost exclusively from the highly productive Ditch IV.

A crew under Juma and Nyapara cleared the trail toward Mchinjiri and Tapaira. Concerned about the lack of news regarding a motor road, Cutler must have been hedging his bets and completing what the administration had originally intended to do in 1924. A contingent of 34 men next moved toward the Noto ridge in order to open the road up the steep plateau.[12]

Of Wyatt and Anderson, Cutler had urgent questions regarding a road:

> Does your road answer as motor road to Matapua at present? Will a road cover the ten or so mileage thence to here, in next two months? Is that my road for my car? Can you give me address of dealers in Ford trucks and, or, Caterpillar trac-

tors in Nairobi? I shall buy my car as soon as I have your or the Commissioner's guarantee that there will be a road fit to use it on. . . . I need to have car in use before October. What would a native driver for car, cost from Lindi, per month?[13]

His pleas for information were repeated to Wyatt, though there is no record of a reply beyond a promise of 80 porters for July 22.

On June 22, 72 of the 80 promised porters arrived.[14] About 20 men from Tendaguru would join this bone safari to make up the difference in men and to carry food for the caravan. Cutler hurriedly prepared letters to businesses in Nairobi and Dar es Salaam regarding the purchase of a Ford truck or tracked vehicle. In a letter to Harmer in London, Cutler passed on what information he had about the price of motor vehicles:

> The motor car can be bought at Mombasa for about Sh. 2,000—a ton Ford truck. A caterpillar, though splendid for the rough roads, would have to come from Europe, be dear to keep and hard to sell. Some of the boxes will certainly need engine traction, too heavy for manpower.[15]

The bone caravan departed Tendaguru on July 25. Boheti led the 108-man safari, as porters slung 26 crates from poles and strode off down the freshly improved trail to the coast.[16] It was also, coincidentally, William Cutler's 47th birthday, the second he had celebrated in Africa. While relieved and proud to see the men off, the strain of his responsibilities was taking a toll, for he had an argument with his new cook: "I discharge cook for his murderous bread."[17]

At this point, 77 excavators were employed by the British Museum East Africa Expedition, and they were unhappy with a change of procedure at Tendaguru. Since little grain had been brought into camp by locals in June, Cutler had begun to pay a weekly stipend of one shilling rather than a grain ration. By July, about 3,200 liters of grain had been stockpiled, and as he had promised, Cutler reverted to distributing cereal rations. Dissent over this among his workers prompted him to seek advice from the authorities:

> their leader . . . stated that they would not take the grain but wanted the money. As my grain store is absolutely filled, this I cannot do, and told them so distinctly. . . . I certainly shall not permit dictation from my labourers.[18]

The incident troubled him enough that he also wrote to F. M. Manning, the district superintendent of the Tanganyika police:

> A patrol would be very welcome here, since my occupation there has never been such a visit here.[19]

Within a few days, most of the men relented and accepted their rations. Cutler was advised to discharge all the men who were arguing over the matter. Just as this difficulty appeared to resolve itself, another developed. Cutler was partially blinded when sand blew into his eye, which he then rubbed too vigorously: "My right eye is very inflamed and can bear no light, whilst writing this I frequently must cease."[20]

Despite all, the fieldwork never languished. With the aid of six young boys who brought in some old German picks, a 1924 quarry was reopened on July 26th. The work force increased to about 90 with the arrival of another dozen Wangoni men at month's end. Twenty-four men and six boys were sent to an unspecified quarry from 1924, which was renamed 3X. Another 45 were assigned to Ditch 2X+.[21]

Cutler returned to the quarries on August 3, the first time since his eye had prevented field work on July 25. He was still in considerable pain, though the haziness of the damaged eye's vision was slowly improving.

When 30 more Wangoni applied for work in early August, there were not enough tools for them. Cutler decided to use the men for another safari of crates. Packing continued for several days, until the 46th crate was ready. Cutler informed Harmer that, in addition to the crates in Lindi, 10 more were leaving Tendaguru with his letter, another 30 were in camp ready to be packed, and there were enough specimens exposed in the quarries to fill another 30 to 50.[22] The first shipment, of 23 crates, had reached London safely in April.

Back in London, much had happened. In May, Harmer spoke at a special fundraising presentation of *The Lost World*. School headmasters had been invited to the screening at First National Pictures' private theater in London. To further raise public awareness, the Museum displayed a few bones from the Tendaguru shipment at the Tanganyika Exhibit of the British Empire Exhibition being held at Wembley Stadium.

In his report to the trustees, Harmer stated that just over £1,000 had been donated to the Tanganyika Fund, and suggested that a special committee be formed to administer the money. Harmer recommended himself, Lord Rothschild, Bather, and Smith as members. Support had come from surprising quarters: the Royal Society and the Percy Sladen Memorial Fund had each donated £100 by mid-June. Louis Leakey wrote to obtain permission to give lantern-slide lectures at public schools in Cambridge, on condition of reserving 50 percent of the proceeds for the Tanganyika Fund.

It was August 10 when Paul, an old hunter, was given four men to reopen a ditch from 1924. It was not identified with a number, but was situated on the right side of the Lindi trail, and may have been Ditch VII.

Cutler plastered in Ditch 2X as well as he could with one good eye on August 19, and then declared the quarry abandoned. An unnamed site was opened that day, north of camp along a tributary creek. Ditch VII rewarded the efforts of the excavators with a large thoracic vertebra. Juma started yet another unnamed quarry near both the Lindi trail and a small German quarry.

The punishing regimen to which he had subjected himself left Cutler exhausted, or worse: "Upon returning home, feeling weak and horrible, found my temperature to be 104°, so went to bed."[23] He remained in camp the next morning. The final entry in Cutler's field notebook was penned the next day, on August 21, 1925.

The next news from the Expedition reached London like a thunderclap. It was a telegram from the governor of Tanganyika Territory to the Colonial Office, dated September 1, 1925, and was forwarded to the Museum authorities:

Regret to report death of Cutler at Lindi 30th August Malaria.[24]

The unthinkable had happened—but how? Commissioner Wyatt, who telegraphed the governor's office on Monday, August 31, included a few more details in a letter to Harmer on the same day:

> Of recent months his health had not been good and he was brought into Lindi from Tendaguru on the 27th instant in a precarious condition. Everything possible was done for him by the Medical Officer but I regret to say that he succumbed on the 30th instant. It may comfort his relatives to know that his end was painless and he passed away quietly after having been unconscious for about 18 hours.[25]

In the death certificate, Dr. Blackwood, the medical officer, listed the cause of death as malaria. In the tropical climate Cutler was buried soon thereafter, in Lindi's European cemetery. A letter from Lachlan reconstructs the last days of Cutler's life:

> He returned from D.S.M. [Dar es Salaam] in much better condition than when he left Lindi. Largely a question of dental +, I think . . . malarial trouble. He returned, however, with the dental trouble underscore(uncompleted)—owing to the fact that, having had so much pain + such a gruelling time with the dentist, the latter either would not or did not complete his task, thinking, perhaps, that Mr. Cutler's health (although better) was not robust enough to stand any more extractions etc.—On the safari up here with me, he complained once or twice of the dentist not completing his work, + on our arrival it was evident from time to time that he was in pain from two or more teeth + told me that he was quite sure that his stomach (which gave him also a good deal of trouble) was receiving poison of some description from his bad teeth. About a month before his death he was working in a far Ditch + returned to the Banda complaining of having got a lot of either dust or mica, from the breeze then blowing, into his right eye + from then onwards his health seemed to give him + myself cause for care. He was, however, a particularly bad patient + would listen to no advice of mine. I said at the time that he should go into Lindi + get the Doctor to look at his eye from which at this time he could see little or nothing. Then—about the 19th of August—he said one morning at breakfast that he had difficulty during the night of getting his breath + remarked that he had had troubles of a similar nature before. On the 23rd—24th ult. his temp. rose considerably + kept rising + falling until on . . . 24th Aug. the Headman + myself told him he must go into Hospital + I added that unless he went in I should send a runner for the Doctor to come up. On the 25th Aug. he was considerably better + able to look after most of the small Safari himself. He would, however, take nothing to eat either the previous night or that morning + altho' I prepared food for the Safari—all I could get him to take with him was a tin of cheese biscuits + some water—he refusing to take anything else. As it was a two-day journey + I had written the Doctor to come out + meet him I felt that he would be quite alright. It appears now that he became delirious about 1 day out + arrived in a very critical condition on the 27th midday—After which I heard nothing until later on the 1st inst when the sad news came to me by runner. To say that I have lost a very good friend + companion in no way expresses my feelings—Although with a great knowledge, far in excess of anything that I possess, yet with it all, he dropped to

my level in conversation + cheeriness + helped me to take a great interest in things of which I had hitherto no knowledge. May God rest his Soul in Peace—I feel his loss more than I can say.[26]

So, prematurely, ended the career of a highly dedicated and skilled field paleontologist. Though it is evident that there was little sympathy in some quarters, his behavior must be placed in the context of his background and the times in which he lived. Self-reliance would have been an essential quality when he emigrated to the sparsely populated homesteading environment of western Canada in the early 1900s, where one was expected to cope by drawing on one's own resources. Further, choosing a physically demanding and financially marginal occupation would have placed a further premium on hard work, perseverance, and independent action.

His attitude toward the Africans was likely no different than that of most men of his era—he instituted a strict hierarchy in which his orders were to be carried out unquestioningly. Health problems, both longstanding and self-inflicted, and an insistence on control over all aspects of the enterprise had adversely affected his performance in the most ambitious collecting program he had ever undertaken. Cutler had enthusiastically accepted enormous responsibilities for the sake of the Expedition, which was the acme of his career. These responsibilities had greatly taxed younger men like Janensch, Hennig, and Reck. For this Cutler had paid the ultimate price. A surprisingly objective characterization came from an unlikely source—Senior Commissioner A. H. Wyatt:

I should like to add a word regarding the late Mr. Cutler's work. In it he had many difficulties to contend with but his keenness and determination carried him through and I am confident that, had he lived, he would have succeeded in adding many valuable specimens to the Natural History Museum.[27]

-12-

1925

A New Recruit

The sudden death of the leader of the East Africa Expedition caught Museum authorities unprepared. Harmer was vacationing at the time and could not easily be reached. As the matter was exigent, Assistant Secretary Smith assumed responsibility. He took the telephone call from the Colonial Office on Tuesday, September 1, notifying him of Cutler's death the previous Sunday. Soon he had a copy of the telegram from Africa. The Tanganyikan government further advised that "Unless successor arriving shortly suggest works be closed down removing stores and equipment to Lindi. Please telegraph wishes of British Museum as to shipping home specimens now collected."[1]

Smith reacted speedily with a flurry of letters and visits. He dispatched a note to C. W. Hobley. In Smith's opinion it was inadvisable to entrust Lachlan with the Museum's funds or the leadership of the Expedition: "we know nothing about him and I rather imagine that he would be quite unfitted for a post of responsibility."[2]

Smith called on the Colonial Office the next day to solicit help with the Tanganyikan authorities. By the third, he had received a copy of a telegram drafted by the office of the secretary for the colonies to Governor Cameron's administration. It stated that Lachlan would not be appointed as leader but should continue assembling crates for field jackets and packing cases for shipment. The local administrative officer was asked to assume control over Museum funds. In August, £300 had been transferred to Lindi: £175 for Cutler's salary and £125 for the purchase of a Ford truck. All too late. The letter informing Cutler of the payment was returned to the Museum marked "Deceased."

Cutler's obituary, penned by Hobley, appeared in the *Times* on September 2. Transmitted by the Canadian Press wire service, it soon echoed through newspapers in Winnipeg, Regina, Edmonton, and Calgary. The high commissioner for Canada in London received an expression of the Museum's regret over Cutler's death, and was asked to extend condolences to his relatives and the University of Manitoba.

Smith sent a message to Dr. Edmund O. Teale, FGS, who was about to depart for Tanganyika Territory in the capacity of director of the Geological Survey.

Teale was asked to assist the Expedition through the advice of his office, and to consider some form of mutually beneficial cooperation in the field.

Smith's next step was an urgent search for a suitable Expedition leader. To this end, Captain G. H. Wilkins was approached. He agreed to join the Tanganyika Expedition, but would only commit to staying for seven or eight months.

George Hubert Wilkins, an Australian, had served on Stefansson's Canadian Arctic Expedition from 1913 to 1916, and on Sir Ernest Shackleton's 1921–1922 Quest expedition to Antarctica. From 1923 to 1925, he had led the Wilkins Australia and Islands Expedition on behalf of the BM(NH). The expedition had collected birds, fish, mammals, insects, plants, minerals, fossils, and human artifacts in remote northern regions of Australia. Wilkins would undoubtedly have made an excellent leader, though his interests were far more focused on polar exploration.

Cutler's last report had indicated there were sufficient exposed specimens to fill another 70 or 80 crates, so quick action was imperative to prevent the loss of material to impending rains. Meantime, one of the most influential men in British paleoanthropology, anatomist Sir Arthur Keith, endorsed a candidate for expedition leader:

> F. W. H. Migeod tells me he is offering his services to the Natural History Museum to go to Tanganyika. I have known him these 10 years; he is . . . the most intrepid + experienced traveller in Africa we now have. He has collected much for us with the utmost care, supplying full + minute + accurate histories with all his specimens.[3]

A recommendation from someone of Keith's stature in the scientific community was not easily ignored. Bather, the Museum's keeper of geology, advocated accepting Migeod as Expedition leader, adding that he felt Migeod would prove a more careful collector than Captain Wilkins. Thus the polar specialist passed from the Tendaguru scene.

Under the impression that Cutler had operated without skilled help, Louis Leakey offered to return to Tendaguru as he had promised at the end of his first field season. Leakey was able to participate for a short time, as he had only completed the first part of his degree at Cambridge. He added an interesting footnote to the life of William Cutler: "I do not . . . believe his parents are alive, but I know he had a brother and an aunt in Canada, and he was I understood from him engaged to a girl there."[4] Cutler's brother had died in England in 1921, and his aunt lived in Britain. The fiancée was probably the E. M. L. of his diaries, who lived with her family, the Lingards, near Wingello, Saskatchewan.

Smith and Bather interviewed Migeod at the Museum on September 9. Migeod accepted the position two days later. The Museum wrote to Leakey, informing him of the appointment and adding that he would be asked to accompany Migeod for about one year.

Frederick William Hugh Migeod was born on August 9, 1872, at Chislehurst, Kent. He received his education in Folkestone and joined the Pay Department of the Royal Navy at age 17. Though he loved the sea, he left the employ of the

Navy nine years later, in 1898. He joined the colonial civil service and, in 1898, voyaged to West Africa, stopping at Nigeria. Migeod spent roughly one year at the confluence of the Niger and Benue Rivers. During this time he became interested in the culture of the local people. In his first book, published in 1908, he compared three West African languages.

The routine of civil service employees was a year on station followed by several months of leave. Migeod returned to England and in 1900 was assigned to the Gold Coast Colony, presently Ghana. He remained in that country for the next 19 years, serving as transport officer. He continued to record information about customs and languages and submitted articles to academic journals. He wrote three more volumes on African languages, as well as a book outlining his thoughts on how primitive man might have coped with his environment, and became a Fellow of the Royal Geographical Society and of the Royal Anthropological Institute.

In 1919, at 48 years of age, Migeod began realizing his dreams of far-ranging and frequent travel. Now semiretired, he returned to Africa to study the ethnology of tribes of equatorial Africa. In the course of 13 months, he traversed the continent from west to east, on foot and via the Congo River. Unable to locate a shipping line that would transport him and his African servants back to the west coast, he resolved to trek across the continent again from east to west.

On Christmas Day 1920, the intrepid band left Dar es Salaam for Tabora by rail. Migeod inspected the remains of the crashed Cairo-to-Cape aircraft, the *Times*-sponsored Vickers-Vimy that had carried Dr. Peter Chalmers Mitchell to Africa. At the end of March 1921, Migeod boarded a ship in Lagos, Nigeria, that carried him back to England after an absence of 18 months. Home was Northcote, or 46 Christchurch Road, Worthing, Sussex, and it was here that Migeod set down a lengthy account of his first journey.

Seven months after his homecoming he set out on a second voyage. In November 1921, he embarked on a trek through Nigeria to Lake Chad in order to record ethnological details and to understand how drought affected human migration. At Yola he stayed with his brother Charles, who was the resident of Yola Province. His party reached Lake Chad in March 1922 and Kano more than three months later, and rode the train back to Lagos on the coast. He recounted this nine-month odyssey in another volume.

At the end of January 1923, Migeod was en route to British Cameroons. He partially scaled the 4,100-meter-high Mt. Cameroons. At the end of March, Migeod's group was proceeding east through Bafut country. Six months later, Migeod once again stayed with his brother Charles at Yola, Nigeria. The party traversed the Nigerian countryside until a halt was finally called. Exhausted, Migeod returned home after 9^1/$_2$ months.

Back in England, he recovered from the effects of his sojourn and began organizing his next book. Migeod was a meticulous planner, having completed several of his earlier treks on the sum of £2 per day, averaged over 10 months. To achieve such economy he constantly adjusted the number of porters employed, monitored the price of foodstuffs in village markets, and informed himself of boat and rail costs.

In late September 1924, Migeod departed for Freetown, Sierra Leone. By early January 1925, he was in Mende country in the southeast of the Sierra Leone Protectorate, where he documented the customs of Mende-speaking people. The voyage home began at Freetown in April, 1925. It had been a six-month tour of discovery, and would yield another book.

In Worthing, Migeod resumed his writing and remained active in the Worthing Archaeological Society. From his books, it is obvious that he was possessed of an insatiable interest in Africa, and was a meticulous observer. He documented his travels exhaustively. His views on Africans were standard for the era: a belief that whites must remain paramount, and that there should be no mixing of the "races." Migeod did not accept indirect rule for Africans, a policy being introduced in Nigeria under which the British would govern through the defeated native rulers. He respected West Africans and felt they were capable of accomplishing more than their counterparts in East Africa. The following passage illustrates the methods he developed through trial and error to render his travels as free of complications as possible:

> I strongly object when travelling to have to give a single order or say anything at all, and require everything to run as automatically as possible. . . . I further find it better never to call a boy, except for some great emergency. I object, too, to be asked if I am ready for lunch, dinner, or whatever it may be, or for a bath. It is for the boy concerned merely to announce they are ready. . . . if you should by chance fall sick, which I usually manage to avoid, you are not stranded with everything broken down because you are too ill to give any instructions.[5]

Years later, upon Migeod's death, William E. Swinton, who joined the Department of Geology at the BM(NH) in 1924, wrote,

> His manner of speech and bearing sometimes gave an impression of arrogance; but in fact he was a kindly and friendly man and an admirable host.[6]

And in a similar vein, he maintained that Migeod

> had something of a reputation for being difficult to get on with but this must have been with those who only knew him slightly. He was a charming host and his nearest friends alone knew of his many acts of friendship and generosity to those in difficulties. . . . For most of his life he was, in the domestic sense, a lonely man, but the abundance of his interests and the delight he found in all educational work and the pursuits of amateur naturalists will long be remembered.[7]

A much less sympathetic assessment was made by the young American paleontologist George Gaylord Simpson:

> He calls it Mee-zhoh, but many's the unsuspecting lad who has addressed him as Mr. My God! This Migeod is a quiet dried up fellah that looks as if he might have a fairly sticky past tucked away somewhere. . . . This little prune has tramped right across Africa on the equator . . . & then not yet sufficiently fed up he tramped right back on the parallel 5° South, which is even worse.[8]

In 1925 the 53-year-old Migeod married Madeleine Marguerite Adrienne Charlotte Banks. Despite a marriage that was said to have been happy, Migeod had been home for only four months when he read of Cutler's death and offered his services to the BM(NH).

In his letter of September 12, 1925, Assistant Secretary Smith defined the terms of Migeod's appointment as leader of the British Museum East Africa Expedition. The appointment would last from October 1, 1925, to December 31, 1926, and could be terminated with three months' notice by either party. Salary was £25 per month, plus £5 per month hospitality allowance. Board and lodging would be paid separately by the Museum. Return passage to London was guaranteed upon termination on December 31, 1926, or when mutually agreed upon.

To avoid independent action of the sort Cutler had undertaken by hiring Lachlan, the Museum permitted Migeod to hire or dismiss Africans, but obliged him to obtain trustee approval prior to any decisions involving Europeans. Migeod was to provide monthly accounts of expenses. The collections were the property of the trustees of the British Museum, and the trustees retained sole rights to publish any results.

On September 14, Leakey drove to Migeod's home in Worthing from Boscombe on his motorcycle. Migeod was favorably impressed by the young man: "it would be eminently desirable to obtain his services. . . . I further suggest his salary might be raised to £15 p.m."[9]

In informing a Museum trustee of the decision to appoint Migeod, Smith made a curious statement: "As he is a man of much higher education than Mr. Cutler, I think he should do even better work."[10]

Leakey met with Smith in mid-September, and was offered employment. On the advice of his Cambridge tutor and his father, Leakey refused the position on September 16. Perhaps he was wary of working with another older man, who at age 53 was, like Cutler, accustomed to operating independently: "I felt somehow from things Mr. Migeod say [sic] and left unsaid, that he did not really want me with him."[11]

On September 25, a letter reached the Museum from Major Thomas Deacon of Waterloo, Liverpool. Deacon offered to help in any capacity and for any length of time on a voluntary basis. With Migeod's agreement, and following a meeting with Museum officials on the 30th, Deacon was appointed.

Thomas Deacon was born on October 9, 1864, in Milton Abbott, Devon. He joined the colonial civil service and was postmaster general of the Gold Coast Colony around 1901. Migeod knew him, having been posted in the same colony at roughly the same time. Deacon spent two years of the war in the East Africa campaign, traveling with the Allied forces from Dar es Salaam to the Rufiji River and Lake Victoria. Following the Armistice he was posted to Kenya and then commanded the King's African Rifles in the Uganda Protectorate for several months. Blessed with a robust constitution that had resisted illness throughout the campaign, Deacon was proficient in Swahili and possessed an in-depth knowledge of East Africa. Being of independent means, he could now, at age 61, volunteer without pay. Deacon was married to Dr. Mary Ariel J.

Stewart, MBE. His employment contract included expenses for room and board and return fare to Africa. He was to provide his services until at least June 30, 1926, and could end his participation with three months' notice.

Migeod familiarized himself with Tendaguru material at the BM(NH). He received offers of help for the Expedition personally, and before long, another offer reached the Museum. Wilfred Hodgson, of Dorothy, Alberta, Canada, wrote, stating that he had been good friends with the late William Cutler and had recently offered to join Cutler in Tanganyika as his assistant. News of Cutler's death had reached the small prairie hamlet via radio. When consulted about Hodgson, Arthur Smith Woodward did not recommend pursuing the matter, and Hodgson disappeared from the picture.

In an update to Harmer, Smith expressed no sympathy for the deceased Canadian:

> Cutler brought this tragedy on himself from his disregard of advice. I found on my visit to the Colonial Office that news of the reckless way he went on had reached them, and they were by no means surprised to hear the news.

In response to an earlier suggestion from Harmer, though, Smith did soften his stance:

> I agree that if it transpires that Cutler had any dependents, the Trustees should be recommended to go beyond their legal requirements and pay a lump sum of say two or three hundred pounds.[12]

In the interim, Senior Commissioner Wyatt and H. L. Lachlan had both written to Harmer regarding conditions at Tendaguru. Wyatt had driven to the site, arriving September 18 and departing the next morning. The round trip over very rough terrain took a day and a half. On Wyatt's advice, Lachlan had inventoried Cutler's personal effects and Museum property. All but 12 men were released by Lachlan on August 31, saving about 1,500 shillings per month in expenses.[13]

Cutler's assistant had not felt qualified to direct any further excavation and sought only to preserve and ship readied specimens. Lachlan organized 20 crates to be carried to Lindi on September 24, leaving enough field jackets in the bone storage hut to fill another 30.[14] Wyatt was to assemble the roughly 250 bearers that would be required for the bone safari. Another 220 specimens remained exposed in various quarries.[15] Half were sitting on pedestals in Ditch 2X+; the remainder were in 2X, 3X, IV, and several unnumbered quarries. Finally, there were another 36 boxes already waiting in storage on the coast. Lachlan included a detailed list of the contents of Ditches II and IV.

Two further points were raised, both awkward. Firstly, Lachlan had received no pay since starting on May 1, despite the promise of an allowance of 300 shillings or £15 per month plus food expenses for six months.[16] His savings were now depleted, and he wished to either renew his contract with the Kikwetu Sisal Estate or accept a job offer in England if the Museum could dispense with his services. Despite loyalty to Cutler, he disapproved of the latter's behavior: "I maintain (+ I have spent also 3 yrs in Sumatra) that no man can go out + work through the heat of the day (as Mr. Cutler did) 1. wearing only a light velour hat + 2. drinking the unboiled water (which the natives take with them into the field) in large quantities."[17] Hearing of Lachlan's desperate finances, Migeod urged an advance of £25 to Lachlan's wife Lucille in London. The outstanding £65 would be paid to him by the administration in Lindi.

Secretary Smith forwarded a telegram to the Colonial Office for Governor Cameron, announcing Migeod's appointment. Migeod left England for France on October 9, his brother and wife seeing him off. Deacon was to stop at the BM(NH) on his way out of the country. His train to France left England on October 21.

More specifics were provided by Lachlan in October. He reported that all cases up to number 54, and a selection from 54 to 88, arrived safely on the coast. This emptied the bone store and Tendaguru camp of all readied specimens. The bones originated from Ditches II, 2X, and IV. The first safari, transporting 17 wooden containers, involved about 130 porters and the second, of 33, required roughly 270 bearers.[18] All remaining exposed bones were covered in thick brown paper and buried under a mound of soil, and their location in the quarries marked with sticks pressed into the ground. A photograph of Ditch 2X was taken by Lachlan to supplement the plan and cross section drafted by Cutler. Notes were left with the administration in Lindi to guide Migeod, should Lachlan be unable to meet him in person. The expense of transporting the crates and Museum equipment to Lindi was assumed by the Tanganyikan government, which would invoice the Museum. The self-effacing Lachlan had served the BM(NH) loyally.

In preparation for a meeting of the British Museum trustees in late October, Assistant Secretary Smith drafted a summary of events. He mentioned that Cutler's sole surviving relatives, his aunts Heloise Watson and Ellen I. Watson and Ellen's husband, all of London, had asked for details of his death. They had last seen him during the war. A copy of the death certificate arrived at the Museum a few weeks later. Cutler had left a large cabin trunk and a tin case containing lantern slides, notes, and private papers in a Museum storeroom.

Despite the fact that neither Migeod nor Deacon had geological or fossil-collecting experience, the Museum authorities felt optimistic that the Expedition had been salvaged. The best possible arrangements had been made, given the circumstances.

-13-

1925-1926

An Expedition Saved

Despite setbacks, there was cause for optimism. The economy in both Britain and Tanganyika Territory was stronger than it had been in the first years of the decade. A decisive new governor was at the helm in Africa and had pledged his support to the Expedition. Lastly, a new Expedition team had been formed, in which Museum authorities appeared to have greater confidence.

Migeod arrived in Dar es Salaam on November 4, and stayed at the New Africa Hotel, the former Kaiserhof of the German era. At Government House, Migeod was introduced to the governor and other officials, and learned that "Poor Cutler seems to have been a very difficult person."[1]

Interestingly, the German claims that Tendaguru specimens had been appropriated had some substance after all: "I saw in the Secretariat the whole of the official file with regard to the dinosaur remains dating back to War Times, and it seemed that South Africa was very anxious to secure specimens for their Museums and that a large shipment was apparently made."[2] This statement caused surprise in London. Harmer had never investigated claims that the South African Museum possessed specimens of *Brachiosaurus* after it had declined to participate at Tendaguru.

It is unclear which Pretoria museum had approached the German colonial authorities in 1910 regarding permission to excavate at Tendaguru. There is no doubt that South Africans had been interested in the site for years. There was a sizable collection of Tendaguru bones in Dar es Salaam when the Allies occupied the city, since Hans Reck's preparators were readying material for the National Exhibition and for a future museum. South African troops frequently moved through the city as well. If Pretorius had arranged excavations at Niongala, transport for fossils could have been provided by South African forces. The South African Field Artillery had operated out of Lindi between late April and early December 1917, and infantry units had fought at Mahiwa and Nyangao. Other than material obtained in exchange many years later, however, there is currently no evidence of Tendaguru fossils in South African museums.

On board a steamer to Lindi, Migeod met a German passenger who had served with von Lettow-Vorbeck, and was a source of information about the ill-fated Bernhard Sattler:

> He was shot by an Askari in 1916 near Kilwa. An askari had broken away and was looting. S[attler] found him in the act + instead of covering him + ordering him to lay down his rifle went up to him to arrest him + was shot dead. The man [Askari] had previously kicked several natives. Other Askaris came up + shot him in his turn + finished him slowly.[3]

Sattler had actually died in late October 1915, although the rest of the story is corroborated by other accounts.

When Migeod and Lachlan twice attempted to obtain an interview with Anderson, the administrative officer in Lindi, they were asked to wait, along with several Africans. This was unacceptable to Migeod, who walked out. Despite a subsequent apology Migeod reported his dissatisfaction with the authorities to Harmer, who annotated this passage "Confidential":

> Mr. Leakey . . . informed me of the unsatisfactory relations existing between the Expedition and the Administration of the Colony. . . . I was certainly surprised at the casual reception accorded me, such as I have never met with in any of my private expeditions in Africa. . . . after an apology by the Officer concerned, for Mr. Lachlan had lost no time in passing around my views on the subject, more cordial relations were established.[4]

The relationship that existed between the administration and Expedition members is difficult to determine objectively. The Territory had suffered grievously during the war and so esoteric a pursuit as paleontological excavation must have seemed inexplicable to government officials. The reputedly uncommunicative Governor Byatt may not have informed his staff in Lindi of the aims of the Expedition, and was not especially supportive of the demands of Europeans. There had been an interval of almost a year between Byatt's departure and Cameron's arrival, during which John Scott was acting governor. Like Byatt, Scott had been criticized by settlers for heading an inert administration and seeing little of the country during his tenure. The supportive provincial commissioner Orde-Browne had long since left Lindi District, and his successor, Commissioner A. H. Wyatt, had only transferred to Lindi the day Leakey arrived. One can imagine that Wyatt and other officials were soon occupied with practical matters of higher priority in this large district.

Further, the terms of the League of Nations mandate did not allow Britain to overly favor her nationals in the Territory. The proposed two-year duration of the Expedition likely did not encourage the extensive local support Cutler had demanded. This was especially true in the case of building roads to isolated sites that had no economic value, or lending something as rare as a motor vehicle. Harmer's subsequent commitment to Dar es Salaam for a longer presence at Tendaguru may not have been communicated to Lindi.

There may have been personal differences between Cutler and government representatives in Lindi, who may not have been convinced of his authority. Any one or all of these conditions may have contributed to Cutler's and Migeod's impression of being slighted as representatives of one of the world's greatest natural history museums.

Shortly after Migeod's arrival in Lindi, Boheti reported to him, being one of the two men retained by Lachlan. The overseer was receiving Salvarsan injections, possibly to treat an infection transmitted by ticks, which caused recurring fever.

Lachlan was placed in charge of shipping the 83 crates in storage at Lindi. When Migeod, a former transport officer, witnessed the impressive bulk of crates amassed in the storage area, he offered high praise: "Mr. H. L. Lachlan has done good work in very adverse circumstances, and the transport of the large cases from Tendaguru was an exceptionally good feat."[5]

Migeod and a string of 22 bearers marched into Tendaguru on November 18.[6] Excavation resumed on November 20. Sites were visible in all directions but marked only by tall sticks in the ground. With Lachlan in Lindi and unavailable to identify Cutler's quarries, Migeod was seriously disadvantaged. He chose an existing pit, which measured six by three meters and was about two meters deep, and designated it M1. From the series of articulated bones concealed under brown paper and loose earth, and Boheti's confirmation, he took this to be Cutler's Ditch 3X. It was situated .8 kilometers from camp. Vertebrae and possibly limb elements soon lay exposed, but the inexperienced Migeod initially mistook the skeleton to be that of a *Plesiosaurus,* a marine animal that had not been found at Tendaguru to date.

With his limited anatomical expertise, Migeod confessed confusion over the number of animals represented in the first ditch. After nine days of excavating, M1 No. 1 consisted of a skeleton with about 24 vertebrae, a humerus, two femora, tarsals, phalanges, pelvic elements, and numerous ribs.[7] The bones were exposed, allowed to dry for an hour, consolidated with shellac, and then plastered. A quarry plan was sketched as well. The excavation was completed and the last jackets were brought into the bone store in camp on November 30. M1 No. 1 was packed into nine bearer loads on December 17, and they were moved to the coast four days later. The articulated vertebrae were placed in sequentially numbered boxes so that the position in which they were found could be reconstructed at the Museum.

At the deepest end of M1 No. 1, and lying at an angle of 120° to it, was what Migeod believed to be a second skeleton. He designated this M1 No. 2. Near the surface, and therefore fragile, were sizable vertebrae. By December 10, it was raining most afternoons, restricting excavation to the mornings. A tarpaulin was stretched over the quarry to allow Boheti to plaster unhindered. A mass of pebbles had been deposited at the south end of the bone accumulation and a rib was bent by the quartz stones. Migeod interpreted this as the action of a stream in which a portion of the animal had come to rest. The quarry measured about 12.5 meters in length. By Migeod's calculations the Expedition's expenses were currently not exceeding £5 per day. At this rate, for a total of 42 days, it had cost about £135 to remove M1 No. 1.

By mid-February 1926, the final tally of bones in M1 No. 2 included two scapulae, two humeri, two coracoids, two ulnae, radii or fibulae, a femur, two tibiae, an ilium, a pubis, possibly an ischium, foot elements, portions of ribs, 30 caudal vertebrae, about 30 other vertebrae including three cervicals, and several other unidentified elements.[8] One femur and one tibia stood nearly vertically in the sediments. The ilium also was lying at a steep angle with vertebrae tight up against it, though these may have been the sacrals. Collecting M1 No. 2 had required about 70 days, for a total of £310 to £325.[9]

Eighty men carried the jackets of M1 to the coast in 54 loads at the beginning of March 1926.[10] Despite pronouncements that the excavation had been completed, Migeod ordered trenches cut across emptied quarries until there were no fragments left. It was not until mid-July that the sections of this large site were photographed, with signposts designating the subdivisions.

Migeod was led to several new and old sites, and proposed to concentrate on groupings of bones rather than isolated, individual elements. He intended to clean all of Cutler's ditches of bone and then return to the German quarries. Not satisfied with merely collecting dinosaur bones, Migeod also gathered botanical specimens with the same passion that Cutler had shown for butterflies. By late April 1926, when the flowers produced by the trees of the area were blooming, his plant specimens totaled 150.

If Migeod was inexperienced in excavation, he was exacting and thorough. Plastering was supplemented by screening matrix and regular mapping of quarry contents. He also took the precaution of excavating another 15 centimeters below bone level.

Migeod described his collecting strategy: "If I am dealing with a whole skeleton I leave not a fragment, but in case only of a selection from miscellaneous bones I naturally only take good pieces + what I am sure about."[11]

Migeod devised a careful system of tracking the field jackets, since they were carried in daily. A slip of paper was labeled with the bone's position or number on the quarry map. This tag was attached to the bone in the quarry and the jacketed specimen was then removed by experienced men. Usually by the end of the morning shift, the labeled jacket was received in the storage hut. The specimen was later cleaned by an assistant. Migeod then glued together as many fragments as possible. A reassembled bone was treated with shellac and splinted, if not already processed this way in the field. The fossil was then ready for packing. Some bones sat around for weeks in the hope that a missing portion could be uncovered and reattached. This practice had the drawback of rendering the storage hut extremely crowded.

With proper glue so difficult to obtain, Migeod turned to boiling a concoction of "gum arabic" extracted from local trees. He admitted it was not strong enough to cement heavy fragments, but had little alternative. In desperation, he ordered methylated spirit and shellac from London, having waited in vain for a reply from the Public Works Department in Dar es Salaam. The materials eventually arrived from the East African capital three months later.

Small wooden packing cases that had originally held kerosene tins were brought to Tendaguru. Two field jackets would be placed in each case, which the carriers would then move to Lindi. Later in the season bones were placed into locally woven oblong baskets, which were then stacked on top of each other in the cases. A list of the contents of every kerosene case was stenciled on its sides. The Equator Saw Mills firm, of Nairobi, was requested to build and ship crates in sections to Lindi, where they would be reassembled, reminiscent of the German method. In town, three kerosene cases were loaded into each of the heavy wooden crates.

According to Migeod, had the larger crates intended for ocean transport to Britain been carried to Tendaguru and filled on site, it would have taken about a dozen men to convey them back to Lindi, slowing progress markedly. This was about four times the manpower than was actually used.

Equator Saw Mills later sent notice that a business in Essex, England, would purchase the Expedition's crates once they had been emptied at the BM(NH). The *Podocarpus* wood would be used to build furniture, so some of the 27-shilling cost of each crate could be recouped by the Museum. Given the expense and trouble of obtaining wood, would it not have been more economical to reuse the lumber, as the Germans had?

Migeod attempted to survey quarry positions, but the vegetation and rolling countryside that had hampered all of his predecessors were never satisfactorily overcome. By sending an assistant up a tree with a bamboo pointer, Migeod was able, from the summit of Tendaguru Hill, to take an imprecise reading of the bearing of quarries to his campsite. Later he had smoky fires lit at ditches as a guide. An aneroid barometer loaned by the Royal Geographical Society was used to estimate elevations of excavations above sea level, and Migeod concluded that the quarries most often cropped out at about 215 meters above sea level. There is no indication that he was clearly aware of which bone-bearing stratum he was excavating.

In Lindi, Lachlan must have felt a sense of pride and relief. On November 23, a Holland Afrika Lijn steamer had slipped out of the harbor carrying 83 crates of Tendaguru dinosaur material. This load weighed 8,270 kilograms.[12] Commissioner Wyatt had written to the shipping company in Amsterdam and London, requesting that "special care" be taken during the transfer.

Migeod was paying a wage of 14 shillings per man, but had no idea how much grain to distribute. Disputes arose immediately. He claimed that "everything is chaotic" due to Cutler's lack of bookkeeping. A few questions to Lachlan, Wyatt, or Anderson would have clarified matters, though Migeod likely had no intention of appearing ignorant. Like Cutler, Migeod chose to shoulder the entire burden of camp and excavation duties personally, without any European assistance.

Migeod did recognize ability in others, and praised Boheti's competent plastering. As promised, Boheti's monthly salary was raised from 50 to 65 shillings, including one shilling per week for rations. Juma's pay started at 30 shillings including rations, and later rose to 42. There do not seem to have been any audiences or shauris, but workers demanded Migeod's medical skills.

Due to the difficulty of operating without a representative on the coast, Migeod engaged the agent for the Holland Afrika Lijn, G. C. Bennett, for £50 per year. Bennett would deal with all shipments of supplies and crates of specimens, including the assembly of packing cases. By the end of May 1926, however, Bennett's lengthy absences from Lindi had reduced his value to the Expedition to the point that the agreement with him was terminated.

For someone not given to hyperbole, Migeod was greatly impressed with the abundance of bone from the outset: "One can pick up bones everywhere. It is one huge graveyard exposed by denudation. . . . There are here enough specimens to supply all the Museums in the world."[13]

Despite the positive assessment, Migeod criticized numerous aspects of his predecessor's fieldwork: "I may be wrong but Cutler was only out to find something sensational in the form of a single gigantic bone + had no use for small things and broken fragments. I find he never used a sieve, + have long since had one made + have retrieved fragments thereby."[14]

Migeod interpreted the piles of bone at Cutler's Nguruwe clearings, some 2.4 kilometers from camp, as evidence of collecting bias: "Cutler as far as I can see only picked the eyes out of a place, and never attempted to complete excavation of a whole animal. . . . Nearly all of Cutler's openings are close to German ditches."[15] Coincidentally, Cutler had made similar remarks about the lack of thoroughness exhibited by the Germans, when he emphasized the amount of material he had been able to remove from an extension to a German quarry. It seems likely that the best specimens had already been collected, leaving fragmentary, duplicate, or nondiagnostic finds rejected as unworthy of the expense of shipment.

Teams from Stuttgart, Berlin, and London had worked the region intensively, generating many quarries and much rejected bone. Unfortunately, no group had left adequate maps that identified the quarries well enough to re-locate all of them in the field. When Cutler had worked close to or extended existing quarries, as Migeod was doing, the area had become confused beyond unraveling. Individual quarry maps, though meticulously drawn by the Germans, had never been published or consulted by the British. There was no accurate overall site plan to guide succeeding teams.

Migeod suffered the consequences of a massive, multiyear campaign by hundreds of excavators at dozens of localities. When this is coupled with his own inexperience, it is not surprising that he could not determine the strategies of his predecessors. Without documentation, it is unlikely that anyone could have made sense of what had been removed from which unit by this time.

While Cutler had worked close to German quarries, it should be remembered that this was the sole strategy of the BM(NH) at the commencement of the Expedition. Presumably much time, effort, and money could be saved by digging near these proven sites, since the Germans had already cleared large areas and opened enormous quarries.

As the field season progressed, Migeod again voiced frustration over his inability to understand the rationale followed by previous excavators:

> The more or less complete dinosaurs are there waiting to be excavated but he seems to have gone on opening more and more ground in all directions, and the bones when unearthed were never treated with shellac.
> ... It is distressing to see at M2 the vast quantity of fine bones excavated by the Germans and left exposed when they had to abandon the diggings. Now they are all perished from so long an exposure to the air.[16]

Undeniably, Cutler had steadfastly opened ever more quarries, but he had done so in an attempt to find complete skeletons worthy of impressive museum displays. The more area opened, he had reasoned, the better his chances. He had been under pressure to provide material in order to justify the expenditure of the Purchase Grant.

Also, there were few exposures at Tendaguru of the kind Cutler had been accustomed to working, and so he had followed the German strategy of excavating large expanses where promising bone accumulations had weathered out on the surface. His fatal mistake had been to carry on relentlessly, badly overextending himself until he had been physically unable to complete the work. Cutler had certainly been capable of determining whether a quarry contained a relatively complete skeleton or not, and abandoned those that had proved unsatisfactory, judged according to his hard-earned Canadian experience.

Migeod next found fault with Cutler's collecting scheme: "He never got out a more or less complete dinosaur."[17] In fairness, it may be that Migeod had been encouraged to bring back complete skeletons, and therefore it became his modus operandi to continue excavating until he could assemble or re-articulate a whole animal. Surely during his extensive travels through Africa he would have encountered incomplete animal skeletons, scattered by scavengers and the elements, and realized the implausibility of unearthing perfect specimens.

Migeod, like others, believed that the Germans had excavated at Tendaguru up to the eve of the First World War, though it is not possible to confirm this. If the "fine bones" really did date to German times, their rejection had been explained by Hans Reck: "There are hundreds of sites at which poor surface finds already seen by us were abandoned, and also many a piece in and near the bone stores was left behind by us that did not seem worthwhile to transport."[18] Migeod was correct that Cutler was unable to supervise as thoroughly as was desirable, or may not have trained or hired skilled men: "Cutler's gangs worked very carelessly and threw many bones on the dumps. They seem to have been left to their own devices. In fact he employed more men than he could attend to."[19]

The perception that earlier teams were guilty of carelessness may also be due to the initial attempts by the Germans, Cutler, and Migeod to retrieve only complete skeletons. Tendaguru rarely had these, but featured many disarticulated accumulations, either in bonebeds or as partial skeletons. Excavation methods were also modified over time, as the area and its stratigraphy were better understood and more experience was gained in fieldwork. Each participant, from Fraas to Cutler, had left some unfinished work, and also clearly stated that the area was far from exhausted.

Yet Migeod had to be vigilant with his own staff as well. Several men took pickaxes to M2 against specific orders. Migeod insisted that excavation must proceed with awls once the bone level was reached. Pickaxes, favored because they removed more matrix more quickly, also badly damaged bone.

Had Migeod freed Lachlan from packing duties in Lindi and brought him out to Tendaguru, some of the confusion over the identification of Cutler's sites might have been avoided. Had there been more consultation with Leakey, more thorough study of the German results, or some communication with the Museum of Natural History in Berlin, greater effectiveness and success might have been possible. Lack of time or inclination for more thorough preparation prior to fieldwork was costing the BM(NH) dearly. However, given the rushed preparations in the wake of Cutler's death and the bitterness of the German Expedition participants, such judgments are all too easy to make in hindsight.

Commissioner Wyatt attempted to drive Deacon out from Lindi by automobile. Despite this extraordinary effort, the vehicle bogged down and turned back to the coast. Deacon continued on foot, and plodded into camp on December 12, about eight weeks after leaving England. Migeod described his appearance as "the worse for wear a little. I had sent the small boy Bohari to meet him with a bottle of beer. He had no equipment although I left everything in store at Lindi for him. . . . He had not even had a cup of tea or coffee all the way."[20] Migeod was grateful for Deacon's presence. Deacon assumed responsibility for laborers, the food and equipment stores, the vegetable garden, and collecting insects. However, once the contract with Bennett in Lindi was terminated, Deacon assumed the role of expediter on the coast, and Migeod remained alone at Tendaguru.

Quarry M2, lying 1.6 kilometers northeast of both the camp and M1, was started December 11. This may have been Cutler's Ditch 2 or the German IX. Each exposed bone was coated with shellac and Migeod placed a tent over the pit to protect the contents from rain. M2 was soon subdivided, and Boheti was reassigned to the quarry. In the various divisions of M2, Migeod felt that he was dealing with a total of six animals. When completed, M2 covered an area of about 24.5 meters by 15.4 meters, with the bone level at 1.2 to 1.5 meters below the surface of the ground in hard sandstone.

At M2 No. 1, two femora, a tibia, a fibula, two possible radii, several tarsals, and rib fragments were packed into five wooden cases. Migeod experienced difficulty in distinguishing elements of the forefeet from elements of the hind feet.

M2 No. 2, bordering No. 1 on the east, was begun on December 12. The hard white sandstone yielded enough elements to fill 10 cases with what Migeod identified as a scapula, a partial pelvis, some vertebrae, three tibiae, three radii, three humeri, three femora, and possibly three ulnae.

M2 No. 3 bordered No. 1 to the southwest and eventually filled 16 cases, according to Deacon. Migeod identified perhaps five radii, two ulnae, a fibula, vertebrae, tarsals, ribs, a pelvic bone, two tibiae, two humeri, and three femora. Partially exposed or incomplete bones were difficult to identify in the field.

23. Preparator Gustav Borchert.
Courtesy of Museum für Naturkunde, Berlin.

24. Preparator Fritz Marquardt and *Brachiosaurus* limb.
Courtesy of Museum für Naturkunde, Berlin.

25. *Kentrosaurus aethiopicus.*
Courtesy of Museum für Naturkunde, Berlin.

26. *Elaphrosaurus bambergi.*
Courtesy of Museum für Naturkunde, Berlin.

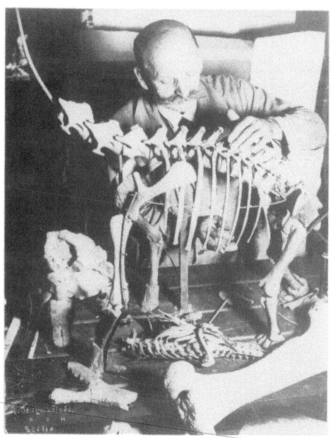

27. ABOVE: *Dicraeosaurus hansemanni.*
Courtesy of Museum für Naturkunde, Berlin.

28. LEFT: Preparator Ewald Siegert and *Brachiosaurus* model.
Courtesy of Museum für Naturkunde, Berlin.

29. Preparator Johannes Schober.
Courtesy of Museum für Naturkunde, Berlin.

30. *Brachiosaurus* mount. *Courtesy of Museum für Naturkunde, Berlin.*

31. Below: *Brachiosaurus* mount, Marquardt. *Courtesy of Museum für Naturkunde, Berlin.*

32. *Brachiosaurus* mount, Schober.
Courtesy of Museum für Naturkunde, Berlin.

33. *Brachiosaurus* mount, Schober (top).

34. *Brachiosaurus* mount.
Courtesy of Museum für Naturkunde, Berlin.

35. *Brachiosaurus* mount, scaffolding.
Courtesy of Museum für Naturkunde, Berlin.

36. Brachiosaurus brancai.
Courtesy of Museum für Naturkunde, Berlin.

37. Preparator Günter Neubauer.
Courtesy of Günter Neubauer.

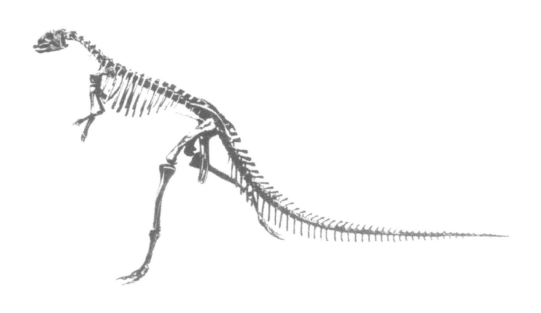

38. *Dryosaurus lettowvorbecki.*
Courtesy of Museum für Naturkunde, Berlin.

39. Above left: Charles William Hobley.
From L. Weinthal et al., The Story of the Cape to Cairo Railway & River Route from 1887 to 1922 *(London: Pioneer Publishing Co., 1923).*

40. Left: William Edmund Cutler.
Courtesy of Department of Geological Sciences, University of Manitoba.

41. Above right: Frederick William Hugh Migeod.
From F. W. H. Migeod, Across Equatorial Africa *(London: Heath Cranton, 1923).*

42. Thomas Deacon. *Courtesy of R. H. M. Stewart.*

43. Francis Rex Parrington. *From K. Joysey and T. Kemp, eds., Studies in Vertebrate Evolution (Edinburgh: Oliver & Boyd, 1972).*

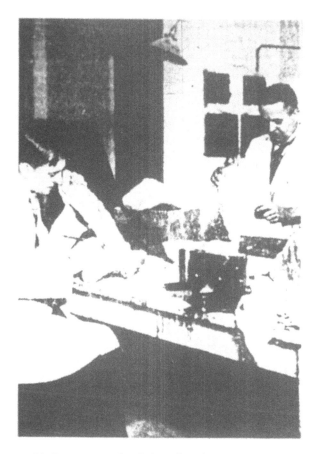

44. Preparators Cyril Castell and Louis Parsons. *From* Times *(London), January 14, 1927, p. 16.*

45. Preparator Georg Wetzel. *Courtesy of Universitätsarchiv Tübingen.*

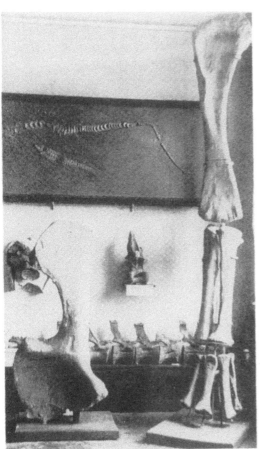

46. LEFT: *Brachiosaurus* cast at Tübingen.
*Courtesy of Institut für Geologie und
Paläontologie, Eberhard-Karls Universität
Tübingen.*

47. ABOVE: Louis Leakey.
Courtesy of Richard Leakey.

48. *Dryosaurus lettowvorbecki* at Göttingen.
Courtesy of Institut für Geologie und Paläontologie, Georg-August Universität Göttingen.

49. William Cutler's headstone, Lindi, photo by author.

50. Tendaguru 2000. Back row, left to right: R. Chami, S. Schultka, H. Kimega, W.-D. Heinrich, S. Kapilima, G. Maier, E. Schrank, E. Msaky, O. Kamanya, O. Hampe. Front row, left to right: J. Mhina, B. Sames, R. Bussert, Y. Kabezi, E. Mgaya, M. Aberhan. *Courtesy of Dr. Oliver Hampe.*

Without illustrations or experience as guidance, it is not surprising that Migeod claimed he found a keeled sternum belonging to a large bird. Earlier, Bather had determined that a fossil Migeod had identified as a tooth was an ungual phalanx.

Juvenile dinosaur remains were unusual at Tendaguru. Migeod believed he had reassembled a tibia and fibula, each of which were only 12.5 centimeters long, if his identifications were accurate. The elements, including a pathological tarsal, were so densely deposited that they overlapped. "It is a mass of bones so that it is difficult to find space to plant one's feet. . . . The site reveals a marvellous prehistoric Armageddon as I might call it."[21]

M2 No. 4 lay to the east of No. 2 and contained nine humeri, eight femora, and seven radii but little else. Migeod thought it might have constituted the remains of four animals. Five crates were filled by this assemblage.

M2 No. 5 lay to the north of No. 2. The partial remains of three animals were buried here, including six femora, three humeri, and masses of almost useless fragments, filling two crates.

M2 No. 6 was located to the south of No. 2, and had originally been designated No. 2. A femur only 36.5 centimeters long appeared in April 1926. Its mate, only half of which was preserved, was found near it under a much larger femur. The site had also produced a 1.1-meter femur, two distal ends, four humeri, two pelvic bones, and two possible radii, plus vertebrae and tarsals. It filled 10 crates.[22]

M2 No. 7 yielded a 1.3-meter femur that was subsequently assigned to No. 3. The interpretation of No. 3 was later substantially revised when the quarry was split into a variety of individuals—two large animals, one medium-sized, and one or two small ones.[23] Five crates were needed to contain the specimens. Bones from M2 No. 7 were filling the shelves and even the floor of the storehouse, and it was felt that the quarry would soon be exhausted.

By August 1926, even the subdivisions of M2 had been further subdivided in an attempt to demarcate the 19 individuals that Migeod now estimated for the quarry: No. 4, for example, became 4a, 4b, 4c, and 4d.

Although Migeod was excavating a bonebed deposit, he attempted to group a mass of disarticulated bones into individual, coherent carcasses, which to him had become intermingled during death throes. While he was justified in trying to associate bones into distinct animals on the basis of uniqueness, size, and so on, it was very difficult in the field.

Migeod theorized that dinosaurs had congregated at a water source during a drought. Dead animals had been trampled into the mud by thirsty and starving survivors. He would return to this theme later in published reports, as it was an analogy to conditions he had witnessed in Africa. Though his interpretation of the depositional conditions of Tendaguru did not find general acceptance, he visualized the environment of the distant past:

> In excavating these dinosaurs and working among their bones . . . One begins to feel oneself actually to be in that remote age. . . . In those Jurassic days the sun rose and set as it does now. There were cloudy days and days on which it rained. The

moon shone at times; and the vault of the heavens with its twinkling stars must have looked much as it does now. Trees and other plant life grew luxuriantly. There were insects, land and sea shells, crustaceans, fish—as well as huge reptiles in the waters. Yet all of these would have made no mental impression on the small-brained and thick-skinned yet mighty dinosaurs that were the highest development of animal life then existing.[24]

Wisely, Migeod restricted his work force to about 25 and used the same men for excavating, packing, and transporting bones. It was his intention to continue at Tendaguru throughout the rains but, as so often in the past, food shortages were becoming critical: "Famine conditions already prevail, and food has to be bought at a great distance. The position will grow worse in the next three months before the first crop of corn is ripe."[25]

Unfortunately, Lachlan's hopes for employment elsewhere were dashed. Migeod exhibited the quietly humane character recalled by Swinton, and wrote Governor Cameron to secure a position for him. Migeod had earlier recommended payment to Lachlan and his wife when both were troubled financially.

Tempted to reduce his work force at the end of December, Migeod found that the matter was settled for him when nine men left. The remainder were only working to raise the annual 10-shilling poll tax that was soon due.[26]

Like that of the Germans, British discipline was strict and unforgiving. An employee named Mfaome had violated camp rules when he refused to join the bone caravan. This infraction could not be ignored: "When I met Mfaome he got it well with my stick. I met him again at 4 p.m. + told him that if he had not been beaten he would assuredly have been sacked. He went away quite satisfied."[27] Even Boheti was not immune to criticism: "Boheti said he wanted a month's leave. I said if he meant to go he need not return. . . . he seems to have thought he would be retained on full pay all the rains + live on his farm some way out of Lindi."[28]

In the face of many frustrations, Migeod was no more tactful than Cutler in the treatment of his laborers. He believed that a firm hand must be taken with Africans, yet this left him severely short-staffed. Oddly, he seemed unbothered by Boheti's departure: "Juma is doing the plastering at M2 very well, + Boheti's loss is largely discounted. I can do without him, though an additional plasterer is useful."[29]

Migeod had no better success than Hennig in curtailing noisy dances once they began. However, his method of dealing with the situation was more drastic: "I gave leave for a dance till midnight last night. At 3 a.m. I could stand it no more. . . . I threatened to burn down Juma's house + made them move everything out. The dance also ceased."[30]

In England, Louis Leakey sought to extend his modest means in several entrepreneurial endeavors. He left over a hundred carved walking sticks with tailors in Cambridge, to be sold in exchange for new clothes. Partly to raise money and partly for the experience, Leakey arranged to give a lantern-slide presentation and talk at a number of schools. Its title was "Digging for Dinosaurs" and it was first given at Leys School, Cambridge, at the end of October 1925. The

BM(NH) stipulated that he seek Museum approval for each venue in advance, and Louis sent half the proceeds to the Tanganyika Fund. By March 1926 he had lectured at over a dozen schools, including Harrow and Eton.

Leakey's boundless energy nearly caused him difficulty. His claims of the distance at which his signaling drum from Tendaguru could be heard were treated with skepticism. To prove his assertions, he climbed onto the roof of St. John's College and thumped a mighty tattoo over the rooftops to strategically placed listeners. The effect was instantaneous:

> A deafening noise echoed all around the courts of the College, and in a minute I could see heads peering out from windows and hear shouts of "What's that?" I realised then that the authorities would be on the war path in few minutes, so I climbed back into my room as quickly as I could.[31]

Louis had received a small grant from Cambridge at Christmas break 1925. With this award and the proceeds from the sale of his motorbike he toured several European museums to examine African bows and arrows. A meeting was arranged between Leakey and Hans Reck at the Museum of Natural History. Leakey excitedly reported that

> This meeting started a very warm friendship between us.
> I had taken with me some of my photographs of the British Museum Expedition to Tendaguru and these I now showed him, and he in turn showed me all of his. We discussed the conditions of life at Tendaguru, and I was able to give him messages from his old native headman Boheti.[32]

This was likely the first face-to-face contact, albeit unofficial, between members of the German and British Expeditions. It speaks well of both Reck and Leakey that they could overcome the animosity of the time and develop a friendship. As recently as January 1925, Reck had written about the British Expedition in the bitter tones adopted by the Germans after the loss of their African colony. He was reacting to Harmer's newspaper reports that the British results were more limited than hoped due to lack of funds. At first Reck had opined why this came to be:

> The Expedition from the outset plainly did not depend on the assistance of its German predecessors and . . . did not consult at all. . . . the success of the new excavation could only be very modest, if the Expedition did not attempt, or if it was not successful in being on intimate terms with the Wamwera living in the Tendaguru region and especially with the trained personnel of the previous excavations.
> . . . It is also possible that in postwar times many people wandered away, died; many a brave Tendaguru man also fell in the War, for example the Patrol Corps of the fallen W. B. Sattler as well as my own was composed almost exclusively of such men. Thus an extensive splintering of our old body of workers during and after the War could have been one of the causes, but certainly not the sole cause, of the limited success of the British Expedition.

Reck's tone became more cynical and nationalistic:

As a scientist one must regret . . . the unquestionable failure of this excavation. . . . How on earth can the accomplishments of Anglo-Saxon science, produced by its most competent representatives, allow themselves to be overshadowed by German successes?

According to the shameless phrases of Versailles . . . it was supposed to be unthinkable that German culture . . . could hold the palm branch of success against British achievement. . . . we Germans can remain proud of past achievements, and in the coming years, as in the previous one, will impatiently look forward with equal interest to those things which British scientific work will produce in our territory. Meanwhile, however, it has still not overshadowed the German [scientific work] yet.[33]

It may be that this provocative outburst was meant for public consumption. Reck had endured a lengthy internment as a British prisoner of war. The effort he had invested in collecting and preparing specimens for the Dar es Salaam exhibition and planned colonial museum had been wasted. Sharing common experiences with Leakey must have overcome some of Reck's anger.

In Africa, Migeod intended to honor the memory of Cutler. Once the rainy season ended and the grave mound settled, a concrete slab with a low headstone could be poured. An inscribed panel, to be sent out from England, would be attached to the headstone. A 40-by-60-centimeter bronze plaque with raised lettering would bear the following epitaph:

To the Memory of
WILLIAM EDMUND CUTLER
Leader of the
British Museum East Africa Expedition
1924–1925
Born in London 23 July 1878
Died at Lindi while on duty 30 August 1925[34]

In September 1926, Deacon wrote to say that the grave plot had been prepared in the European cemetery in Lindi.

Upon investigation, the administrator general in Dar es Salaam determined that Cutler had died intestate. The £375 in his account would go to his aunts, Ellen and Heloise Watson, and an uncle after deduction of liabilities.[35] Between June and September 1926, an ever-lengthening correspondence took place as law firms attempted to settle Cutler's estate. Higgs and Warris, Solicitors, were variously referred to Tanganyikan authorities, the Colonial Office, the British Museum, and the University of Manitoba.

Another law firm, Burnie and Coleman, announced that they possessed Cutler's will and were acting on behalf of interested parties. The will referred to "scientific collections" in Canada and England. Cutler's cabin trunks in storage at the Museum had been opened and contained notebooks listing specimens stored in Calgary, presumably the scientific collection. Here the trail of letters ends; it is not clear how the situation was resolved.

A large area had been opened up by the Germans and Cutler south of the trail to Lindi. A site here was designated M3 on January 15, 1925. It was another of Cutler's ditches, .2 kilometers from Tendaguru Hill. Migeod cautioned that the bones were in such poor condition that it might not be worthwhile to ship them.

On February 5, Juma and five others were ordered to reopen a Cutler ditch about .8 kilometers from the bone store. Migeod dubbed it M4. A second femur, only 31.5 centimeters in length, was found in M4 No. 2. Work on the quarry was halted on May 28.

The grain grinder ordered by Cutler finally arrived and was proving popular with the men, though most of the women still preferred to pound corn kernels in their mortars. Food was still in short supply. Migeod was approached by several men wishing to travel further afield for provisions. Juma, who suffered from a deformed palate, occasionally used notes written in Swahili to communicate more comprehensibly, and conveyed the men's intentions in this manner. Before he could translate the note (with the assistance of a dictionary), Migeod endured an angry tirade from Momadi Ngoni. He understood nothing of this, yet another occasion when insufficient language skills hampered the Expedition.

Daily frustrations were growing, just as they had for Cutler. Migeod was enraged when Juma and his crew quit work at 10:30 instead of 11:00 A.M. An alarm clock had been supplied to help crews maintain a daily schedule: "I took the clock + hove it at a tree. This did not break it so I had it fetched + hove a second time which finished it. Juma then understood I was displeased."[36]

Scorching heat, monotonous diet, bad water, communication delays, and the stress of dealing with excavation and camp duties were taking their toll physically as well. Migeod's spleen was enlarged, so he began a course of regular quinine doses, but was plainly losing his sense of proportion. His temper flared when workers asked for grain on credit, even though he was selling millet at a reasonable 15 cents per liter.[37] "If they spend all their money at once they must go hungry. . . . If all this country cannot support my weekly average of 15 men in the camp it must be a mighty poor one. It is not as if I had a hundred men."[38] The weather and the crop yield varied considerably from year to year in the southern part of the country. In some years, rain was plentiful and the harvest bountiful, and in other years famine ruled. In German times Lindi District, known as the breadbasket of German East Africa, had often failed to produce enough food to sustain its own sparse population.

Migeod blamed an inadequate diet for the exhaustion he had felt in the last weeks. The Expedition operated on a precariously narrow margin. The capacity of the countryside to sustain population concentrations was limited. Migeod's years of experience in West Africa had taught him that small, mobile groups could easily fend for themselves with relatively minimal expenditure. This made him impatient with the unreliable food supply at Tendaguru. As the Expedition was permanently stationed in one spot, it placed greater stress on local resources.

Pumpkins and maize would become available in another month. Millet had to be ordered from the coast since local crops would not ripen for another three months. This added to the cost of the Expedition. Sweet potatoes were planted locally, and were available to supplement the camp fare. Migeod's chicks, from several hatchings, were falling victim to predators or dying from parasite infestations. He had hoped that they would supply eggs to supplement his diet. Migeod noted that the large land snails of the genus *Achatina* were never used as food in the region. None of the British Expedition members appear to have used the shells to mark potential quarry sites as the Germans had.

Migeod was still obsessed with chickens in late August, since their meat and eggs relieved the monotony of his diet. When Juma was discovered with hens, "I said I wanted four fowls for myself + he was off duty until he got them. . . . I do not encourage competition with me in fowls."[39] Migeod, like Cutler, felt he was not receiving what he requested in foodstuffs due to the deliberate intransigence of the locals. While they preferred to sell to coastal traders, they did not enjoy a rich diet themselves, especially during times of poor crops.

Migeod, like Leakey, was suspicious that Lindi postal workers were tampering with his letters, despite reassurances from the Museum that all correspondence had arrived intact: "there is undoubtedly a thirst for information—not scientific—on the subject of the expedition among some of the Indians. Either they are trying to get information for the Germans, or they suspect something else besides bones."[40]

With a shorter wet season this year, Migeod belittled the practice of halting work during the rains: "Cutler closed down from January to the end of May, which also the Germans did. . . . The truth is the people this side [of Africa] do not know what rain is. They ought to sample some parts of West Africa. I am glad I did not allow myself to be choked off."[41]

Six small cervical vertebrae were uncovered on April 24 at a new site, M5. It was located 1.6 kilometers northwest of Migeod's house. Preliminary excavation began about 18 meters from the cervicals, at the location of surface finds. The quarry extended into a steep hillside about six meters above the streambed. Over the ensuing months, the excavation was pushed back 12 meters from the baseline into the hillside, and was three meters deep where it ended in the hill. There was little to show for such an expenditure of time and effort, and the quarry was closed on August 1. Fifteen samples of matrix were taken from the wall of M5 to aid in understanding the strata.

One of Migeod's keenest scientific desires was doomed to certain disappointment: "I am still hopeful I am near a skull. I do not feel I have done anything until I get a head."[42]

When Boheti returned to Tendaguru on May 29, Migeod relented and re-hired him. Eight new Wayao were taken on, which meant there were 40 workers at Tendaguru.[43] A fresh site called M6 was located 650 meters west-north-west of Migeod's house on Tendaguru Hill. Boheti found a belemnite and a few bones here. By late October, the unit presented a 1.2-meter femur, designated M6 No. 2. M6 No. 1 was a few meters away, and contained unidentified disarticulated material.

On June 5, M7 was started, a cutting into the side of Tendaguru Hill at its northwest end. Boheti and a crew of six were in charge of the quarry. An incomplete, 1.84-meter-long humerus began to appear. Using the steel arm balance ordered by Cutler, which Migeod had disparaged as an expensive extravagance, a 1.7-meter-long femur in M7 No. 2 was found to weigh about 195 kilograms. Ribs between 1 and 1.5 meters long and a vertebra measuring 45 centimeters in diameter were found. A fossilized tree trunk, almost 13 meters long, was uncovered in the same quarry. M7 was thought to contain seven animals by the end of August, and Boheti also found belemnites below the bone layer.

A zone pockmarked with pits .8 kilometers south of Tendaguru Hill indicated extensive excavation by the Germans and Cutler. Migeod decided to try his luck nearby, at a spot that lay at a bearing of 340° from his house. Bokari and his crew were sent to this site, M8, on June 11.

In addition to a scapula, a humerus, a radius, an ulna, and a femur, 3.4 meters of articulated caudal vertebrae were uncovered. By June 22, the skeleton measured eight meters from the scapula to the last caudal. There was a curious break at the base of the tail. Three vertebrae were found lying vertically above one another and the next three were on a level surface about 61 centimeters lower than the main series. To Migeod it seemed that after the animal's death, something had pressed down on the tail where it joined the body, and pushed several vertebrae deeply into the substrate. The remains sloped at an angle, with the posterior end lying almost a meter higher than the anterior end.

Migeod believed he had the skull of a ceratopian in M8. This was improbable, given the geological age that the Berlin Expedition claimed for the sediments, the preponderance of sauropod bones, and the fact that horned dinosaurs had at that time only been found in North America and Asia. In the early days of July, Migeod visited productive sites and recorded the stratigraphy as seen in profile on quarry walls.

A shipment of 53 cases of bone left Lindi on June 11, bound for London by way of the Cape. The total weight of the consignment was 4,527 kilograms.[44] In the crates were jackets from the first seven subdivisions of M2, or Cutler's Ditch II.

Deacon sent news that another stockpile of 21 crates had been shipped on July 9. Seventeen contained field jackets, one contained the dried botanical collection, and three contained specimens that had been chosen by Deacon from the mass of bones purportedly abandoned in Lindi by the Berlin Expedition. They may have been those excavated by Pretorius or Hichens, and were undoubtedly part of the same collection rejected by Cutler. One hundred bones were shipped for the Museum to use for exchange or sale.[45] This consignment weighed 2,187 kilograms, and represented more subdivisions of Quarries M2 and M3.[46]

Fires were now often visible, and one even swept up to the top of Tendaguru Hill. Disaster was only narrowly averted on July 28. Flames were observed nearby, and intermittently the entire complement of laborers was summoned to combat the conflagration as it threatened the campsite. Migeod removed the contents of his house and the provision store, but the menace appeared to recede. Suddenly, the winds shifted. Fanned by a strong northeasterly, a wall of flame more than .8 kilometers broad raced directly toward them. Once again huts were hastily emptied. A store at the base of a small valley ignited, and then the privy. This was a mere 15 paces from Migeod's house. The inferno whipped up the summit of the hill and down the south side as far as the path.

Migeod blamed himself for not saving more supplies. He was pleased with the cooperation of everyone but the cook, who was publicly chastised for not helping to fight the fire. Migeod surmised that such criticism was completely lost on him and most locals: "They are very very dense these East African natives."[47] That evening an assistant was ordered to sleep up on the hill with Migeod, to give additional warning of a flare-up.

Drugs, photographic supplies, and most of the brown wrapping paper were lost. Incinerated tool handles could be replaced. Neither the plaster of Paris nor the calcium carbide for the lantern were affected. As the bales of burlap were unrolled it was discovered that much could be salvaged. The loss of photo developing materials was not critical, as Migeod made no use of the Expedition plate camera. He never constructed a darkroom like Cutler, but instead used his own range-finder camera and developed the negatives in a tank. A new latrine hut was erected the next day.

After the fire had denuded the hill, Migeod located what might have been a German dwelling at its northwest end. Its placement did not appeal to him, as he thought it was a far windier spot than the one he occupied, about nine meters further down the hill. Hennig and Janensch had welcomed the breezes that carried away noise, dust, mosquitoes, and smoke.

An old friend of Cutler's in Lindi, a police inspector named Revington, called at Tendaguru on July 31, en route to an elephant hunt: "He rather confirmed what Cutler + Leakey complained of + I recognize too that the natives were long ago told by an Administrative Officer of the District that this expedition is of no importance + they need not trouble about it."[48]

On one of his rare visits to Tendaguru, Deacon took offence when Migeod implied that Deacon could easily be replaced. Migeod's account is reminiscent of the relationship between Cutler and Leakey:

> Then he objected to his work being treated as only "small boy" work, + said he had better go home if that was his position in the expedition. . . . I do not want another European with me as I prefer to work the natives direct. . . . If he did scientific work it would be different. But D. does nothing. I could easily arrange somehow to do his work at Lindi which consists of getting a carpenter to make cases + then putting the <u>already</u> packed boxes inside.[49]

Two elephant hunters, G. W. Parlett and W. Kershaw, sent greetings to Tendaguru on August 22. Parlett was a one-armed man Migeod had met in British Cameroons in 1923. Migeod trekked northwest for about three hours to their camp near Nakihu. They lived in a grass hut on the banks of the Mbemkuru. The three Europeans agreed about the intellectual capacities of East Africans: "P.[arlett] & K.[ershaw] have both been in W.[est] A.[frica] + we agreed how inferior in intelligence is the native on this side + how useless in comparison."[50] The two hunters had spent more than a year in East Africa, and several months in the Tendaguru region. They planned to move on to Portuguese East Africa after they shot their limit of elephants in Tanganyika Territory.

Migeod recruited the pair to look after Tendaguru while he returned to England for several months. They would continue to excavate, and collect insects and fish from the Mbemkuru. They were free to hunt elephants, but one man was to remain at Tendaguru at all times. No agent would be retained at Lindi, as one of the two hunters would make any necessary trips. Migeod engaged them at £11 per month each, which was to pay for food, drinks, and a servant.[51] Parlett would receive £50 per month from the Standard Bank of South Africa in Lindi, to cover labor and other expenses. Boheti was paid £3 per month, and the laborers about £1 each. The headman Juma earned just over £2.[52] The two hunters were to come to Tendaguru at the end of October, in order to receive several weeks of training in fieldwork prior to Migeod's departure. If all went as expected, Migeod would return for the following season.

At the BM(NH), Bather had pronounced himself pleased with Migeod's meticulous excavation methods, the adoption of sieving, and especially the drafting of detailed quarry maps. Bather also alluded to a burgeoning problem: "We are having a great job, even to find room for the cases, much less to unpack them."[53]

On August 24, the first mail in 31 days arrived, giving rise to both joy and disappointment. Bather again wrote to complain that the flow of material reaching the BM(NH) had caused severe congestion:

> we have come to a standstill for the present, since all our available space is occupied by material unpacked from the earlier boxes and we cannot empty any more boxes until we have proper storage accommodation.[54]

Assistant Secretary Smith sounded almost defensive:

> do not construe my silence to denote any lack of interest on my part in the success of the Expedition . . . everything connected with the Expedition has been going on so well that there has been no . . . call for a letter.[55]

Given the steady flow of crates this was probably true, especially after the embarrassing delay in Cutler's shipments.

Bather sounded unconcerned about Migeod's osteological difficulties: "I have asked Dr. Swinton to make one or two drawings of Tarsals and Carpals such as may be of service to you; but I don't know that it matters very much at the present stage."[56] This is surprisingly indifferent, considering Migeod had repeatedly admitted his inability to distinguish skeletal elements. Perhaps Bather thought that since Migeod had done so well there was little point in providing much support. It was true that Migeod was not accustomed to asking for help, but his letters clearly stated his difficulties.

What had happened to the publications Cutler had brought with him? Although the *Archiv für Biontologie* volume dealt mainly with stratigraphy and geology, it did contain some illustrations of bones. And by 1926, both Hennig and Janensch had published well-illustrated papers on Tendaguru stegosaurs, sauropods, and theropods in *Palaeontographica,* from which tracings could have been made. There was also no lack of illustrations from papers by Marsh and other American paleontologists, dating back to the 1890s. Why had Migeod not been provided with renderings before he left, or early on? He made it clear that such guidance was necessary: "I am most grateful for the sketches of the hand and foot bones. It is so trying to work in the dark with no guide whatsoever."[57]

What a far cry from the lengthy, placating notes penned by Harmer and others to Cutler. Migeod likely projected an aura of confidence regarding Africa, and his detailed monthly progress reports and meticulous excavation methods had produced an uninterrupted flow of specimens. This combination, plus a change of Museum keepers to Bather and Swinton, may have left the impression that there was a minimum of problems. This is not the feeling gained when reading Migeod's field notes. The parallels between his frustrations and those of Cutler are remarkable and were clearly stated.

Absent from Worthing for more than 10 months, Migeod was ready to return home. As stipulated by his contract, he had given ample warning for a December departure, and also informed the Museum of his plans to engage Parlett and Kershaw. Numerous letters and telegrams were generated in an attempt to clarify terms for both himself and Deacon, leading to considerable aggravation for both men in the field. Part of the confusion stemmed from the fact that Migeod had been writing to Harmer for the last 10 months, yet it was Smith who replied. This situation was compounded by lengthy delays in mail delivery. Letters could take five to six weeks each way, and tersely worded telegrams allowed ambiguous conclusions to be drawn.

Deacon experienced similar confusion over the terms of his agreement. Like Migeod, he was happy to return after a few months' vacation, but did not receive a reply for months, and when he did, it was hardly what he expected. A letter from Smith arrived November 12, in which it was implied that Deacon was obligated to stay until December.

The Expedition was disturbed by such miscommunication or misinterpretation on several occasions. The chronic uncertainty of steamer schedules aggravated the situation. Plans for shipments or personal travel could be formed with the best of intentions and then collapse when a ship failed to appear.

Unknown to either man in the field, discussions regarding the coming season had been taking place in London between Bather and Harmer since late July. In Bather's mind, there was uneasiness about the leader of the Expedition:

> Migeod is, as you say, doing his work in a very business-like fashion, but his recent letters have contained remarks indicating a lack of geological knowledge and experience which may, I fear, land us in difficulties. Should he find anything entirely new, such as a bird or early mammal, we should have to consider the necessity of sending out a trained geologist to make sure of the horizon.[58]

Harmer concurred that a geologist should be sent to Tanganyika, and Bather suggested W. E. Swinton, who would study the Tendaguru specimens. Harmer approved, pending a medical examination. Presumably Swinton was amenable, for almost exactly one month later a memorandum was submitted to His Majesty's Treasury. It requested that special leave be granted to allow Swinton to spend three weeks in Berlin studying Tendaguru specimens at the Museum of Natural History.

William Elgin Swinton was born in Kirkcaldy, Scotland, on September 30, 1900. In 1920 or 1922, he joined the Scottish Spitzbergen Expedition. In 1922, he received his B.Sc. from the University of Glasgow. The British Museum (Natural History) appointed him curator of fossil reptiles in 1924, the same year Bather replaced Arthur Smith Woodward.

Bather intended broadening the field of investigation to other East African sites. C. W. Hobley was consulted regarding suitable localities. Dinosaur bones had been discovered to the west of Tendaguru, on the shores of Lake Nyasa, just a few years earlier by A. Holt of the Veterinary Department of the Nyasaland government. The specimens—caudal vertebrae, fragments of femora, humeri, tibiae, and parts of a scapula and ilium—had been forwarded to Dr. Frank Dixey, the government geologist of Nyasaland in mid-1924. A few weeks later, Dixey and Holt had returned to the site, Mwakasyunguti, 7.5 kilometers south of Vua. They collected additional bones and found more at Florence Bay and Deep Bay.

The fossils had been shipped to paleontologist Sidney Henry Haughton of the South African Museum in Cape Town. Haughton later described the partial carapace and plastron of the turtle *Platycheloides nyasae,* and assigned some of the dinosaur bones to the sauropod *Gigantosaurus dixeyi.* The weathered and incomplete condition of the bones made a closer comparison with German illustrations of Tendaguru material inconclusive. Dixey recognized that the sediments in which the bones were found represented a new and undescribed sequence, deposited in small rift valleys that ran north to south, parallel to the shore of Lake Nyasa. He returned in 1925, and described them as the Dinosaur Beds.

Bather, aware that both Dixey and the governor of Nyasaland were on leave, in France and England respectively, broached the idea of the BM(NH) operating in Nyasaland to the government geologist, who welcomed the collaboration. Dixey and Assistant Secretary Smith communicated with the governor of Nyasaland, Sir Charles Bowring, who offered full cooperation.

A letter from Louis Leakey arrived at the BM(NH) at the end of August, and presented the germ of another idea to Museum officials. Leakey was planning his own expedition to East Africa, and had mentioned to Assistant Secretary Smith a month earlier that he hoped to examine some archaeological sites about 8 to 10 kilometers from Tendaguru. Now that funding was in place he renewed his offer to look for natural history specimens for the BM(NH) and promote the Museum Expedition if feasible: "I . . . hope to pay a visit to Mr. Migeod if he is there. . . . I am taking out a cinematograph which I have been lent by a London firm, + would like, if you would give permission, to film the work at Tendaguru."[59] The groundwork was thus laid for an expedition in 1927 that would be quite different from those of the preceding years.

Assistant Secretary Smith explained the plans for future fieldwork to Migeod. Swinton, who would describe the specimens scientifically, would assess the site geologically for a few months. Leakey would join Migeod after completing six months of archaeological fieldwork, to collect fossil mammal remains in caves and prospect for additional dinosaur sites. Migeod's response was positive:

> I am glad to hear Mr. Swinton will accompany me out next April. One does not always arrive at much discussing a doubtful point with one's self alone, and his geological knowledge will be invaluable.
>
> I shall welcome back Mr. Leakey, whose previous brief service here will have been a useful experience.[60]

Leakey had spent eight months in the field, at this point only two less than Migeod, hardly "brief service."

The Tendaguru work force stood at 43 on September 3, which allowed for more excavation, but strained finances. Migeod was irked, as the Museum account had been overspent by £76, and daily operations were funded from his own salary and a bank overdraft. Yet he had asked London for more funds in early June. Three weeks later, word was received of a £500 deposit in the bank account at Lindi.

Another quarry, M9, was opened near a German site past M7, to the north of Tendaguru Hill. Juma began excavating on August 25. It seemed to Migeod that the German Expedition had left the locality without taking the exposed bones with them. On September 18, the efforts at M9 were halted. Ten days later all its contents, including a scapula and a femur, were waiting in the bone store.

A new site, M10, was opened on September 16. It was, as always, close to German quarries, near the Mtapaia trail about 275 meters past M9, to the north of Tendaguru Hill. M10 had reached a depth of 1.3 meters, descending into a sandstone layer, when work ceased on the 21st.

It was also on September 21 that a start was made on M11. It was located between M9 and M10, on high ground between two dry streambeds north of Tendaguru Hill. Juma was in charge. Excavators were moved up the slope the following day as no bone showed. An odd outcrop, a few meters in area, appeared. It reminded Migeod of a layer of broken eggshell, and was 2.5 to 5 centimeters thick. A 60-centimeter-square portion was plastered and sent to London, but there was no final identification. By October 22, a humerus, femur, fibula, and radius and pieces of what might be pelvic bones had come out, along with several belemnites.[61]

A new locality, known as M12, was opened on October 20. It was an extension of one of Cutler's larger ditches, on the trail from M7 to M9, M10, and M11. In surface prospecting, two humeri and a vertebra were found.

Baggage belonging to a group of guests who were en route to Tendaguru in automobiles was carried into camp on October 1. The party consisted of Mr. C. H. A. Grierson, the acting provincial commissioner, two other men, also government officials, and their wives. The visitors had all arrived by late afternoon. It was only the second time motor vehicles had reached Tendaguru.

Grierson warned Migeod that his expenses would soon rise. Millet had increased to 20 cents a liter out of Lindi, a situation about which Migeod had complained to London in the past. All the guests departed on October 4.

In Lindi, another major consignment left the port October 5: 63 cases of fossils, weighing 6,310 kilograms.[62]

Some small boys in camp impressed Migeod by building a toy vehicle with seats and an awning, inspired by the automobiles that had come to Tendaguru recently. Another pair of boys caused trouble in camp, and relatives of the boys' victims demanded 15 shillings each in compensation. Instead, Migeod punished the boys with "a dozen each," presumably strokes with a cane.[63]

In London, Bather prepared a document for the British Museum trustees. It outlined the progress made and recommended future directions. As of October 21, 250 crates of specimens had been received from Tanganyika Territory, only 25 of which had been opened. Storage space at the Museum was becoming critical. Bather predicted that Tendaguru could be excavated for another 50 years before the accumulations were exhausted, but saw little likelihood that new dinosaurs would be discovered. He suggested that the Expedition explore to the west:

> 4) . . . if the course of this ancient river were followed westwards, other deposits, possibly with a different class of reptilian remains, might be found. . . . It is therefore proposed that the expedition should trek across the Jurasso-Cretaceous rocks of Tanganyika Territory to Lake Nyasa, passing down the Ruhuhu Valley. . . . Those sections should be carefully studied, in order to correlate them with the sections examined by Dr. Dixey on the other side of the Lake.

> . . . 5). The deposits in the Oldoway Gorge are of extreme importance for the study of the early Mammalian fauna of Africa including Man. . . . should be investigated as soon as possible.[64]

Parlett and Kershaw arrived at Tendaguru on October 31 as planned. They were toured around the area the next day, and Kershaw began mapping at M12, which had yielded part of a small animal after initial disappointment.

Rain poured down, flooding the quarries. Migeod roamed 6.5 kilometers along the Mtapaia path in search of prospects, hoping to discover some worthwhile sites to keep his replacements occupied. A femur, about .4 kilometers beyond the trail that led from the water hole to the jumbe's farm, was deemed worthy of further investigation and dubbed M13. The site was abandoned on November 11, after the femur was removed.

M1 was reopened and tentatively renamed ?M14 on November 16.

Workers presented Migeod with an ebony walking stick and honored his imminent departure with a noisy sendoff. The final consignment of the year, another 20 crates, left Lindi that day. Seventeen cases contained jackets from M6, M7, M8, M9, and M11. Three more crates held botanical, zoological, and geological specimens as well as Cutler's notebooks and rifle.

Frederick Migeod left Tendaguru behind him on November 17, 1926. He had labored at the dinosaur graveyard since November 18, 1925—one year exactly. He had not visited Lindi or Dar es Salaam during that period and had only taken a one-day trip to the Mbemkuru. It had been an impressive feat of endurance, and had saved the Expedition.

Had the British Museum East Africa Expedition been canceled or postponed, it might have never resumed, considering its initial modest success and precarious funding. The Museum's credibility might have been questioned in England and Tanganyika Territory. Certainly many specimens lying exposed in the field would have weathered beyond recovery.

Instead, the majority of Cutler's hard-won finds had been brought safely to the BM(NH). A second campaign had spanned the rainy season, operated for a longer uninterrupted stretch than any previous field effort, and yielded a large haul of specimens. The number of crates shipped to London had increased more than tenfold over the previous year: about 289 in 1926, versus 23 in 1925. Roughly 245 had reached the Museum by year's end. If one subtracts from this total the 83 stemming from Cutler's efforts, another 17 as the collections of invertebrate fossils he had made at Dar es Salaam, one more of his fossil and geological specimens stored at Lindi, three crates of German "leftovers" from Lindi, and a few containing his butterfly collection, then William Cutler's excavations had yielded about 110 crates of dinosaur bones weighing 10,228 kilograms. The cost of the first field season had been £1,787.10.9.[65]

With Cutler's figures subtracted, Migeod had filled roughly 180 crates with his own excavation results in 1926. Although weights were not provided for all of his boxes, one can estimate from available figures that his efforts had produced about 17,450 kilograms of bones. This collection represented 431 loads carried to Lindi by 530 bearers.[66] Migeod calculated that he was sending remains of about 30 dinosaurs.[67] About four hundred botanical specimens had been shipped to the BM(NH), as well as geological and soil profile samples, invertebrate fossils, and preserved modern fish and reptiles. Thirteen quarries had been opened and their elevation determined. An attempt at surveying had yielded a useful map of quarry locations. Crew size had ranged from 17 to 49 per month, and men had served as porters when possible.[68] This approach had proved much more manageable for a single European.

Administratively, proper accounts of expenditures had been kept and regular progress reports mailed to London. More careful methods of fieldwork had been adopted. The contents of each quarry had been precisely recorded, and the units mapped. Triangulation had been attempted to pinpoint quarries, and individual sites had been identified with a sign on a post. Matrix had been sieved for small fragments and samples taken of quarry wall profiles and unusual deposits like the "eggshell pavement." Migeod had adopted archaeological excavation techniques on a paleontological dig. Like Cutler, Migeod had photographed the progress of the Expedition and had kept a daily journal and notebooks. He had also recorded meteorological readings. He had exercised the greatest possible economy, though Cutler's expenditures could not be considered extravagant.

Both men had been beleaguered by similar problems: a limited budget, lengthy communication delays with the Museum, the alleged indifference of local government officials, difficulties in procuring provisions and obtaining supplies, language problems with a constantly fluctuating number of workers, poor diet, personal health troubles, heat, insects, fires, and rain. All these factors had placed a great deal of pressure on a single European and led to much frustration. Both men had insisted on working alone to maintain independent control.

The personality of the individual magnified the effect of the hardships. While Janensch and Hennig tended to have disagreements over how to best run the Expedition's daily business, they appeared to be quite adaptable. These two, and Hans and Ina Reck as well, seemed more flexible in their dealings with their laborers, and in private and public writings openly admired them. Perhaps this was because they were younger and had fewer preconceptions than their British counterparts, though they lived in an era when paternalistic colonial attitudes prevailed.

Cutler and Migeod, on the other hand, experienced many more confrontations with a far smaller work force. Personalities were not the sole cause. The aftereffects of the war and a different administration had changed the relationship between blacks and whites a great deal. As Reck surmised, key trained overseers may have been killed or relocated by the war. The implementation of indirect rule by Governor Cameron may have made administrative officials less sympathetic to European demands, but the Englishmen, and Migeod especially, had also been hindered by their expectations of European and African behavior. Leakey had not harbored the same suspicions of the locals, partly because of his age, but largely because of his upbringing in Kenya. He understood and respected Africans.

Cutler had rigidly tried to maintain the field methods that had served him well in Canada. He had been a practical man of action with little patience for bureaucracy. Migeod, the consummate recorder, though showing much more attention to detail and documentation, had been impeded by a lack of field experience and unfamiliarity with anatomy and geology. Where Cutler seems to have fallen out of favor with Museum officials because of his personality quirks, Migeod had lost credibility because of his technical weaknesses. Still, Migeod had loyally answered the call when the Museum was in desperate straits. All of the men of Tendaguru had selflessly given their utmost.

In Lindi, Migeod discussed the construction of a rest house for Museum staff with a carpenter, to avoid having to rely on government facilities in town. While waiting for a ship, Migeod typed a 17-page manual for collecting fossils, titled *Tendaguru: Notes on Excavation of Dinosaurs*. He met W. Fryer, the new district officer, possibly Major Wyatt's replacement. A letter from Assistant Secretary Smith informed Migeod that the London office of the British Broadcasting Company had asked for an interview.

Migeod left Lindi on December 15, 1926. At a dinner with Governor Cameron in the capital, he learned of the myriad changes Cameron was instituting. The former 22 German administrative districts were re-formed into 11 provinces, and each province assigned a provincial commissioner. These individuals were assisted by district officers. The hierarchy was completed with administrative officers and cadets. A Geological Survey had been created in 1925 or 1926.

Perhaps the change for which Cameron was best remembered is the introduction of indirect rule. By way of the Native Authority Ordinance, Africans were granted certain powers to administer their tribal affairs, such as the collection of taxes. British administration became more decentralized as a result of Cameron's reforms, and provincial officials had a wide range of responsibilities to Africans. Migeod agreed with many of these reforms but did not support indirect rule.

Deacon, who had celebrated his sixty-second birthday in Lindi, arrived in Liverpool on December 20. On January 16, 1927, Frederick Migeod was reunited with Madeleine, who was disappointed that they had been unable to spend Christmas together. Her husband had been absent 15 months, and turned 54 at Tendaguru. It was time he took up the regular thread of his life.

-14-

1926–1927

Berlin in Chaos

Life in Germany had improved as the nation's currency stabilized with the introduction of the rentenmark in the fall of 1923. Banks in the United States were once more supplying desperately needed credit. Loans worth billions of marks were invested in factories, which also reduced unemployment. The French army was evacuated from the Ruhr in 1925 and the Rhine around Cologne in 1926. In 1926, Germany was admitted into the League of Nations.

Many consider the period between 1924 and the 1929 stock market crash Germany's "golden years." Creativity flourished, especially in Berlin. Film production was centered in the capital. Theater life was vibrant. Nightlife became almost notorious for the cabaret. The mid-1920s were the beginning of an era of Max Reinhardt, Bertolt Brecht, Fritz Lang, Marlene Dietrich, and Peter Lorre. Both the Dada and Expressionist art movements, which were established in the capital, and the Bauhaus architectural design school, founded in Weimar, had powerful influences on Berlin.

The city had modernized as well. The largest power plant in Europe was constructed in Berlin between 1924 and 1926. Telephones were installed in the homes of the wealthy, and radio stations began broadcasting in 1923. Transportation systems were upgraded as city and suburban rail lines were electrified. Trams, trains, motor buses, and taxis were making horse-drawn vehicles a rarity. When Berlin's western docks were completed in 1923, the capital became the second largest river port in the country.[1] A modern airport was under construction at Tempelhof.

Many jobs were created with public works initiatives that included canal excavation, road building, and the construction of recreation facilities and exhibition halls. Public housing projects came into existence after 1923. They were urgently needed, for the city's population had increased by an average of 80,000 per year since 1918.[2] But no amount of housing relieved the overcrowding prevalent in the working-class districts of eastern Berlin. Living conditions in the tenement flats were appalling.

In this atmosphere of increasing prosperity, artistic expression, and greater economic and political stability, another success was recorded by the Museum of Natural History. In late 1925 or early 1926, the second Tendaguru dinosaur was mounted and placed on public display. Quarry dd, in the Middle Saurian Bed near Kindope, had yielded a medium-sized bipedal animal that Werner Janensch had named *Elaphrosaurus bambergi* in a brief paper published in 1920. Expedition patron Paul Bamberg was honored in the species name. The same productive quarry also contained remains of the sauropods *Dicraeosaurus, Brachiosaurus,* and *Barosaurus.* Janensch's full description was delayed due to the difficulties of the postwar years, and the work did not appear until 1925.

A second series of Tendaguru monographs was published in the prestigious journal *Palaeontographica,* to coincide with Wilhelm von Branca's 80th birthday. Major themes included descriptions of reptiles and discussions of invertebrates, stratigraphy, and geology. Later Tendaguru papers in *Palaeontographica* benefited from a generous grant from the Prussian Ministry of Science, Art, and Public Education. The sum of 8,500 marks was made available in early 1926 for printing costs and the production of plates. As with Hennig's *Kentrurosaurus* monograph, the cost of producing the figures for Janensch's *Elaphrosaurus* paper was borne by the Prussian Academy of Sciences.

The final count of *Elaphrosaurus* skeletal elements from Quarry dd was 16 presacral vertebrae (initially defined as seven cervicals and 10 dorsals),[3] a sacrum composed of five sacral vertebrae and one fused dorso-sacral, 18 caudal vertebrae, two partial ribs (an anterior and a mid- to posterior dorsal), one chevron, a right humerus, metacarpals (possibly I and IV), both ilia, both ischia, the left pubis, and a partial left hind limb consisting of a femur, a tibia, a fibula, an astragalus, metatarsals II, III, and IV, two phalanges of the second digit, and one of the fourth digit.[4]

Janensch was confident that the remains were derived from a single individual. Many of the bones from dd were encrusted in a light-colored impervious calcite, which also filled cracks in the bones. Preservation was generally good, though most vertebral processes were missing. This made it impossible to articulate the neck, spinal column, and tail to determine the degree of flexibility that the animal possessed when alive. As the caudal vertebrae did not represent a complete sequence, Janensch referred to the literature on *Struthiomimus, Ornithomimus, Ornitholestes,* and *Compsognathus.* He estimated that there might have been 42 vertebrae in the tail of *Elaphrosaurus.*[5]

Mounting must have been completed early in 1926. Since many bones were missing, they had to be modeled in plaster. Mirror images of ribs were created from the few fragments in existence. The remainder were based on those of *Struthiomimus altus.*

Janensch believed the animal would have possessed abdominal ribs in life but declined to add them. Wherever gaps existed in the sequence of vertebrae, replacements were modeled according to adjacent vertebrae. In the case of more sizable gaps, the replacements were based on those of other genera that most closely resembled those of *Elaphrosaurus.* The atlas and axis were produced entirely in plaster, as were the skull and lower jaws. It was felt that a headless mount would appear unfinished so Janensch made the attempt, basing a reconstruction on *Compsognathus* and *Ornitholestes*. He admitted that the model was bereft of any scientific value.

About 150 carnivore teeth had been recovered from the Middle and Upper Saurian Beds.[6] Janensch assigned teeth of a certain size and morphology to large carnivores like *Allosaurus, Ceratosaurus,* and *Megalosaurus,* which were represented by a few skeletal elements at Tendaguru. There were also rare remains of possibly three species of theropods smaller than *Elaphrosaurus.*[7] When these were discounted there remained a selection that would be suited to an animal of the size of the new mount, so jaws were modeled with teeth. In 1928, Janensch became aware that Franz Nopcsa assigned *Elaphrosaurus* to the subfamily Ornithomimidae, a group featuring toothless beaks. In addition, Henry Fairfield Osborn described *Oviraptor* as a beaked coelurosaur, so *Elaphrosaurus* may indeed have had the same jaw anatomy. However, at the time the African dinosaur was mounted Janensch added teeth, believing that this older and less specialized animal may have possessed them.

To achieve the vertebral formula of 23 presacrals, as in *Struthiomimus, Ornithomimus,* and *Compsognathus,* plaster vertebrae were inserted in positions 8, 16, 17, and 18. Similarly, caudal vertebrae in positions 3, 4, 8, 9, 12, 13, 15, 16, 18, 20, 22, 24, 27, 30, 31, 33, 34, 36, 38, and 39 through 42 were artificial reconstructions.[8]

The shoulder blades were also based on ornithomimids. After the mount was completed, more field jackets from Quarry dd were prepared and two scapulae and coracoids were recovered. They proved to be wider than those in the theoretical skeletal outline, but as the glenoid cavity matched the humerus so well they were assigned to the same animal. Most of the bones of the forelimbs were modeled, but again new elements came to light as more field material was prepared. A left radius from Locality RD in the Upper Saurian Bed had been excavated under Hans Reck's supervision in 1912. It verified that the reconstructed limb component was accurate. Quarry dd also yielded a proximal fragment of what appeared to be the first right dorsal rib and a portion of a right rib shaft. Both bones confirmed the accuracy of the plaster replicas.

The missing end of a pubis was also rebuilt in plaster. Since there were no muscle attachment scars on the existing metatarsals to indicate more digits, the hind foot was restored with three toes. Janensch cautioned that rudimentary "splints" could not be ruled out.

In order to distinguish the man-made reproductions from the natural material, the surfaces of plaster elements were given a rough texture before being painted. O. Bügen and senior preparator Ewald Siegert created the plaster replicas. Another preparator, Johannes Schober, faithfully realized Janensch's concept of a very swift, lightly built animal by mounting the skeleton in a running pose. To reduce the amount of metal armature visible, the small and fragile vertebrae were not cored and strung on a central rod, but rather were suspended from metal wires. The pectoral girdle, sacrum, and pelvis were all supported by short, horizontal metal braces attached to the wall panel behind the skeleton. Finally, a framed glass case was installed around the mount. In its upright pose the skeleton measured 5.69 meters in length and stood 2.31 meters tall.[9]

In Berlin, Hans and Ina Reck each published a book dealing with their experiences in German East Africa during the war. Ina's reminiscences, elegiac and nostalgic, appeared in 1925 under the title *Auf einsamen Märschen im Norden von Deutsch-Ostafrika,* or "On solitary marches in the north of German East Africa." The book recounted happy days in Ufiome as she traveled the countryside, and her escape and eventual internment in Dar es Salaam. Hans's account, *Buschteufel,* or "Bush devil," narrated a series of incidents from that long and hard-fought campaign and was published in 1926. Another book of his was in preparation that same year, but was never printed. He called it *Masr el Kahira: Die siegreiche aegyptische Erinnerungen,* or "Masr el Kahira: The victorious Egyptian reminiscences." It recalled his imprisonment outside of Cairo.

Preparation of Tendaguru material had advanced to the extent that duplicate specimens were available by the mid-1920s. Before the war, von Branca had enumerated five individuals to whom he felt deeply obligated for their generous financial support of the Expedition. Correspondence from 1911 shows that von Branca had promised backers dinosaur remains for museums and institutions of their choice. Among these patrons was Duke Johann Albrecht, regent of Braunschweig, who headed the Tendaguru Committee. The Braunschweig Polytechnikum maintained a natural history museum that was to receive bones.

Richard von Passavant-Gontard, the Frankfurt proprietor of Gebrüder Passavant, Seidenwaren, had asked for duplicate bones for the Senckenberg Natural History Museum in 1909 and 1911. August Roechling, of the Mannheim coal and steel conglomerate, had donated a total of 12,000 marks, and asked that bones be sent to the Mannheim Geological Collection (Geologische Sammlung). Von Branca was keenly aware of the debt he owed Eberhard Fraas of Stuttgart. Had Fraas contacted another German museum with his exciting news in 1907, the Berlin institute might never have amassed the impressive collection it now owned. In recognition of this debt, Pompeckj, as Museum director, had promised the majority of duplicate ornithopod material to Stuttgart in April 1920. In 1927 or 1928, Janensch sent *Kentrurosaurus* material, including a humerus, a hind limb, three spines, and three vertebrae, to Stuttgart.[10]

In appreciation of the freight discount offered by the German East Africa Line, the Mineralogical-Geological State Institute in Hamburg was to be presented with Tendaguru bones. Following discussions between Janensch and Georg Gürich in the fall of 1926, Janensch offered a plaster cast of the right forelimb of *Brachiosaurus.* This would supplement a humerus already sent to Hamburg, and when assembled would stand five meters tall. Janensch also offered a cast of a *Brachiosaurus* rib and original stegosaur material. The forelimb was eventually sent.

Friedrich von Huene was perhaps the first German paleontologist to visit the British Museum (Natural History) to view the results of the British Museum East Africa Expedition. In a letter to Janensch written in late March 1927, he gave his impressions, though it is likely that they were influenced by the dearth of prepared specimens. The mystery of bones missing from the 1914 Dar es Salaam exhibition remained unresolved:

Also spoke with Migeod who last excavated at Tendaguru. . . . What was discovered there looks bad and is relatively sparse. Migeod does not have the slightest notion of palaeontology, is an anglicized Frenchman. He brought back two large crates of well-prepared bones from the Berlin excavations, that were there somewhere in the country; but the articulated forefeet mentioned by Reck are not among them. Bather and Swinton are of the opinion to return these to Berlin, as they do not wish to keep stolen goods. . . . Migeod considers it likely that the Boer Pretorius has set aside still more good material from the German excavations. . . . The British now want to attempt to obtain these items as well from Pretorius, then they would likely also somehow find their way back to Berlin; but P.[retorius] would certainly not return them as gifts, if he even has them. Possibly, the articulated hand will be found among these things.[11]

In London, British Museum trustees approved Bather's suggestions for fieldwork at their sitting of November 27, 1926. The *Times* published Bather's description of the new field season, along with another plea for public support. It was admitted that the Treasury was unlikely to provide more funding.

Leakey was to offer "general assistance" to Migeod, and would receive £15 per month salary plus expenses.[12] In late December, Louis Leakey's fortunes had changed, and he was anxious to follow up personal opportunities independently rather than join another field program in a subordinate position. Leakey went on to say that he still hoped to meet the BM(NH) team at Olduvai Gorge.

Deacon confirmed his willingness to rejoin the Expedition in 1927, which provided some continuity. Migeod was permitted to lecture on Tendaguru to the Royal African Society in London in February. He used the same lantern slides he had prepared for a similar talk to the Worthing Archaeological Society. Smith asked Migeod to contribute an article with photographs to a popular journal that the Museum published—*Natural History Magazine.*

Planning for the upcoming field campaign seemed to be proceeding well until a second change was forced upon the Museum. Swinton had been advised not to go to Tendaguru for unspecified health reasons. The loss of the second Expedition member, its geologist, had a domino effect. Harmer described it to the trustees on February 25:

> Mr. Migeod had done his work admirably, but it appears that he has not the special knowledge of Palaeontology which is required to make the work a complete success.
> . . . it will be best to advise the Trustees to place the leadership in the hands of Dr. John Parkinson, whose employment as an expert Geologist was approved at the last meeting of the Standing Committee. . . . This would involve dispensing with Mr. Migeod's services; and although it would be going too far to say that the gentleman entirely welcomes the proposal, he appears to be willing to admit that he is not a Palaeontologist.[13]

In rapid succession Leakey, Swinton, and Migeod were out of the picture, and a "competent field geologist" had been selected. If Deacon accepted the changes, he would most likely remain at Tendaguru while Parkinson carried on to Nyasaland. Uncertainty now troubled the Expedition.

How John Parkinson entered the scene is not clear from the East Africa Expedition archives. He was notified on January 13, and his terms were accepted by Museum trustees. Conditions of his employment included the usual restrictions on publishing, and a salary of £66 per month. Parkinson could engage in private work while in Tanganyika, provided the Museum received half the proceeds, less expenses.[14]

John Parkinson was born in London in 1872, and studied engineering and geology at University College, London. In 1894, the 22-year-old visited Australia on a geological field trip. Three years later, he married Eleanor F. Whitlock. They eventually had one son and one daughter.

During 1897, Parkinson undertook field trips in Great Britain and northern Italy. The focus of many of his papers was petrological. Geological studies took Parkinson around the world in 1900. He visited Sri Lanka, India, and Burma (now Myanmar). This was followed by a trip to the Canadian Rocky Mountains and Yellowstone Park in Montana.

In 1902 and 1903, Parkinson lectured in geology and geography at Harrow in London. Admitted to St. John's College, Cambridge, as an advanced student, he received his B.A. by dissertation in 1903. Later that year, he was appointed leader of the Imperial Institute Mineral Survey of Southern Nigeria, and spent the next three years in West Africa. Parkinson next accepted the position of geologist with the Liberian Development Company. He published several anthropological papers written during his tenure. Parkinson continued to be active in West Africa, specializing in economic geology, particularly the search for hydrocarbon resources.

Among his hobbies he counted writing, and successfully published his first novel, *A Reformer by Proxy,* in 1909. Two years later his second novel, *Other Laws,* appeared. It was likely a fictionalized account of his travels. Africa drew him back in 1911, this time to the east coast. He investigated metamorphic rocks in the East Africa Protectorate, in collaboration with the Magadi Soda Company. This was hot work in the arid, lava-covered stretches between Lake Magadi and the Kapiti Plains, just north of the border of German East Africa.

During the First World War, Parkinson's scientific expertise was recognized by the Colonial Office in London, which sent him to the Northern Frontier District and Jubaland of the East Africa Protectorate. This assignment also took him to Abyssinia, currently Ethiopia. He undertook a water reconnaissance for the Magadi Soda Company in 1914 and 1915, which would prove valuable to any troop movements in the region.

Prior to the end of the war, Parkinson became chief of geological staff in both Trinidad and Venezuela. By 1920, he had spent five years in Trinidad. He had earned his M.A. and, in 1923, his Sc.D., and was a Fellow of the Geological Society and the Royal Geographical Society.

Parkinson unquestionably had a wandering lifestyle, in which his wife and family probably rarely participated. As a field geologist with three decades of experience, he was accustomed to the rigors of the tropics. This made him an excellent choice for the British Museum East Africa Expedition.

Meanwhile, the elephant hunters Parlett and Kershaw were maintaining their post at Tendaguru. In November and December 1926, quarries M3 and M4 were still producing bone. Trenches added to M3 and M1 proved barren. M14, about 90 meters east of M1, was extended with little result. Insects, especially butterflies, were collected. Boheti left for his farm at the end of November and the labor force was reduced to 20.[15]

A month later, the rains were regular and heavy, causing leaks in Parlett and Kershaw's dwelling. They spread their tents over the roof to prevent water seeping into their quarters. The new storehouse was nearly complete, while several of the more dilapidated huts in the laborers' camp were torn down. It had proved difficult to keep within Migeod's budget allowance.

A new quarry was opened on January 1, 1927, opposite M3 on the east side of the Lindi trail. About a dozen men were available for excavation.[16] Rain had fallen almost continually for 28 days in January 1927, and was calculated as the heaviest precipitation in six years.[17]

Another site was selected on February 1. At Nguruwe, Parlett reopened the large Cutler quarry, which again yielded bones in good condition. Other work was suspended due to daily downpours. Reconnaissance trips to the north and west were canceled as the Mbemkuru was in flood and trails were impassible.

Parlett rented a house in Lindi adjacent to Edward de Souza's shop at 70 shillings per month, rather than having one built as intended by Migeod. It had six rooms, a cooking area, and a fresh coat of whitewash. Samuels of Smith, Mackenzie & Co. suggested that an iron shed be erected nearby to store field jackets.

Parlett spent several weeks in Lindi packing jackets and replenishing supplies. He was suffering from malaria and treating himself with quinine. He also inquired about the price of a wireless set. This would have immeasurably improved communications, had it been capable of transmitting to and receiving from Lindi. Unfortunately, at £62 it was considered too expensive.

The main dwelling at Tendaguru underwent badly needed repair after the north end collapsed under the rains. Parlett admitted that this absorbed time and staff that could have been used for excavation but insisted that refurbishing was essential. Native huts had fallen down and the old bone store was in imminent danger of disintegration.

By mid-February 1927, Kershaw had made a more detailed assessment of Nguruwe. In the past, a large area had been cleared down to two meters below surface level. Subsequent rains had carried soil back into the pit, so laborers were ordered to resume excavation. The bones exposed by the most recent effort were in very good condition and it was hoped that there might be a single skeleton. All other quarry work was placed on hold. Tendaguru must have been thoroughly miserable: leaking and rundown dwellings, few laborers at hand, flooded quarries, a monotonous diet, and rain constantly interrupting excavation.

At the end of March 1927, Parlett was still at the coast when he received the announcement that Migeod would not return to Tendaguru. His disappointment was profound, as the men had hoped to show Migeod the new buildings and the excellent prospect at Nguruwe:

> He [Kershaw] expects to get a complete skeleton from Nguruwe, but now I think our enthusiasm is declining. . . . I think the BM authorities very unwise . . . in abandoning Tendaguru and area. . . . I am half inclined now to get into touch with the Metropolitan Museum, New York and see whether they would not like to carry on this Nguruwe field. It is a pity the BM did not get on with the assembling at home if they needed a show specimen so badly.[18]

Migeod's reaction to the changes cannot be traced. Possibly, feelings were bruised. In a rare letter, Harmer expressed gratitude for Migeod's efforts:

> The whole expedition under your direction has in fact been dealt with in a thoroughly satisfactory and capable way. Moreover, we do not forget that you came to our aid at the time when the Expedition was in such a difficult position owing to Mr. Cutler's sudden death.

The next passage could be interpreted as an attempt by Harmer to justify the Museum's course of action:

> We are constrained to make our money go as far as possible, in order not to be obliged to close down our operations too soon. . . . Full advantage cannot be taken of our opportunities unless we have the assistance of an experienced geologist, and you will understand that this has been our sole motive for asking Dr. Parkinson to help us.[19]

Museum officials were undoubtedly disappointed by the lack of a complete skeleton or remains of dinosaurs different from those collected by the Germans. Considerable sums had already been expended, yet it seemed the work could go on indefinitely without producing the desired results.

The BM(NH) underwent another staff change in March. Director Sidney Harmer retired at age 65. His replacement was Charles Tate Regan, the keeper of zoology. Migeod, curious about the state of the crates he had labored to bring home, was disappointed when Swinton informed him that the sole preparator, L. E. Parsons, was occupied with other specimens.

Born January 12, 1889, Louis Emmanuel Parsons won a scholarship to attend the Westminster Art School in London, where he studied drawing and sculpture for five years. In 1908, at age 19, he was employed by the BM(NH) as a workshop attendant in the Department of Geology. Quick to recognize his talents, Museum keeper Arthur Smith Woodward put him to work preparing and mounting a partial *Iguanodon* skeleton in 1909. Between 1910 and 1914, Parsons completed for exhibition in the geological gallery the skeletons of the Miocene camel *Stenomylus,* the Pleistocene bovid *Myotragus,* and the Jurassic ichthyosaur *Ophthalmosaurus.*

Because of his experience and abilities, Parsons was appointed preparator in the Department of Geology in 1921, without sitting for the civil service examinations. He had also painstakingly freed another dinosaur skeleton from its enclosing matrix: the ankylosaur, *Scolosaurus cutleri,* collected by William Cutler in 1914, in Alberta's Steveville badlands. The specimen was placed on public display in 1925, when Parsons was 36 years old.

John Parkinson was contacted by a London publishing house, H. F. G. Witherby, at C. W. Hobley's suggestion. The firm requested a popular account of dinosaurs in Africa, so Parkinson wrote to Smith to discuss the matter.

By late April, the rains were beginning to subside at Tendaguru, though the paths were still impassable. It took 4¹/₂ hours to reach the old hunting camp at Nakihu, and Parlett was forced to float across the Mbemkuru on a log. A Canadian geologist by the name of Eltz had visited Tendaguru the week before, in search of coal deposits in the region. He was likely in the employ of the Geological Survey under E. O. Teale, and hoped to meet Parkinson at Kilwa.

Little excavating had been done. Quarry M7 was dry, although Nguruwe was flooded. There was now a good collection of jackets waiting to be transported to the coast, but they were left at Tendaguru for the new leader to ship. Parlett and Kershaw appeared resigned:

Tendaguru is now like a cemetery—quiet + no life, and we shall not be sorry to get away on our hunting again. . . . were it not so lonely, it would be very attractive. . . . The boys are very sorry the place is closing down I think.

Neither man was as close to Deacon as they were to Migeod:

We both miss you here very much. . . . However we shall have the pleasure of old D's company soon I suppose. I am sure Fryer will be glad to see him back again as he was, I believe, quite a source of income to the Fryer family.[20]

Fryer, an administrative officer, was coming to Tendaguru in May.

Parlett and Kershaw sent a telegram to the Museum on April 28, warning that they would cease operations unless they received more money. Believing that funding had been arranged in advance by Migeod, Assistant Secretary Smith ignored the request. He expected Parkinson to be on site by the end of May.

Upon arrival in Mombasa Deacon sent Smith a brief note regarding Parkinson: "I have no doubt we shall prove congenial company. . . . I think he is inclined to be unduly nervous and anxious about his personal health in the tropics."[21] Deacon and Parkinson disembarked at Lindi on May 23.

The 55-year-old Parkinson soon set off down the coast for Mikindani. A two-wheeled cart was purchased and used locally to transport equipment and specimens. Deacon cleared equipment through customs and prepared loads for the interior. On June 2, Parkinson received word from two sources that bones had been discovered near Newala on the Makonde Plateau, 120 kilometers southwest of Lindi. A report from the same locality had been investigated in 1910 by the Germans. After a week's journey, Parkinson was disappointed to discover that the "bones" were fragments of asbestos weathering out of a seam of the mineral.

During this time, Deacon packed leftover jackets from Migeod's fieldwork into nine crates. Parkinson discarded some of the contents, mostly fragments, or bones lacking provenance data.

On June 26, 1927, Parkinson and Deacon stood at Tendaguru Hill. Parlett turned the accounts over to Deacon while Kershaw toured the quarries with Parkinson. The elephant hunters packed and left Tendaguru for their old camp at Nakihu. Parlett penned the last progress report to Migeod from Nakihu on June 29. The men had spent about eight months at Tendaguru, mostly under uncomfortable conditions. From Parlett's letter, it appears Migeod's decision to hire them had not met with the Museum's approval: "D[eacon] says that the Museum authorities only expected you to run a native caretaker at Tendaguru, and the impression I gleaned was that our appointment was more or less a misguided step. . . . Deacon has given up the house in Lindi. He says its [sic] too near the Indians?"[22]

Parkinson was impressed with the work undertaken at Nguruwe, and planned to deepen M7 in the search for a geological fault he believed the Germans had overlooked.

Fryer reached Tendaguru on July 11. Since Cutler's demise, the Lindi administration had twice made an effort to reach the site. Parkinson soon commenced excavation. Jackets originating from Nguruwe were all discarded. The Museum was advised that the five Nguruwe crates shipped in June would also likely be of little value. In compensation, an articulated series of bones appeared at a second, lower bone level at Nguruwe, presumably the find that had pleased Parlett and Kershaw. Parkinson placed the site at 1,800 meters south-southwest of Tendaguru Hill, at a bearing of 18°. This lower horizon was exposed in an area measuring 9 by 12 meters.

Efforts were renewed at M2, a group of quarries about 1,500 meters east-northeast of the hill, near the German site Quarry IX. Teeth, fish scales, belemnites, and other invertebrate fossils were collected. The pit opened at M2 measured 930 square meters. About 22 meters to the west, a partial pelvis, vertebrae, and other remains from a single animal were pulled from the ground. There was little else coming out of any other quarries, though no details are available on localities, dates, or crews.

In an attempt to obtain representative samples of other dinosaurs, the stegosaur site St (EH) was reopened at Kindope after 14 years. Parkinson's workings extended over 279 square meters, versus the German total of a thousand, though he excavated deeper, to 9 to 15 meters below the surface. At Ig (WJ), a tarpaulin was stretched over a portion of the sun-baked quarry and ornithopod bones were removed in blocks. Time was also spent at nearby Quarry dd.

Deacon was in charge of equipping the excavation crews and improving the campsite while Parkinson directed the scientific work. Huts for laborers were rebuilt and overgrown paths were freed of grass that stood almost two meters tall. About 40 locals were rehired at an average monthly wage of £1 per person.[23] This included individuals engaged in construction, excavation, transport, and camp duties. Parkinson had a higher opinion of the locals than Migeod, and greatly respected Boheti, who had returned to Tendaguru.

Parkinson was eager to understand the stratigraphy and clarify the age of the sequence by recovering invertebrate fossils. He familiarized himself with the relevant German publications on the southern region of the country, and emerged as the most informed representative sent out by the BM(NH) to date.

Exposures were examined in streambeds, but as the Germans and Migeod had found, the low-relief and vegetation-choked landscape offered few opportunities for correlating locally exposed sequences. The geologist soon concluded that a topographical survey was essential. He placed flags on Tendaguru Hill and distributed eight others about the countryside to enable triangulation and traverses. These white cotton flags were affixed to bamboo poles tied to the tops of trees. Intersecting bearings were taken with a 4.5-centimeter prismatic compass from the hill to the quarries and from one quarry to another. Intervening distances were calculated by pacing and further use of the compass. Once the quarries were pinpointed, a protractor was used to produce a map at a scale of 1:50,000. An aneroid barometer was used to determine elevations.

Dissatisfied with this imprecise method, Parkinson insisted that a surveyor was required for a period of six months:

> Drs. Hennig and Staff state that heights were determined by aneroid and a few by boiling point thermometer. For close work at these altitudes such methods cannot be considered satisfactory, but . . . the Germans employed them, after finding that the beds with which they had to deal were nearly horizontal, sketching in the "outcrops" on the outline thus produced.
> . . . To trace a bed by following its outcrop is for all practical purposes impossible.
> . . . The German workings certainly do not appear to have been sufficiently deep to demonstrate the relations between the fossiliferous horizons found in the latter and in the bone beds.[24]

Bather did not expect this development, which would inevitably involve additional expenses and complications:

> I thought Parkinson had enough surveying knowledge + experience to do this work himself. . . . The German work is unsatisfactory, and Parkinson was chosen because he is a practical field geologist.[25]

Assistant Secretary Smith contacted the Royal Geographical Society and the Colonial Office to inquire whether they knew of a surveyor who could reach Tendaguru quickly and cheaply. The Department of Surveys in Tanganyika was too short-staffed to assist, but advised that there were several freelancers in the Territory.

Parkinson specified that the surveyor was obliged to supply instruments for a three-month assignment at Tendaguru. A salary of £300 per year was proposed, plus the cost of travel, meals, and accommodation.

Contrary to Parkinson's understanding, the German Tendaguru Expedition had attempted geological mapping by means of an aneroid barometer to determine elevations, and a prismatic compass for triangulation and traverses. Lengthy and deep trenches had been excavated to establish stratigraphic succession. Parkinson's inability to find evidence of survey beacons or obtain additional details from Boheti is not surprising, given the amount of subsequent activity in the area. No other aspect of the Tendaguru Expeditions received such consistent attention from researchers as the dating and stratigraphic succession of the sediments.

German conclusions had been challenged on several fronts. Firstly, the paleogeographic reconstruction and depositional environment of the sediments had been interpreted differently by American and British researchers. Secondly, dating the deposits and establishing the stratigraphic succession around Tendaguru were contentious issues that led to a sharp exchange in the scientific literature between 1926 and 1937. It was so blatant that paleontologist W. J. Arkell labeled it nationalistic. Thirdly, correlations of strata at Tendaguru with areas to the north were revised by subsequent British fieldwork, as were geological structures such as faults and depressions.

Some of the disagreements stemmed from the difficulty of placing the Jurassic-Cretaceous boundary, which at that time was debated around the world. Participants in these discussions employed different approaches, lithologic and paleontological. They offered a number of geological mechanisms to explain the patterns and succession of deposition: subsidence, faulting, land and sea level oscillations, and disconformities. The value of different invertebrates, from bivalves to ammonites, in dating the sediments was emphasized.

As early as 1914, Charles Schuchert at Yale had examined German scientific results, hoping to refine the dating of the famous American Morrison Formation. He presented a paper at the end of 1916 that was published in mid-1918. Schuchert construed the mass burial of ornithopods and stegosaurs as evidence of drowning in rivers that were in flood. Carcasses had been carried downstream and deposited closer to the ocean, where decomposition and disarticulation had taken place. The alignment of bones was the result of stream currents, not wave action. Similarly, sauropods had moved through coastal marshes that were brackish to freshwater. Rivers had transported dinosaur cadavers into these marshes or lagoons.

The American paleontologist also took issue with the German assignment of the Upper Saurian Bed to the Lower Cretaceous. Supported by British paleontologist S. S. Buckman, Schuchert argued for the inclusion of the Upper Saurian Bed in the Upper Jurassic, and accepted as correct the placement of the *Trigonia schwarzi* Bed within the Lower Cretaceous. He also suggested a break between the *Nerinea* Bed and the Middle Saurian Bed.

Questions on geological dating were next raised by Finlay Lorimer Kitchin. This British paleontologist knew W. D. Lang, the keeper of geology at the BM(NH), so his views may have stimulated the Museum's search for a geologist to resolve stratigraphic issues at Tendaguru. Kitchin specialized in Mesozoic stratigraphic paleontology, published on bivalves in India and South Africa, and became especially familiar with the genus *Trigonia*. By reputation, he was meticulous and experienced. In a 1926 paper, he charged German researchers with serious errors in the conduct of their fieldwork and museum interpretation. He pointed out that the results published in *Archiv für Biontologie* had not been widely read by British researchers. This is not surprising, considering that the papers he questioned were published shortly before World War I.

At issue was an assertion by Erich Lange that several genera of geologically short-lived bivalves, that were obviously Upper Cretaceous, were found not only in the *Trigonia schwarzi* Bed but also in the *Trigonia smeei* Bed. Kitchin asserted that it was impossible for one of the genera, *Trigonia smeei,* a gregarious, ornamented invertebrate, to have survived without change for such a long period of time. Lange had described specimens that did not occur in Jurassic beds anywhere in the world. Kitchin suggested a geological unconformity, whereby the Cretaceous *Trigonia schwarzi* Bed lay directly atop the Jurassic *Trigonia smeei* Bed. This meant the Upper Saurian Bed was entirely Lower Cretaceous. His choice of words was unquestionably provocative, and elicited swift responses from Dietrich and Hennig.

Dietrich's rebuttal appeared in 1927 and stated that the entire invertebrate fauna of the *Trigonia smeei* Bed (ammonites, corals, gastropods, bivalves, and echinoderms) was of Jurassic age. Further, the faunal suite could be traced at numerous localities, and the cause of the confusion was Lange's unfortunate use of Krenkel's Cretaceous specimens. In fact, these specimens were not even from Tendaguru, and Lange had never claimed that *Trigonia smeei* persisted unchanged through long periods. However, in an important revision, Dietrich placed the Jurassic-Cretaceous boundary above the Upper Saurian Bed. He based his opinion on the strength of the complete Upper Saurian Bed fauna: dinosaurs, fish, a mammal, pterosaurs, and invertebrates. Now all three dinosaur-bearing horizons were considered Upper Jurassic.

Hennig reiterated the care taken to label specimens with ink as soon as they had been collected. He pointed out that the earlier assumptions of Bornhardt and Fraas had been overturned by the German Tendaguru Expedition, so revision was not unthinkable. More recent work in the American Morrison by George Gaylord Simpson strengthened the argument for a Jurassic age, and the faunal similarities between the Morrison and Tendaguru Beds were another factor supporting an older date for the African site. Hennig suggested that if Kitchin had such serious reservations about the German work, he could have asked for an explanation rather than publicly raising such a noncollegial controversy. The matter was still not settled by the end of 1927, and high hopes were placed on John Parkinson's geological expertise.

The partial skeleton discovered in M2 in July had been plastered and the quarry closed by mid-August. When exposed, the skeleton, which included a pelvis, measured about 6.5 meters in length. Deacon considered it more complete than anything collected by Migeod. After the bones had been removed, trenches had been dug on either side, but produced nothing. Other exploratory pits had been opened nearby. Quarries M9, M9a, and M11 to the north of Tendaguru had been enlarged and deepened but were barren of bone, as had been the case in M3 to the south. Three more quarries near M1 had brought only fragments to light, and the third new unit, No. 15, had been abandoned when it attained dimensions of 7 by 2.4 meters and a depth of 2.1 meters. No. 15 was located about 32 meters from M9 and M9a.

The efforts at Kindope had yielded few rewards—both St (EH) and Ig (WJ) had been reopened and bones had appeared 3.7 meters below the surface. Parkinson theorized that the skeletons from the Middle Saurian Bed were distributed along the channels of an ancient lagoon network or were accumulated, transported, and redeposited by postmortem erosion. Invertebrates had been collected at a number of localities.

The trek to Nyasaland had been canceled, since December was the rainy season in that country. Instead, Parkinson proposed that he and Deacon travel to the Amani Institute in northern Tanganyika. Its director had promised them the use of a room and laboratories where the season's notes and maps could be prepared.

Advertisements for a surveyor had been placed in local newspapers and the Kenya *Official Gazette,* eliciting five replies. Four candidates were so distant that they could not reach Tendaguru before the rainy season, but Parkinson sent a letter to the fifth applicant, W. O. Millington Rees, at Kilwa. Rees had not replied by early October, so Assistant Secretary Smith ordered instruments to enable Parkinson to complete the work himself. The geologist's unilateral promise of £300 per year salary for a surveyor had alarmed Smith, who cautioned that such expenses would jeopardize future field seasons.

Parkinson apologized to Bather for the time-consuming Newala excursion. Bather accepted the reasons but also admonished him, "Don't spend too long at Tendaguru. We have to remember that in order to satisfy the Trustees bones have to be got, but stones, if in your opinion of real scientific interest, would also be acceptable."[26] Bather and Swinton appeared to be losing interest in the dinosaurs of Tendaguru, given that this fieldwork had drained budgets and created a shortage of space at the Museum for several years.

On September 14, Parkinson and Deacon set out for Niongala, a village composed of a few huts. Kitchin had written to the BM(NH) the previous March, urging that special attention be paid to the stratigraphic sequence in this area. Parkinson checked two sites and Deacon and African assistants visited a third. Localities at which invertebrates had been found in the past were mapped with a prismatic compass. Hennig believed the area to be a region of subsidence, bordered by two geological faults. Parkinson was only able to locate one. He felt that the structure of the area resulted from a geological disconformity, or erosion followed by subsidence.

Even Parkinson, who was familiar with the scientific results of the German expeditions, claimed that the large bones exposed in the quarries at Niongala were abandoned by the Germans due to the outbreak of the war. It is unlikely that he knew of the wartime excavations of Pretorius and Hichens. Several men were set to work at Niongala digging both at and near to the old quarries. Fragments were found, but the only things of value unearthed were a few teeth.

Parkinson then continued to Mtapaia and Kinjele, another tiny settlement, and minutely examined the geology. The old German sites of Aa and Locality 15 were deepened by BM(NH) crews. Aa had provided the Germans with 18 articulated vertebrae and a sacrum of *Brachiosaurus,* complete with numerous marine shells in the body cavity. Shallow exploratory trenches were cut in the Kinjele Valley but no bones were found. The German quarries to the northwest of the village were also enlarged by a crew of eight. No traces of the lowest bone-bearing horizon could be discovered in the valley. At Nautope, a deserted village, Parkinson found no evidence of the faulting proposed by Hennig.

As September ended, the work force was reduced by 30 men, leaving about 50. Bearers carried 150 jackets into Lindi. A small team of Africans was retained at Tendaguru to continue excavating at M2 and Kindope. The British initially attempted to extract individual bones from the hard sandstone at Ig (WJ), in addition to removing blocks of matrix, resulting in breakage of some specimens. Other bones were considered "too imperfectly preserved" to send home.[27] At the close of the field season, around the end of September, Parkinson stopped at Tapaira, 14 kilometers southeast of Tendaguru. Invertebrate fossils were collected at two sites. Already the geologist was hinting at the possibility of a geological disconformity between the *Trigonia schwarzi* Bed and underlying strata.

Upon reaching Lindi during the first week of October, after three months at Tendaguru, Parkinson collected Tertiary fossils along the coast. During the rainy season, he proposed to work south of Leakey's sites in Kenya Colony, as the East Africa Protectorate had been renamed in 1920. Parkinson hoped for help with customs formalities and transport. Charles Regan, the new Museum director, wrote to the Colonial Office, which in turn passed the plea for cooperation on to Governor Grigg in Kenya.

Major Deacon, who was lodging with the Fryers or the Griersons, arranged for Smith, Mackenzie & Co. to ship 29 crates in late October. Fourteen of the boxes held the contents of M2. Deacon explained that the skeleton had been found 1.2 to 1.5 meters below the limb bones collected at M2 in 1926, but that he was sure the limb elements belonged to the skeleton being shipped this year. Deacon included photographs and a scale diagram of the quarry. Another six crates were filled with specimens of the small ornithopod bones from Ig (WJ) at Kindope. Eight boxes held jackets from Nguruwe, and the final container was packed with Parkinson's fossil invertebrates from Niongala and Mtapaia.

Kershaw and Parlett returned to Britain in October. On Parkinson's recommendation, Assistant Secretary Smith promised to look for some position for Parlett at £3 per week. He arranged for Parlett to observe museum methods of skinning small mammals in the Zoology Department. Smith also wrote him a letter of introduction to Peter Chalmers Mitchell of the Zoological Society, stating that Parlett hoped to return to Tanganyika and was offering to collect specimens. With this, both men, like H. Leslie Lachlan, pass from the story.

Other developments had taken place in London since mid-October. C. W. Hobley suggested that the region around Tendaguru be designated an official reservation to protect the fossils. He cited as precedents the American fossil reservations in Utah and South Dakota. This same step had been taken at Tendaguru by the Germans, but with the adoption of British and Indian statutes, previous legislation no longer applied. Hobley suggested that excavation be sanctioned only with a permit from the governor, that only approved scientific institutions be granted access, that the governor have the authority to reject permit applications without providing a reason, and that unsanctioned removal of fossils be declared a punishable offense.[28]

The BM(NH) trustees approved the suggestion on November 7, and advised the secretary of state for the colonies to "direct" the Tanganyika Territory government to enact the declaration. Dr. Parkinson would assist by outlining the geographic area.

A lapse in communication had occurred, which irritated Bather. He had heard nothing of Parkinson's plans for Kenya. Though all of Parkinson's interim reports were addressed to Bather, Assistant Secretary Smith appears to have assumed responsibility for the administrative support of the Expedition. Swinton and Lang, the deputy keeper of geology, were both away in late October, leaving few people to attend to Parkinson's reports. Somehow, a breakdown took place, which elicited an aggrieved response from Bather. Smith later advised Parkinson to address all correspondence to the director of the BM(NH).

Parkinson completed a trip north to Kisimbani, near the Kikomolela Plateau, between October 27 and 31. He intended to continue further north to Matapua but rain fell, rendering roads almost impassible to vehicles. Invertebrate fossils were collected at Kisimbani, Nandambiti, and Namgaru.

Parkinson made a final visit to Tendaguru around mid-November. On each trek, he examined outcrops, paced off distances, and took bearings. Quarry Ig (WJ) at Kindope had yielded even more bones since the end of September. Deacon was preparing to ship them, and work was halted at WJ when it was considered unlikely that much more of value would be found. An exploratory trench was underway, 185 meters from M2, providing well-preserved small bones.

Parkinson and Deacon were in Lindi again at the end of November, and the rains came anew. The plan was to leave Lindi around mid-December. Deacon prepared a final shipment prior to their departure. Twelve crates were packed by December 13: 11 containing the results of Ig (WJ) and one holding Parkinson's invertebrates. Two weeks later, after spending three months in Lindi, the men boarded the steamer heading north. John Parkinson's first field season at Tendaguru was over.

In all, 50 crates had left Lindi. Nine represented the last of Parlett and Kershaw's efforts. Twenty-nine contained dinosaur bones. Two cases bore the invertebrates that were crucial to resolving longstanding stratigraphical controversies. Wherever possible, their locations had been correlated with maps published by the German expeditions. Parkinson noted to Bather that it was difficult to collect from every level of an exposed sequence, because exposures were rare, the grass obscured topography, and many specimens were weathered out of their original positions. In London, some of the previously forwarded invertebrates were unpacked and identified by Leonard Frank Spath, a specialist in such fossils.

At least two sketch maps had been produced from Parkinson's surveying efforts. Geological reconnaissance had been undertaken between Lindi and Tendaguru, as well as south to Newala.

The crews employed had been comparable in size to those engaged by Migeod a year earlier. In May, there had been 20 on Parlett's payroll. June was busiest, with 30 bearers hired for the entire month just to carry equipment and supplies between Lindi and Tendaguru. The crews increased to 40 men in July, 70 in August, and 81 in September, and then reduced to 54 in October, 52 in November, and 40 in December. The supervisory and camp staff had included Boheti as overseer (at a pay rate of 65 shillings per month), two headmen (paid 50 and 42 shillings respectively), a cook (earning 50 shillings per month), and two personal servants at 30 shillings per month. Laborers received 12 to 18 shillings each per month plus 20 cents per day subsistence allowance.[29]

Deacon admitted that transport costs had been high with so many men hired to shift equipment and jackets. To justify the expenditures, Deacon pointed out that a single government officer was allowed 26 porters for safari. Clearing undergrowth for Parkinson's surveying traverses had also required much labor. Parkinson estimated expenses at about £50 per month.[30] A headman and four laborers were kept at Tendaguru over the rainy season to prospect for new fossil sites and maintain the camp buildings.

A promising start had been made. Stricter geological control had been brought to the British Museum East Africa Expedition, and new sites had yielded different animals than those previously sent to London.

-15-

1927–1929

Geology at Tendaguru

John Parkinson reached Mombasa on December 27, 1927, while Major Deacon disembarked at Zanzibar to visit friends. Public Works and Nairobi Museum officials offered advice on various archaeological sites that had been investigated by Louis Leakey. The BM(NH) men were able to reach a number of localities in a rented Ford automobile.

Parkinson and Deacon inspected sites around Koru Station in early March, then traveled between Naivasha, Nakuru, and Kericho in a Chevrolet car they had purchased. In early April, their investigations took them to Molo and Londiani, north of Nakuru. Porters were hired to carry gear for more isolated fieldwork. By mid-month they were near the Suk Hills, almost two hundred kilometers northwest of Nakuru. They spent several days in the Cherangani and Chemorangi Hills, a region dotted with extinct volcanoes. The furthest point they reached was near the Ugandan border, about 120 kilometers southwest of Lodwar. The region between this site and the southern end of Lake Rudolf, now Lake Turkana, did not concern Parkinson geologically so they headed south.

In London, the Expedition's finances were so depleted as to cause great alarm. A multipronged fundraising campaign was initiated, as the continuation of fieldwork was now in doubt.

The committee, originally established by Harmer in 1924 to administer the Tanganyika Fund, renewed its public appeal. By 1928, members responsible for what was now called the British Museum East Africa Fund consisted of Museum trustee Lord Rothschild; director C. Tate Regan; W. D. Lang, the keeper of geology; and Assistant Secretary G. F. Herbert Smith. An appeal notice was drafted and a banker's order attached. These notices were distributed to selected members of the Geological Society. Annual subscriptions in any amount could be paid to a London branch of the Westminster Bank.

On April 24, Regan and Smith approached the Colonial Office for advice on contacting governments of other East African territories administered by Great Britain. They explained that annual Expedition costs ranged from £2,500 to

£3,000. Although the £2,000 grant from Parliament to purchase museum specimens had been supplemented by donations of £1,100 to the East Africa Fund, these monies would be exhausted by September 1928.[1]

The Colonial Office forwarded an appeal to the governments of Uganda, Kenya, Tanganyika, and Nyasaland, accompanied by a summary of the Museum's activities. Lang strengthened the case for continuing the work by raising the threat of intervention by other countries: "It is imperative that if our knowledge of the Dinosauria is to advance, these deposits should not be allowed to remain unworked, and it seems a pity that we should now abandon them to the Americans."[2]

The tremendous success of the American Museum of Natural History's Central Asiatic Expeditions had eclipsed British efforts in Africa. Without more money the Expedition would have to cease operation by the end of December 1928.[3] Assistant Secretary Smith communicated this gloomy news to Parkinson on May 3, alluding to the Museum's dissatisfaction with the specimens in its possession, most of which had not even been unpacked:

> The Geological Department appear to be a little disappointed with the results of the Expedition so far as they have investigated them, but they are hopeful that the fossils you are sending may be in better condition and of greater interest than those previously received.[4]

In early June 1928, Assistant Secretary Smith wrote to a member of Parliament using Lang's argument of the threat of foreign interests as an inducement:

> If the Trustees have to close down the Expedition at the end of December . . . it is unlikely that the Museum would take up the work again. In that event the Americans will probably step in; a position, I think, humiliating to British prestige.[5]

If all this was not sufficiently worrisome, further unwelcome news arrived from Africa. In early June John Parkinson cabled the following: "Doctor very strongly advises six months leave."[6]

Smith must have groaned at this development, which would add delay and cost at an awkward time. The auditor general had recently raised several issues in a report on Expedition expenditures. Now a request for a passage home to London had arrived. More importantly, what had gone wrong in Africa? There was little Smith could do but accede.

Parkinson was staying with friends north of Nairobi. He had contracted amoebiasis, for which monthly injections had been prescribed. Treatment would end in September, when Parkinson intended to return to Tendaguru for two months before the annual rains. He communicated the news to Deacon, who had left for Lindi on June 1.

Parkinson explained his situation more fully to Smith:

> I do not at all desire to return home but in view of the Doctor's wishes and the necessity for this monthly treatment, it would have been risky to do otherwise and impossible to work at Tendaguru unless half of each month were spent in travelling to and fro.[7]

The geologist suggested that he might return to England and examine the invertebrates that had been collected.

Rees, the surveyor, finally wrote to Parkinson and was advised to contact Deacon in view of Parkinson's illness. It was hoped that traverse lines could be run from the summit of Tendaguru Hill to the village, to M2, to Ig (WJ), to Nguruwe, and to the Maimbwi Valley. Smith approved Parkinson's return to England, but in view of the uncertain financial situation, instructed the geologist not to engage a surveyor.

Then Deacon announced he was without funds, and asked for direction in Parkinson's absence. The major was instructed to commence excavation at Tendaguru with a limited crew, and £100 was transferred to Lindi. He was to report at once if unable to uncover good material.

Parkinson and Deacon were surprised at the Museum's insistence that more specimens were desired, especially in the light of complaints about lack of space and technicians to prepare material: "for we both thought we had sent you more than you could deal with at the moment—and anything of value found this year."[8]

The struggle to keep the Expedition in the field continued. At the end of June, Regan drafted a memorandum to the Treasury, asking that further sums be approved out of the Reserve Fund. The Museum first had to prove that specimens equal in value to the amount expended had been received. Then the auditor general and Treasury officials demanded assurances that the Museum had received Treasury approval to spend sums in excess of those originally allocated.

Assistant Secretary Smith raised the fear of American intervention with C. W. Hobley, who was never a passive supporter. Hobley replied with four suggestions for promoting the Expedition: ask the *Times* to back an appeal; consult a "publicity expert" such as Charles Higham; approach a prominent scientific society and ask that they include the appeal leaflet in an issue of their journal; and finally, encourage a BM(NH) trustee to take the lead. Hobley had the Prince of Wales in mind. A letter from the prince's private secretary to "magnates" known to him should encourage others to contribute.

Smith wrote to a number of newspapers, including the *Times,* the *Morning Post,* the *Daily Mail,* and the *Daily Telegraph.* The editor of the prestigious journal *Nature* also offered to print a notice. Two subsequent newspaper reports announced that over five hundred cases of specimens had already reached the Museum. The figure was inflated; according to most BM(NH) reports, the maximum was closer to 350.

The response to Smith's appeal was modest throughout July, with no single donation exceeding £10. Several men of substance indicated that they would not become involved. Edward, Prince of Wales, contributed £10.

When Parkinson reached England, he was advised to discuss the invertebrate fossils with Swinton. Swinton had visited Germany in 1928, and had written to Janensch regarding the Tendaguru collection. Informed that the jackets from M2 were not uncrated yet, Parkinson urged that this be done, presumably to ease fears of low-quality specimens. Parkinson's sincerity about rejoining the Expedition was not in doubt, as he offered to pay his own return passage to Africa.

Gloom pervaded the fundraising, and Smith confided to Parkinson that there was no hope whatever of convincing private sponsors or East African governments to contribute enough to extend the fieldwork. Ultimately, it was agreed to send Parkinson back to Tendaguru for a few months to complete his topographical survey, collect additional invertebrates, and close out the effort with Deacon.

Parkinson also conferred with Sir Arthur Keith at the British Museum in Bloomsbury. When Keith heard Parkinson was returning to Tanganyika Territory, he urged the geologist to explore Hans Reck's sites at Olduvai. Intrigued, Parkinson wrote to Regan with this suggestion on August 23. Lang, too, believed it would be "most desirable" for Parkinson to collect fossil mammals at Olduvai if such work did not interfere with Tendaguru. Given the Expedition's financial crisis, the practicality of this plan was questionable. Parkinson would be lucky to be employed by the Museum after another three months in Africa, and could hardly expect to mount another remote expedition.

The Expedition had been at Tendaguru under European supervision since late June, when Deacon had returned to the site after a six-month absence. When the major learned of the geologist's intention to work through December, he cabled his disapproval. The small crew left at Tendaguru had "met with small success."[9]

Parkinson urged the major to increase the size of his labor force and rehire Boheti. By July 9, Deacon was prospecting with 20 men. Results from the month of July appeared meager: nine quarries had produced little. Only one locale convinced Deacon to recommend that the Expedition continue another month. He raised a point that had been obvious to both the Germans and Cutler from the start: "there are certainly quantities of fossil remains in the area, but . . . before good complete specimens may be expected excavations will have to be carried on on a large scale."[10]

Deacon suggested that a few men remain at Tendaguru once excavation was halted, to maintain the dwellings, at a cost of about £5 per month. Perhaps he had not abandoned hope for a renewal of the Expedition by some means, or had not been informed of the imminent shut-down.

At the end of August, Deacon reported that 16 men had been occupied throughout the month carrying jackets to the coast.[11] The site that had been discovered at the end of July proved productive and may have been the source of many of these loads. Deacon predicted the porters would be kept busy for another two or three weeks. His estimated expenses for September were modest at £56, including the cost of timber and carpenters for crating specimens.

Deacon dispatched 44 crates to London on October 3.[12] Twenty contained remains excavated southeast of M2 and were marked M2/D. Deacon called them "correlated specimens," presumably because he considered them part of the same animal. Another three crates held "unrelated bones" and were labeled MTP/MISC, likely from Mtapaia. Thirteen boxes marked NG/D originated from Nguruwe. Five more, designated NG/MISC, were "unrelated specimens" from around Nguruwe. The last container was filled with Migeod's books.

The Museum's request for reduced freight rates involved a multitude of parties, including Museum shipping agents F. Stahlschmidt & Co., the shipping line Gray Dawes & Co., and the British India Steam Navigation Company. A 50 percent concession was granted by the carrier, as in the previous year.

Fully recovered from his amoebiasis, Parkinson returned to Tanganyika Territory in early October. The governor of Uganda, Sir William Gowers, was aboard the same ship. When Parkinson broached the subject of assistance to the Expedition, Gowers offered no hope.

In London, Assistant Secretary Smith wrote to the Treasury to bolster support for the Museum's latest application. He emphasized that the acquisition of other collections had made great demands on the Purchase Grant, and that the recent public and private appeal had raised a mere £70. His frustration was obvious: "It seems quite impossible to interest our wealthy men in this sort of thing."[13] Smith asked if the Treasury might vote in favor of an application to increase the Purchase Grant by £2,000 solely to support the Expedition. The threat of American entrepreneurs was raised yet again.

In early November, and with no warning, disturbing news reached the Museum via another telegram: "Been hospital fortnight | Wish leave early | Deacon."[14]

This terse message must have sent a chill through anyone who remembered the untimely end of William Cutler. Around the third week in October, Deacon had entered the hospital in Lindi suffering from an unspecified illness. His early return to London was approved, and he made his way home via Zanzibar, Lourenço Marques, and Cape Town. Deacon was in Southampton by December 21, 1928. It is a tribute to his constitution that he was able to remain active in remote Africa at age 64. Parkinson was most likely at Tendaguru collecting fossils. He had initiated a quarry near M3, but the matrix was so friable that little of value was collected.

In mid-November, the East African governments finally replied to the Colonial Office. Uganda declined to support the Expedition, and Nyasaland could only contribute £50 because of tight budgets. Tanganyika had not replied. The greatest hope came from Kenya, whose governor had suggested that the Legislative Council should allocate £1,000 in the 1929 budget. The governor also asked that specimens be donated to the new Coryndon Memorial Museum in Nairobi.[15]

A telegram reached Parkinson from the Tanganyika Geological Survey in Dodoma, approving an expedition to Olduvai Gorge. The sum of one thousand shillings was available to pay for labor. The grant offer was too little too late for Parkinson, who turned it down. A London mining firm had contracted him to search for coal deposits in the Ufipa region of the Territory, commencing January 1929.

Around mid-October, Governor Cameron replied to the Colonial Office in London regarding a fossil preserve around Tendaguru. No special legislation was required to create such a designation. Pending formal declaration, the controller of mines and the land officer were directed to pay special attention to any applications for land or mineral rights in the region.

Parkinson's description and sketch map of the fossil reserve were forwarded to the district officer in Lindi. They specified that Africans around Tendaguru were to be compensated if fossils were found in their fields or villages. Restrictions requested by the BM(NH), such as limiting work to approved scientific institutions, were to be addressed by the authorities in Dar es Salaam as "questions of policy."[16] Parkinson sent his final suggestions for the boundaries to Provincial Commissioner Grierson on December 10, incorporating Deacon's practical suggestions of using geographic features like rivers, streams, and roads.

The reserve was defined as follows:

> From the Kitumbini or Mikadi bridge over the Mbemkuru River on the Lindi—Kilwa road southwards to the Matapua branch road, thence via Matapua and Mangalingali along this road to the beacon bearing S.E. from Tendaguru Hill and known as Lipugilo [Lipogiro to the Germans], thence northwards up the Tendaguru road to the side path leading to the Maimbwi River following this to its junction with the Mbemkuru River thence following the course of the latter to the point of departure at the Kitumbini bridge.[17]

Parkinson was obliged to conclude the Museum's affairs by himself. Two men would stay at Tendaguru from January to April 1929. They would be paid 18 and 16 shillings per month respectively, plus 6 shillings per month subsistence allowance until the end of May.[18] Boheti would be retained at half pay, or 32.5 shillings per month, from January through April, and was instructed to search for pterosaur remains at M2. Any specimens collected would be left at Tendaguru.

The superintendent of police accepted a 12-gauge shotgun and a .423-caliber Mauser rifle, to be kept in police stores. Expedition gear was transferred to the warehouse of Smith, Mackenzie & Co.

Maps, sections, and a report were drafted by Parkinson during the first three weeks of December. The plans included a sketch of quarry locations, with invertebrate fossil discoveries marked.

Another profile illustrated Parkinson's interpretations of the stratigraphic sequence. He posited a disconformity between the *Trigonia schwarzi* Bed and the underlying bed, and felt there was no marine layer between the Middle and Upper Saurian Beds. The nearly horizontal inclination of the deposits and lack of outcrops made it difficult to correlate the strata. This was one of the reasons he ordered quarries deepened to expose profiles. Yet he admitted not being entirely successful in unlocking the relationships:

> Extensive and systematic pitting combined with careful levelling are necessary to determine the conduct of the beds of this Series; laborious work which, however desirable, under the circumstances could not be carried out. In this manner alone could a disconformity be proved.

Despite this frustration, the geologist maintained that if he could not conclusively demonstrate the stratigraphical relationships, then the Germans were in no better position to do so:

> it appears to me doubtfully wise unhesitatingly to accept the vertebrate fossils from certain pits as belonging to one of the Saurian horizons, as shown in the

sketch-map "Die Grabungsstellen am Tendaguru," when by slightly shifting a Geological boundary at "S" for instance, or, by inserting a minute dip the bones could be placed in the other horizon of considerably greater or lesser antiquity.[19]

In a paleogeographic reconstruction, Parkinson described the region in Mesozoic times as a coastline whose margin consisted of sand and mud flats. The arms of a river ran through the coastal flats to form a delta, like the modern Niger River. The delta fanned out into expansive shallow lagoons that were isolated from the ocean by barriers of sand. This gave rise to an estuarine environment. Periodic flooding swept river mud, silt, and dinosaur bones into the lagoons. The coastline subsided tectonically at the same pace as the sediments were deposited, allowing marine sediments to be carried into the lagoons.

One of the first people to hear of the Tanganyika government's agreement to fund the Expedition for 1929–1930 was Parkinson, in late December. He hurriedly passed the excellent news to Smith on Boxing Day. While Dixey, the government geologist of Nyasaland, supported the Museum's plea to his government, he was doubtful of success, due to the poor price being realized for the protectorate's tobacco crop.

The BM(NH) authorities were asked to provide a full report to the Tanganyikan authorities by November 1, 1929. This request was made in order to prepare for the 1930–1931 fiscal year; government authorities hinted that the £1,000 grant might be renewed another year. Parkinson mailed the final accounts and labor payroll book to London on the last day of the year. A total of 1,369.65 shillings was left in the Expedition account in Lindi.[20]

Parkinson was in Dar es Salaam by January 10, 1929. Though he had only spent about six weeks around the Tendaguru and Lindi areas in 1928, he had collected and accurately located numerous invertebrate fossils. Accessible exposures in the region had been examined by an experienced field geologist. There was now support for a stratigraphical interpretation that differed from that of the Germans. The fossils of the area had been protected thanks to the map for a reserve, drawn up by the two Museum representatives.

Deacon had been on site for roughly four months in 1928. Altogether, 63 crates of dinosaur bone and fossil invertebrates had been shipped to London that year, bringing the total to 378, according to Swinton's report.[21] Even though the ambitious plans for the 1927 and 1928 seasons had not been realized, a clearer understanding of the geology around Tendaguru would allow a more accurate interpretation of the events that led to the preservation of the dinosaurs.

-16-

1929

Migeod Returns

A ten-page report was prepared for the Tanganyikan authorities by W. E. Swinton. Again, it was claimed that the Germans had employed a small crew to excavate around Tendaguru up to the outbreak of World War I, though no evidence of this has been found in German museum archives. Reck may have sent Boheti or a team to the site in 1914 while he prepared for the National Exhibition. Krenkel had never reached Tendaguru before hostilities were declared. In the section of his report titled "Scientific Results of the Cutler and Migeod Periods," Swinton wrote of the difficulties of assessing the Tendaguru material:

> the Geological Department has had . . . neither space nor preparator staff for
> speedy treatment of the collections. . . . an attendant was subsequently appointed
> and he has been almost continuously occupied with the Tendaguru bones. . . . the
> process of unpacking and giving the bones preliminary treatment has been unde-
> sirably slow. Only a comparatively small portion of the collections have been ex-
> amined although the work is now proceeding much more quickly.[1]

The newly appointed preparator was Cyril Philip Castell. Born June 30, 1907, Castell passed the examination for London University in 1926, when he was recommended to the BM(NH) by a school teacher. The 19-year-old was hired as a workshop attendant in the Department of Geology in December 1926. One of his first assignments was the preparation of specimens from Tendaguru. In the 1928 annual report on Castell's performance, Lang, the keeper of geology, indicated that Castell performed the "monotonous" work "ploddingly and cheerfully." Castell would later disarticulate and remount the Carnegie *Diplodocus* cast with technician F. O. Barlow.

In late February 1929, Assistant Secretary Smith informed John Parkinson that the BM(NH) appeal to the Treasury for additional funding had been rejected. Consequently, Parkinson's employment with the Expedition could not be extended past December 31, 1928.[2] Yet there was more afoot than Smith alluded to. Almost simultaneously, Swinton asked Migeod if he would consider returning to Tendaguru. One suspects that Parkinson was not asked to return primarily because of the limited funding available for salaries. He was also now on contract with a private firm in the Ufipa District, about 120 kilometers southeast of Lake Tanganyika.

The 1929 field season called for the utmost thrift. Migeod was asked whether he could operate at Tendaguru without a European assistant, and whether he had sufficient confidence in his geological and anatomical skills to continue excavating without a geologist.[3] The terms were spartan; not even Deacon was invited to volunteer. Migeod agreed to the arrangements, and steamed out of Marseilles on March 23, 1929.

While Migeod was en route, the governor of Kenya expressed his regrets that the Legislative Council did not support the allocation of £1,000 to the BM(NH) Expedition.[4] Assistant Secretary Smith's answer to the Colonial Office illustrates the desperation to which the Expedition's financial status drove the Museum: "I believe Nyasaland offered £50 as a gesture of goodwill. Will that be forthcoming? Every little bit helps when there is so little to spare."[5]

Parkinson was unaware of the dire financial situation in London, and of his termination, when he wrote the Museum in late February, asking if his services were required at Tendaguru in April. As late as mid-April, he was still hopeful of being sent to Olduvai somehow, offering to forfeit his paid passage home. Unaware that Migeod had been hired, Parkinson continued expediting Museum business from Ufipa, directing the bank in Lindi to keep Boheti on half pay until April "in order that he might visit Tendaguru from time to time and report results of any work done there."[6] Parkinson sent the Museum a draft version of the Tanganyika government's fossil-protection legislation, *An Ordinance to Provide for the Preservation of Objects of Archaeological and Palaeontological Interest*. It passed the Legislative Council, and Governor Cameron assented to it on April 10.

In anticipation of the Tanganyika Agricultural and Industrial Exhibition, to be held in Dar es Salaam September 2–6, Parkinson proposed a show of Tendaguru specimens to stimulate public interest, and the organizers informed the BM(NH) of the cost of a display area.

Parkinson ran afoul of the strictly enforced publication policy of the BM(NH) when confiding his skepticism about Migeod's approach:

> Between ourselves, I do not think the Tendaguru sequence and consequent age of
> the deinosaurs will be materially advanced by Migeod, and . . . I have suggested . . .
> a paper of some 3,500 words being in substance composed of my last Report. . . .
> I have signed a contract for a book on the Tendaguru deinosaurs to be ready late
> this year or early the next. I fear the step was a rash one.[7]

Assistant Secretary Smith bluntly replied that all rights were reserved by the trustees, and that he was to formally apply for permission to publish.

Meanwhile, Migeod arrived in Dar es Salaam in mid-April 1929. Governor Cameron asked whether the BM(NH) would continue excavation at Tendaguru indefinitely, and whether "all the existing output [had] been studied yet."[8] Migeod assured the governor that there was a great deal of potential left untapped at the site, and "that it was preferable that the British Museum should have the first opportunity before Americans or Germans applied to dig."[9]

On April 17, 1929, Migeod was back in Lindi, having been away for 16 months. He engaged 24 carriers and a cook. On the morning of his safari to Tendaguru only half the carriers reported. On the following day, he sent the cook and 11 men in advance, abandoning half the loads in Lindi. Migeod stood at the foot of Tendaguru Hill on April 26.

Fourteen men were hired immediately, at the rate of 12 shillings per month wages plus 6 shillings for subsistence.[10] A small garden was started, a wise move given the state of food supplies.

Boheti was back as head overseer, and for the first week, the men cut the tall grass around the hill and built new huts for the work force. As the remaining loads had not arrived from Lindi by the end of April, several men were sent to the coast to retrieve the most urgently required items.

Migeod was ready to start excavating at the beginning of May, but, as in 1926, was hampered by lack of direction. Neither Parkinson nor Deacon was available for consultation, though Migeod may have spoken to Deacon in England. Migeod saw only a portion of the notes made by the geologist, and relied on the African overseers who had participated in the 1927–1928 field season. Boheti would have been invaluable, but it is not known how well Migeod was able to communicate in Swahili.

M1 was selected and a team of 17 men was assigned on May 1. In Cutler's day, the quarry had covered an area of about 46 square meters and was almost a meter deep. When another dozen locals applied for work there were 36 men in camp including Boheti, the cook, a camp assistant, and two water haulers.[11] A crew cut a trench from east to west across M1 on May 7, and, in time, opened a second ditch at right angles to it. The bones were poorly preserved and abraded by tumbling through water.

By May 30, Trench 1, which had extended the old M1 No. 3, was discontinued. Trench 2 was pushed 18.5 meters north, then east, but was also barren. When it was reopened once more, there were still no results, so it was finally abandoned. A large bone was found on the surface 46 meters past Trench 2. After six men spent a week excavating to a depth of 1.2 meters, this site was shut down.

In early May, Migeod visited Kindope and three men were set to work. Seventeen men were assigned to excavate at Nguruwe, 3.2 kilometers south of Tendaguru, on May 9. The site was designated Trench 1/1929. There were 38 men in camp between May 5 and May 11, with about a dozen assigned to building houses; another 15 at M1; and about 17 at Nguruwe.[12]

Elephants, always rare visitors, left their footprints at Nguruwe in the early hours of May 16. Trench 2/1929 was commenced at the site on the following day. While collecting plants along the Mtapaia Stream, Migeod encountered some exposed bone. He put about 10 men to work developing the site on May 20. Only a vertebral centrum emerged. The ditch was barren and work ceased ten days later. An exploratory cut made a few hundred meters upstream also foundered.

Regrettably, the camp food supply again caused contention. Although he had been skeptical of Cutler's complaints in his first year at Tendaguru, Migeod reacted promptly in 1929. He had a much better relationship with Acting Provincial Commissioner Grierson than with Anderson. Migeod outlined his grievances, asserting that the Germans, with up to four hundred men employed, had had fewer complaints about the willingness of local chiefs to sell food. Inexplicably, he also referred to their "large European staff," though there were never any more Germans at Tendaguru for lengthy periods than Britons: "Trouble began in Cutler's time, when . . . an Administrative Officer told the chiefs the expedition was not a matter of any account; and naturally they are only too pleased to observe to the full any hint against a European."[13] Commissioner Grierson reassured Migeod that the local jumbes had been asked to cooperate fully. He also promised that the tall grass that choked the trail all the way from Lindi would be cut, though there was little hope of reopening the motor road.

Parkinson sounded surprised that Migeod was back at Tendaguru, and apologized for not being able to meet and discuss the fieldwork. Again, Migeod does not seem to have been thoroughly briefed on results from Tendaguru, despite the concern about gaps in his osteological and geological knowledge. He admitted seeing a few of Parkinson's reports and taking notes but had no copies with him, nor had he seen Parkinson's survey maps of the region.

Part of this lack of coordination can be traced to the slowness of the mail. Correspondence between Parkinson and the BM(NH) took many weeks each way, and then passed through several hands before it was available to an Expedition successor in the field. Events moved forward relentlessly in the meantime—such as the hiring of Migeod as leader for the next season, and his departure for Africa.

To the director, Migeod reiterated his conviction that Parkinson and Deacon, despite deepening the excavations at several sites, had been unable to find more bones because the bone-bearing layers were thin. Migeod discussed the occurrence and abundance of bone at Tendaguru, which he believed to be a highly localized deposit, extending no further than 1.6 kilometers beyond the hill: "it is not unlikely that the output of bones will decline, as seems to have been taking place in the last two years."[14] There is little doubt that the British Museum East Africa Expedition was experiencing a decline in the abundance of bone. Given the numerous seasons of excavation in a relatively small area, the rate of erosion may not have been sufficient to uncover fresh finds in a region with few exposed strata.

The Germans had done more prospecting than their successors. Since Migeod's arrival in 1925, the British work force had been perhaps 10 percent the size of the largest German crew. There were fewer bodies assigned to the prospecting, and the task was carried out neither as intensely nor as far afield. Teams had been sent out in sweeps by Janensch and Hennig, and reports had been received regularly from distant areas.

A smaller work force in British times also meant that fewer locals had been traveling to and from Tendaguru, giving them less opportunity to move through the countryside and notice exposed bones. With the myriad logistical details that demanded attention daily, and the fact that the British stayed close to German quarries, it is natural that Migeod was convinced that bone abundance was declining.

John Parkinson did not share some of Migeod's views. On May 11, he posted a report to London. He believed the area was far from depleted, and insisted that it was critical to understand how widespread the dinosaur deposits were and to determine the time range of deposition. It was equally important to discover how the deposits had been formed and the bones brought into the area. This would explain the conditions in which the dinosaurs had lived and died. To this end, he recommended quarries be enlarged and deepened, so that strata could be correlated at various sites.[15]

In Parkinson's paleogeographic interpretation, water channels had swept animal remains from the interior into lagoons. Flooding periodically broke through sand barriers and carried carcasses further out, as at Kinjele. In his opinion, the main channel of the lagoonal system would supply abundant specimens, if it could be located. His conclusions were the opposite of Migeod's. German quarries at Kindope were far deeper than those at Tendaguru, and extremely rich. It was therefore imperative to continue at Kindope, and he specifically listed St (EH), Ig (WJ), and dd (HR) as worthy of further investigation: "At the present time the honours of Kindope rest entirely with the Germans and it is impossible not to feel that the results of the British Expedition are being critically compared with those they themselves obtained."[16]

Parkinson later drew an analogy with the Epe Lagoon, part of the Niger River delta near Lagos. Migeod replied that his model was the Lorian Swamp in Kenya Colony, where many animals had perished due to drought. There was no malice in Parkinson's words, as he offered Migeod his best wishes: "I do hope you will have a successful time and that the Goddess of Chance may smile, for really She has a lot to do with the Tendaguru bones."[17]

Migeod did not revise his paleoenvironmental reconstruction, despite another note from Parkinson on October 11. The geologist pointed out that his lagoon analogy was based on the sea's breaching the barriers to the lagoon and depositing ammonites, belemnites, and marine sediments at M2 and Kinjele or Mtapaia.[18] Migeod's interpretation was based on the effect of climatic change that he had observed personally; his analogy referred to an inland geographic feature that had no connection to the ocean. Parkinson's model was based on the correlation of geological evidence and used a delta as an example. None of the Expedition participants was able to spend more than the shortest time discussing their results with successors, or even examining each other's notes. Ultimately, BM(NH) officials accepted Parkinson's theory.

In 1929, F. L. Kitchin returned to the debate on Tendaguru dating. He regretted the fact that since 1924, there had been no further clarification forthcoming from the British Museum East Africa Expedition, and felt compelled to respond to Dietrich and Hennig's rejoinders to his 1926 paper. Kitchin admitted the improbability of his suggestion that the marine *Trigonia schwarzi* Bed had to lie atop the Jurassic *Trigonia smeei* Bed. It was now obvious that the Upper Saurian Bed and Middle Saurian Bed contained an almost identical dinosaur fauna.

Far from stymied, Kitchin maintained that dinosaur species, like invertebrates, could not have survived the considerable time span between the two beds unchanged. He concluded that both Saurian Beds were of Lower Cretaceous age. He sought support among the invertebrates. Kitchin classified *Modiola* and other genera from the transitional sand that graded into the Upper Saurian Bed as Cretaceous types. In addition, the similarities between *Trigonia smeei* at Tendaguru and the same bivalve in supposedly Cretaceous horizons in India sealed its fate as a Cretaceous genus for Kitchin. Once more, the vexatious suite of bivalves originally described by Lange was resurrected. They were, according to Kitchin, unequivocally Cretaceous—if they appeared in both *Trigonia smeei* and *Trigonia schwarzi* Beds, then both beds must be Cretaceous.

Corals described by Dietrich in 1926 were dismissed as insufficient evidence of a Jurassic age for the *Trigonia smeei* Bed. Even the ammonites that Josef Zwierzycki used to date the bed were, according to Kitchin, derived from other levels. Kitchin insisted they had been swept into Cretaceous sediments from eroding Jurassic sediments. Further, he suggested that the invertebrates from the *Nerinea* Bed, *Modiola* and *Mytilus,* were Lower Cretaceous when compared with Indian and South African genera. John Pringle, also of the Geological Survey of Great Britain, confirmed these determinations. The deposition of the strata was thought to have been quite rapid. This was Kitchin's last published comment on the matter.

On August 1, Assistant Secretary Smith completed the long-awaited report of the Expedition for Governor Cameron in Tanganyika. It repeated, almost word for word, sections of Parkinson's account to the BM(NH) of May 11. It also expressed disappointment with the results of over five years of effort and expense: "there is no indication that the Museum has gained a complete dinosaur, and the signs of skulls are disappointingly few. . . . The remains received . . . come from practically all the types of dinosaur found by the Berlin collectors."[19]

The Museum administration concentrated on closing out Expedition accounts. In response to a query from the Treasury Smith explained salary differences among the leaders. Cutler and Migeod had each been paid £300 per year, and Parkinson £800 per year.

> Mr. Cutler was a collector only and otherwise not an educated man, and Mr. Migeod is not a geologist and, being in receipt of a pension, was content to receive a nominal salary.
> Dr. Parkinson was . . . a thoroughly competent geologist . . . and £800 was the lowest rate which he could accept in view of his family commitments.[20]

When Boheti requested leave at the end of May, he was refused. The head overseer then decided to take the next two weeks without pay and departed. By June 4, with a total workforce of 41, a new attempt was made at Nguruwe, this one labeled Nguruwe Bluff. It was a steep bank at the end of a ridge along which the path to Tendaguru ran. At mid-month, the Nguruwe crews ranged in size from 27 to 30 men.[21]

Around mid-July, Nguruwe was officially divided into NG central, NG south central, and NG Bluff. Work at the first subdivision was terminated. Migeod installed a sundial here to enable the laborers to tell when to break off their daily work, an improvement over his experiment with an alarm clock. The site was too far away from Tendaguru camp for them to hear the signal, which was a crowbar, struck with a piece of steel. Migeod had the crew dig almost another meter deeper in a vain hunt for a pelvis. South central was concluded on July 12.

The bone layer at NG Bluff was approximately 7.5 meters higher than at NG south central. Ribs, 1.25 meters long, were exposed here. Boheti reported a scapula, which Migeod believed was a pubis. One of the large vertebrae had processes over 60 centimeters long. The large "pubis" from NG was brought into camp in sections on August 3, and after some cleaning, Migeod revised his identification and admitted that it must be a scapula. NG Bluff occupied about 26 men on a morning shift. By late October, it measured approximately 40 meters in length.

NG also produced nearly complete large cervical vertebrae, of which Migeod was particularly proud. About two weeks in August were spent cutting into the top of the hill to widen the ledge on which the bone layer was exposed. Once this surface was laid bare, a so-called bird skull appeared. Migeod initially described the fossil as 40 centimeters long and resembling a pterodactyl skull. It is difficult to identify a skeletal element in the field, since it can be partly exposed, incomplete, poorly preserved, and lying at a confusing orientation. Migeod later revised his earlier claim of a bird or pterosaur skull, conceding it was more likely to be a dinosaur vertebral process. By mid-September, NG extended over an area measuring 27 meters in length.

Another extension to an old quarry was commenced on June 22. Six men under Juma worked outward from M8, and called the new site +M8. At M8 No. 2 a mass of "associated" bones were exposed about three meters from the skeleton that had been unearthed in July. A pelvis and a curving arc of vertebrae lay in the sun. All were badly fractured, with mud filling the cracks. On August 23, M8 No. 2 was cleared and fragmented bones were brought into camp piecemeal, where Migeod repaired and reconstructed them one at a time.

The work force in late June stood at 41, with an average of 26 at Nguruwe and 14 at M3 under Boheti. Having produced little, the northern trench at M3, 23 meters long and 1.5 meters deep, was closed on July 10 due to the poor condition of the bones.[22] In a subsequent report, though, Migeod announced that a femur 2.1 meters long had come from the site. About 90 meters north of M3, he created +M3 or XM3, and at a depth of 1.2 meters the bone level finally appeared. About eight men worked here under Kibwana's supervision.[23] Operations were terminated on July 15, again due to poor bone preservation.

Finally, it was possible to burn off the thick covering of grass and begin prospecting. Migeod had been forced to spend the first seven weeks of the field season enlarging existing quarries. No obvious prospects had been identified from the previous year, and without a copy of Parkinson's May report, he was unaware of the geologist's advice to concentrate on the trio of German quarries at Kindope. Burning was always an unpredictable operation and, as in 1926, flames threatened the camp when the wind shifted suddenly on June 28. A firebreak prevented its spread.

Migeod decided to experiment with another method of shipping, in which small boxes bound with iron strapping would be sent to England. These cases, which had previously held kerosene tins, would be more portable than the crates built previously. With no one in Lindi to oversee the assembly of crates, the change was obligatory. One wonders how the shipping arrangements were made without an agent in the port, since Migeod himself never accompanied the loads to the coast. He calculated that his monthly expenditures of £70 would not only allow him to remain in the field until the end of October, but also leave sufficient funds for future work.[24]

The first bone shipment of the year left for Lindi on July 2. Of a work force of 48, 10 went to Lindi with loads. By September 7, 48 loads had reached the coast. Some limited prospecting was undertaken by the third week of July, where grass had been burned.

A new quarry, situated 1.6 kilometers along the trail to Lindi, was designated M18 on July 10. The jump in the number sequence between this quarry and Migeod's last 1926 quarry, M14, was due to three sites that had been opened by Parlett and Kershaw, but not officially named. Eight men were assigned to the new M18, which reached a depth of 1.2 meters but only yielded significant finds on the side of the trail closest to Tendaguru. M18 was closed on August 5, after about three weeks of excavation.

The lack of suitable illustrations again left Migeod irritatingly unable to identify some of the bones. Parkinson had taken most of the German publications with him. Neither Migeod nor Museum officials appear to have remembered the lessons of 1926, when the Expedition leader complained of being in the dark with regard to osteology.

At +M2, begun July 13, Migeod worked on the northern border of the main M2 quarry, to the west of a dry streambed. Vertebrae appeared at a depth of 1.2 meters. At the end of July, +M2 was closed after about two weeks of work.

A mass of what appeared to be fish scales was collected from the central area of M2. Despite the best efforts of eight or nine men, M2 was closed on August 30, as finds were disappointing. On August 27, Migeod set some men to opening a pit 46 meters east of M2, near an old exploratory ditch.

M19 was situated on a hilltop, about four hundred meters north of M2. Since its inception on September 2, it had yielded several vertebrae. Sixteen to 18 men excavated for a month, reaching a depth of 1.85 meters, before work ceased.

Migeod gathered botanical samples as eagerly as in 1926, and used the numbering scheme from his last stay. He also sent a package of quartzite flakes to London, which he believed were the product of fires. These were human artifacts like those that Janensch and Leakey had discovered. On August 9, Migeod celebrated his 57th birthday. The labor crew numbered 46 during the week of August 4–10 and Migeod planned to excavate until November, weather permitting.

Confusion about who was supplying material for the Dar es Salaam exhibition was finally cleared up, with Migeod agreeing to collect suitable specimens. The selection included a 1.07-meter femur from NG south central, an 83-centimeter humerus from M18, a vertebral centrum from NG Bluff, a metatarsal from M1, a terminal phalanx from M3, and a claw from M8. They were carefully placed into four boxes on August 12.

Several experienced men left at the end of August. Migeod refused to supply a certificate attesting to their employment with the Expedition, since they left partway through the season. He also felt that if they returned in the following year they should not be paid more than the standard 18 shillings per month, though most of them were earning 20 to 22 shillings when they left.

Boheti and two others were sent to Mtapaia and Kijenjire to prospect, and returned on September 12. They saw several old German quarries but found nothing promising.

On September 18, Migeod assigned 10 men to a site around a native farm near the gaping hole where Skeleton S had been excavated. The quarry was labeled M20, but rewarded a week-long effort with only a rib head. Another new site was opened on September 21, about 180 meters east of M4.

At the end of September, Migeod informed the Museum that the governor's office had asked for a report on the 1929 season. The primary reason for the early termination of the Expedition this year was the mass exodus of workers after payday: "They have earned enough to pay their taxes and buy a piece of cloth and are satisfied. It is these departures in a body, and the necessity of waiting till new men drift in for a change from somewhere else, that makes the labour conditions so unsatisfactory."[25]

Migeod hoped to return to England from Dar es Salaam on December 15. As predicted, 22 Wangoni gave notice on September 30. Calamity could be averted, as some were willing to carry loads, and another seven boxes were hauled to the coast. Only 17 men were excavating between October 6 and 12, from a total of 37.[26]

Word reached London that the Tanganyikan government hoped to make another £1,000 contribution for 1930–1931. Assistant Secretary Smith wrote to the governor of Kenya in November, asking him to consider a grant as well, and offering duplicate specimens for the Nairobi museum. An identical offer was made to the Tanganyikan government, in the long-awaited report of the British Museum East Africa Expedition. However, in 1929, there was no Tanganyikan museum in which to display specimens, and the BM(NH) was not even able to unpack and store its Tendaguru material. The same space problems that had plagued the Berlin Museum of Natural History were obvious in London. Worse, the manpower level was lower, with one preparator and one attendant.

Migeod sent a nine-page report, dated October 7, to the chief secretary in Dar es Salaam. He outlined the Expedition's efforts since 1924, admitting that his account was not detailed, as he had no references at hand. Again, the lack of communication that had hindered the Expedition was stressed:

> In the absence of any notes or detailed digging reports from either Dr. Parkinson or Major Deacon, I went over the ground with different workmen who had actually done the digging, and viewed such trenches and pits as had come into existence since I was there last.

Perhaps sensitive to Parkinson's urging to dig deeper, Migeod highlighted his method of addressing this concern:

> In a number of my old diggings, therefore, small excavations were made to a depth of five or six feet below my lowest level. Nothing more was found.

The degree of confusion caused by the many unmarked quarries Migeod encountered in 1926 was also reiterated to Dar es Salaam:

> In time I became able to identify a few of his [Cutler's] sites, but of most of them I never succeeded in tracing the connection between diggings and notes, for apparently he never made a map.[27]

Migeod concluded with an account of his practical arrangements. Laborers received the local rate of 18 shillings per month, with a few experienced and long-term men receiving 20 to 24. Boys earned less. Boheti was paid 80 shillings per month. Laborers worked a 48-hour week, usually five hours in the morning at distant sites and three hours in the afternoon at quarries that were closer to Tendaguru Hill. An experienced excavator was placed in charge of a quarry from start to finish whenever possible, and Boheti moved amongst the important finds as they appeared. The biggest changes were made as economy measures, namely placing jackets in kerosene cases and using his excavators to carry them to the coast. There was little choice, with no one to assemble crates in Lindi.

In between other duties, Migeod was still prospecting locally. On October 12, he noted a possibility near M9, calling it +M9. It was abandoned on October 19, and the crew relocated to M9a, on a hill on the east side of M9. M9a was an old German quarry.

Another site, M21, was opened between M12 and M7, on a ridge northwest of Tendaguru. Eight Wangoni were taken on, which eased the labor shortage somewhat by bringing the total up to 26 men. Up to 17 men were still at Nguruwe in the mornings and about the same number at M21.[28] M21 was also closed on October 19, due to disappointing results after about three and a half weeks.

Migeod, again annoyed by his inability to purchase fowls, complained to the administration in Lindi. The response from District Officer Bampfylde could only have further aggravated Migeod: "As regards foodstuffs, vegetables are practically unobtainable in this district, as every person, who lives in Lindi knows to their cost."[29] As in 1926, the food problem was most likely caused in part by the varying capacity of the countryside to support a stationary population of any size, especially when reduced rains yielded substandard crops.

Migeod's discontent was voiced in his October progress report to London. He recounted a litany of unsatisfactory conditions, including the unpredictable labor supply, his forced diet of tinned beef due to lack of alternatives and the alleged intransigence of local headmen, and saline water.[30]

When 10 more men departed on October 31, the work force was reduced to 16, including Boheti, the cook, and the cook's helpers. Then the kitchen boy left and there were only seven men available for quarry duty.[31] Nguruwe was closed down and its crew of six sent out to prospect. Migeod toured the sites to ensure they were completed to his satisfaction. One of his few remaining pleasures seemed to be a healthy baby in camp.

Past a ridge where Cutler had abandoned iron-encrusted bones, Migeod was pleased to uncover most of the elements of a foot. This became M22, and the specimens were packed in box 84, which was brought back to England by Migeod. Boheti traveled to the village of Kipande to recruit for the upcoming safari to Lindi, but came back alone.

It was with little regret that Frederick Migeod walked away from Tendaguru on November 9. He had been at the site continuously for seven months. As in the previous season, two men were to remain at Tendaguru and maintain the camp. They would draw wages of 20 and 22 shillings monthly, while Boheti was engaged part-time at 20 shillings per month.[32]

Migeod compiled lists of equipment still at Tendaguru and supplies placed in storage in Lindi. The bank was informed of the arrangements with the caretakers and provided an account of the balance remaining in the Museum fund. Smith, Mackenzie & Co. was contacted regarding surveying equipment that Parkinson intended to ship to Lindi. A list of equipment required for the next field season included formalin, zinc oxide, boric powder, carbolic crystals, bandages, an eye bath, and an ear syringe.[33]

Migeod reached Dar es Salaam in late November. He met the chief secretary at Government House, who informed him that another £1,000 had been placed in the 1930 budget estimates to support work at Tendaguru. With about £400 left over from the 1929 season, a reasonable sum would be available in 1930.

John Parkinson returned to London in November or December 1929, having completed the draft of his popular Tendaguru book on the journey home.

Museum officials appeared more concerned with closing the accounts and answering the challenges of the auditor. Careful husbanding of financial resources was a paramount concern, and the smallest concessions were relentlessly pursued. Smith again telephoned Stahlschmidt, the Museum agent, to obtain a reduction on the freight charges for Tendaguru specimens.

Major Deacon was disappointed with the sum he had been allowed for his return fare in 1928. Whereas he had been granted £100 in 1926, he had been reimbursed only £80 in 1928. In late November, Wooddisse, the Museum accountant, with Regan's support, presented Deacon's case to the trustees and was successful in obtaining additional money: "as Major Deacon gave his services to the East Africa Expedition without payment the Trustees will wish to treat him generously and remove, if possible, any sense of grievance."[34] According to a report prepared by Assistant Secretary Smith, the total sum expended between January 1924 and March 1929 was £10,680.[35]

The accomplishments of the 1929 season were not insignificant when one considers that there was only one European in charge of an average of 40 men per month. Later in the season a labor shortage had severely restricted the modest quarrying efforts in the immediate vicinity of Tendaguru. Prospecting for additional sites further afield was curtailed. The food supply was unsatisfactory, though water sources held out and the weather did not restrict activities.

Five new quarries, M18 to M22, had been opened. At least six old quarries, M1, M2, M3, M6, M8, and M9, had been enlarged and deepened by an array of trenches. The most productive site of 1929 was an enlargement of the German and British quarries at Nguruwe. These had become the focus for the year. Other areas of investigation had included Kindope and Mtapaia. A plan had been drawn of quarry positions relative to one another, and several survey bearings and elevation readings had been taken.

A total of 84 wooden kerosene boxes had been shipped to London. In the past, two or three of these had been placed in a larger crate, so one could estimate that in 1926 terms, these 84 boxes represented anywhere from 28 to 42 crates. Approximately 260 botanical specimens had been collected in addition to those from 1926, bringing the final running tally to 658. A variety of zoological specimens had also been sent to the BM(NH).

Because the utmost economy had been exercised, there was money left in the bank for 1930. It was also possible that another grant would be forthcoming from the Tanganyika government.

-17-

1930

Migeod and Parrington, Tendaguru and Nyasaland

Across the ocean from Tendaguru, London, and Berlin, events were unfolding that would plunge the world into an economic crisis of unprecedented duration. On October 29, 1929, the New York stock market crashed, ushering in the Great Depression. Burdened by intricate international trade and payment systems, enormous war debts, reparation payments, and tariff barriers, Great Britain and Europe soon felt the effects of the American collapse. East Africa was not immune. Tanganyika Territory relied heavily upon agricultural exports to generate revenue, exports that were to become much more difficult to sell overseas. Decreased revenues forced the government to slash expenditures on all projects. It was a stroke of luck that £1,000 had already been committed to the British Museum East Africa Expedition by the Tanganyikan government.

In January 1930, John Parkinson, now living in Worthing, submitted his Tendaguru paper to the Geological Survey of Tanganyika. A draft of his popular book was also nearing completion. In mid-February, Museum authorities returned the manuscript to him with the qualifier that he should not indicate he had permission of the trustees: "undesirable because awkward questions might be raised as to how far the Museum could support some of your—no doubt justifiable, but nevertheless debatable—speculations."[1]

F. W. H. Migeod was approached to command the Expedition once more. He produced a preliminary estimate of expenses and a plan that included the long-desired reconnaissance of Nyasaland. The cost for a ten-month campaign for one man was estimated at £1,050 and for two men at £1,560. This would cover return ship passage but allow no salary for the second participant or freight charges for transporting specimens. The Museum's increasing space constraints are evidenced by Migeod's suggestion that "further development there [Tendaguru] might be brought down to a nominal amount until it is possible to reduce the undealt-with material in the Museum."[2]

In late 1929 or early 1930, W. D. Lang, the keeper of geology, held discussions with the faculty of biology at the Cambridge University. He requested the secondment of F. R. Parrington to the Expedition, to accompany Migeod.

Francis Rex Parrington was born at Bromborough, Cheshire, on February 20, 1905. His father passed away when Rex was only two years old. In 1924, Parrington enrolled in the natural sciences at Sidney Sussex College, Cambridge, where he greatly enjoyed zoological lectures by Clive Forster-Cooper. This eminent zoologist, along with D. M. S. Watson, had written to the *Times* in support of the BM(NH) Expedition in 1925.

In 1928, Parrington was named Strickland Curator of the University Museum of Zoology at Cambridge. He had previously curated the collection of amphibians and reptiles, and now began teaching zoology courses. According to Parrington's obituarist, Alan J. Charig, the BM(NH) had invited him to join the East Africa Expedition in 1929. Due to the ill health of Forster-Cooper, Parrington was obliged to remain at Cambridge and teach. The offer was renewed and accepted in 1930.

Parrington's travel expenses would be covered, and meals and camping gear would be supplied. He was free to make zoological collections for Cambridge University, while the Tendaguru fossils went to the BM(NH). In addition, Parrington was permitted to publish on his collections within three years of returning to England.[3]

The division of specimens compensated Cambridge for the loss of Parrington's teaching services. No salary was mentioned, as Cambridge agreed to continue paying him a partial salary. Parrington accepted the terms with the provision that he be able to return to England via South Africa if he assumed the extra cost personally. This last suggestion originated with zoologist D. M. S. Watson, who encouraged Parrington to collect Karoo-age fossils in South Africa following the close of the East Africa Expedition. Watson had initially been suggested as a candidate for Expedition leader in 1920, but the suggestion had been overruled by A. S. Woodward, who promoted William Cutler.

Just 25 years old, Parrington was the youngest member of the Expedition since Louis Leakey. Perhaps this is why Migeod had some misgivings, which he expressed to Assistant Secretary Smith: "I hope Parrington will be able to stand the climate and the long march."[4]

A substantial African trek was far more to Migeod's taste than another lengthy sojourn in the insalubrious environs of Tendaguru. The no-nonsense Migeod would likely have been an unforgiving safari leader, disparaging of shortcomings in any companion.

Nyasaland government geologist Frank Dixey visited the BM(NH) in January 1930, offering to guide Migeod to promising sites, and clarifying logistical questions. The Tanganyikan authorities were asked to provide aid on the march from Tendaguru to Liwale and Songea, and Nyasaland officials to procure porters. Departure was set for March 22 from Marseilles. The effort at Tendaguru was to last five or six weeks. Crews would then trek overland to Manda on the eastern shore of Lake Nyasa. A steamer would carry them to Florence Bay on the western shore. Here they would need about 40 bearers for about seven or eight weeks of fieldwork, and would then return to Manda around the end of October.

Migeod ordered equipment and provisions, including a galvanized bath and a case of burgundy from the Army & Navy Co-Operative Society, Ltd., for a total of £141.[5] He stipulated that £12 of his monthly salary be paid directly to his wife.

On April 16, 1930, Migeod and Parrington stepped off the *Dumra* at Lindi. Rain had fallen for a month beyond the average season, making the journey to Tendaguru a trial. The tired band dropped their loads at the hill eight days later.

The first quarry, No. 1, was reopened at Kindope on April 26. It was located on the main north road below a hill, and Boheti bin Amrani was placed in charge. The trusted overseer was entering his 11th year of service at Tendaguru. It is unlikely that anyone else could boast such a continuous record of excavation at the site. Work at No. 1 stopped five days later, on May 1, due to lack of bone and the waterlogged condition of the quarry.

In rapid succession, 11 quarries were opened in the immediate vicinity of Kindope. In light of Parkinson's suggestions, it seems likely that Migeod had been directed to operate here.

No. 2, about 5.5 meters north of No. 1, was operated from May 1 to 6 under Boheti. A femur in poor condition was recovered.

No. 3, about 11 meters south of No. 1, was operated from April 30 to May 8 under Kibwana. Only surface fragments were recovered.

No. 4, on a plateau to the southwest, was operated from May 2 to 7 under Juma. Cut down to a depth of almost 2.5 meters, it yielded nothing.

No. 5, up the hillside between No. 1 and No. 3, was operated from May 5 to 7 under Momadi Ngoni. The quarry was also barren.

No. 6, at the summit of the escarpment south of the roads leading down the hillside, was operated from May 6 to 10 under Boheti. No finds were reported.

No. 7, near top of a hillside next to No. 5, was operated from May 7 to 10 under Momadi Ngoni. No finds were reported.

No. 8, near the main road, was operated from May 7 to 10 under Juma. No finds were reported.

Work at No. 9, to the northeast of the old German Quarry dd in thick undergrowth, commenced on May 8. Major Deacon had worked nearby, and with a few surface fragments showing, it was decided to investigate further. The vertebral column of a large dinosaur soon appeared and the quarry was named M23.

No. 10, on the Kindope road, was operated from May 10 to 14 under Seidu Momadi. Only surface fragments were recovered.

No. 11, near the first quarries, was operated from May 10 to 11 by Momadi Ngoni. Only fragments were recovered.

By May 14, crews were concentrating exclusively on M23. The remaining quarries had produced so little that they were all abandoned. The big skeleton was an exciting find, made within two weeks of the commencement of excavations. A series of articulated vertebrae, soon to stretch over a distance of 3.7 meters, and a hint of massive hind limbs were visible by May 9. Migeod first thought the tail was uncovered, but when he treated exposed surfaces with shellac on May 14, he identified the bones as part of the neck. Confusion reigned for several more days, until he finally decided that the head would be located to the north, with the neck descending into the ground.

The sacrum and dorsal vertebrae could be identified, and 8.6 meters of skeleton lay in the sun by mid-May. Migeod tentatively identified it as the remains of *Dicraeosaurus,* based on vertebral processes. Invertebrate fossils were collected from a marine layer above the dinosaur-bearing horizon. Migeod declared this skeleton superior to M1 No. 2 and M8 No. 1. In an unpleasant introduction to Africa, Parrington had come down with malaria; he had been ill for six days and was unable to share in the excitement of the find.

The posterior section of M23 had become disarticulated toward the tail while ribs had been pulled in the opposite direction, over the neck. The quarry was enlarged on May 17, when a femur began to appear. Toward the end of May, it seemed that the last of the cervical vertebrae had been exposed, and the first bones were pulled from the ground on May 29. At this point, the skeleton had been under excavation for three weeks. A trench was driven across the quarry, but work was halted on this extension on June 2.

On June 3, Migeod wrote that the series of cervical vertebrae lay at right angles to the dorsal vertebrae. The final neck elements and skull were missing, and a scapula had drifted in among the cervicals. Nine caudal vertebrae were in storage at Tendaguru, along with part of the sacrum, which had suffered from erosion and root damage.[6]

The animal was of impressive dimensions: the scapulocoracoid measured 2.23 meters. A femur, once completely uncovered, would be 1.23 meters long, and the fourth cervical vertebra was one meter long. An ilium appeared on June 7. At mid-June, a carnivore tooth over 16 centimeters long was pulled from between the humeri and vertebrae. Crews were plastering vertebrae and carrying them into camp daily. Ribs and "cervical tendons," most likely the very long and thin cervical ribs, rested on shelving in the bone store by June 18. One rib was 2.5 meters long, with a head that measured 24 centimeters long by 15 centimeters wide. The main shaft was only four millimeters in diameter. Sketches were made of individual elements in situ. An ilium was hauled into camp on June 24, followed by a scapula one day later.

A right ilium and two humeri had been uncovered by the end of June. One of the humeri was 1.4 meters long, an ilium was just under a meter long, and a dorsal vertebra measured about one meter from the center of the centrum to the tip of the neural spine.[7] The final portion of the huge scapula was brought into camp on July 5, after a week-long effort to remove it from the ground. A 2.37-meter-long rib came out of the quarry on July 12. For the first time, Migeod was prepared to identify skeletal elements. He had filled a notebook with sketches of dinosaur bones, copied from German publications.

By July 31, six cervical vertebrae had been entirely encased in plaster jackets, and two of them could only be moved through the combined effort of 12 and 16 men respectively. Two others were even larger, but been split into manageable sections. The dorsal vertebrae had also been divided into three or four parts, each of which was then plastered. Even so, a single dorsal had demanded a team of 10 men to shift it. A series of relatively complete ribs had been found almost two meters beneath the vertebral column, while those from the other side of the body were closer to the surface and scattered.

Cement was used to produce field jackets at the end of June. The entire supply of plaster had been used up in an effort to protect the heavy but fragile cervical vertebrae. The use of cement as a plaster substitute was controversial among the men. Boheti and Juma in particular were dissatisfied and reverted to gum arabic and gauze. Did no one remember the jackets of German times, which were produced with local clay, coir, and strips of cotton?

On June 21, a caravan struggled out of camp as far as Matapua, hauling 12 boxes of field jackets from M23. Throughout July, bearers undertook the Sisyphean task of carrying loads out of camp. There was no labor shortage, as Migeod records a group of men leaving every two or three days throughout July and early August.

Ribs and vertebrae of M23 were sometimes split apart by Migeod because they were so tightly articulated. Two of the largest vertebrae in the spinal column, at the junction of the cervicals and dorsals, were proving to be a challenge. Both were enormous, and would have to be divided. For the time being, the exposed surfaces had been plastered and a tent erected over them.

Migeod prospected for the first time from August 8 to 11, moving along a distant ridge to the northwest. His main goal was to follow marine horizons and collect invertebrate fossils that could help clarify stratigraphical questions.

After three months at Kindope, Parrington had been transferred to Lindi around July 21 to expedite shipments. He hired four carpenters to assemble crates, an expensive undertaking. Thirty-two planks were purchased at five shillings apiece, and the wages of the carpenters over two full days added £1 to the cost.[8]

A one-ton truck and a driver were hired from C. W. Carnegie-Brown in Lindi and driven to Matapua, about 80 kilometers away. Carnegie-Brown and his partner G. M. Sibold were the proprietors of the Tanganyika Transport Company—Mechanical Transport Service. The firm held a government contract for transport between Lindi and Songea. Parrington accompanied the truck to and from Matapua. Here, field jackets were placed in crates in the truck under Parrington's supervision. The casual manner in which Migeod mentioned motor vehicles in his reports illustrates the changes in the country since German times.

The journey was a precarious one, for although the road had been cleared from Matapua to Tendaguru, Migeod was wary: "I did not feel like undertaking the risk of the bones going by lorry on this section. Mr. Parrington reports that even on the main road the lorries with half loads had great difficulty at some of the hills, and a capsize was imminent at times."[9]

Parrington's misgivings were twofold: safety and cost. A smaller truck could comfortably carry 12 to 16 half loads, and the larger one-ton truck could safely manage 20 to 25. To transport the truck's maximum capacity, around 40 loads, "would spell disaster."[10] Also, a round trip, over 160 kilometers long, was costly. Despite these concerns, the change from previous years, when Cutler had pleaded with officials to build a road and lend him a vehicle, is striking. The soundness of Cutler's proposed method was vindicated.

Grass was cut in Lindi to cushion the jackets and the crates were stored in the compound of the Tanganyika Transport Company. The first shipment of the season, 12 crates dating back to June, was probably already nearing England. Another consignment of 15 steamed to London on July 12, though it is not clear who expedited these first two lots. Soon a stream of loads from M23 began flowing into Lindi. By August 5, another 51 cases had arrived. The pace was frantic. Migeod recorded that shipments had left for London on July 28 (22 boxes), August 4 (2 boxes), and August 11 (27 boxes).

Parrington was informed that all Indian merchants in Lindi were sold out of plaster of Paris. In desperation, he visited the hospital to procure some, but was turned away empty-handed. The young Cambridge zoologist was experiencing other frustrations as well. Though few of his letters are dated, he likely sent the following to Tendaguru in late July: "I feel a dreadful slacker hanging around Lindi while you still wrestle with the 'body.' . . . It is very trying to do business in Lindi! This year—next year or any old time + they always contrive to make you feel that it is really your own fault for being in a hurry."[11] Migeod was probably not bothered by Parrington's absence from Kindope, preferring to work independently.

Plans to march the entire distance to Songea on foot were canceled when late rains prevented burning of grass and prospecting. In addition, Migeod estimated that M23 would demand at least two more months of work.[12] He expected that Rex Parrington would reach Songea by truck around mid-August and arrange for the onward safari. It was now hoped that both men could leave Songea around the end of August.

Preparations for the upcoming Nyasaland leg were complicated by uncertainty regarding shipping schedules, availability of bearers, and when M23 might be closed out. The district officer at Songea was apprised that supplies would now be sent in advance on trucks, and was asked to procure 80 porters for the leg from Songea to Manda.[13]

Migeod's attempts to secure a timetable for the Lake Nyasa steamer were time-consuming. He had received no word through the Colonial Office in London or the chief secretary in Nyasaland. A reply from the superintendent of marine transport at Fort Johnston, Nyasaland, had taken 25 days to reach him. Finally, he wrote to the Union Castle Mail Steamship Company, Ltd. in Beira, Portuguese East Africa. This office in turn obtained a schedule from the treasurer of the government of Nyasaland and forwarded a copy for Migeod to Lindi.

Migeod wanted to rendezvous with Dixey, the geologist, on schedule, but only one steamer stopped at Manda, the 318-tonne *Guendolen*. This made connections uncertain. It was possible to hire an Arab dhow to cross the lake, but the district officer warned that this was an uncomfortable way to make the journey.

Migeod wrote to the director of public works at Zomba, Nyasaland, for assistance with plaster of Paris in the protectorate. There was none to be had as far away as Dar es Salaam, so the Nyasaland official was asked to ship 90 kilograms to Manda. The Expedition would substitute an equal quantity of cement if plaster were unavailable.

Parrington ordered, scrounged, and packed supplies for the Nyasaland safari. Two trucks would leave Lindi ahead of him, and a final truck would be waiting for Migeod in Lindi once he had completed M23. Much of the scheduling had to be provisional, as transport services were still in their infancy. The vehicles, possibly war surplus, were punished on poor roads and were prone to mechanical breakdowns. Drivers suffered from fevers and other tropical illnesses.

After three weeks of frenetic activity in Lindi, Parrington rattled out of the town on August 11. Two days later, he arrived at Songea. The passage was not without incident: "We got in yesterday after a rather hectic journey enlivened by the whims of an antediluvian Chevrolet."[14] Expedition gear was distributed among the 80 carriers who would proceed to Manda. On August 28, two weeks after arriving, Parrington walked out of Songea. One contingent of porters had preceded him, and another group accompanied him.

Several weeks previously, Tanganyikan Survey geologist Gordon Murray Stockley had been mapping Karoo beds roughly 70 kilometers north of Songea, to trace coal-bearing outcrops. He had uncovered a bonebed and informed the Museum Expedition of its location. Stockley later published on similar finds in the district. Four days and 80 kilometers out of Songea, Parrington passed a cairn of bone fragments left by a bearer near Stockley's site. Parrington too found bone, identifying it as the partial skull of a dicynodont, a herbivorous therapsid reptile. After five days of trekking, his group met with the advance team of porters at Manda on the eastern shore of Lake Nyasa. Here the men awaited Migeod's arrival.

August 15 marked the final entry in Migeod's excavation notebook for several months. He decided that he had to leave Tendaguru in order to carry out the remainder of the field program, although a portion of M23 was still lying in the quarry. The site would be reopened after the Nyasaland investigations. The lack of proper materials at Lindi, including cement, contributed to his decision. Two caretakers would remain on site and inspect the quarry weekly until Migeod returned.[15]

Camp was closed and most of the remaining laborers were dismissed. On August 21, after four months of work, Migeod walked out of Tendaguru. Another consignment of crates was assembled in Lindi, bringing the total since June to 94 boxes.

The truck provided by Carnegie-Brown slowly bounced along the trail out of Lindi on the morning of August 27. The passengers included Migeod, the cook Michael, Ali, Awusi, and the indispensable Boheti bin Amrani. Seven hours later they reached Masasi and camped. After eight more dusty hours the next day, they rolled into Tunduru and stopped for the night at Carnegie-Brown's depot. Sibold drove the final 10 bone-jarring hours to Songea. The six-hundred-kilometer journey from Lindi had taken four days.

By September 5, bearers had arrived and the safari left camp outside of Songea. Migeod picked up dicynodont dorsal and caudal vertebrae, a jaw fragment, and a humerus near Munya Maji Stream, where Parrington had found the bone cairn. Lake Nyasa glistened in the distance on September 8, and shortly after noon they met Parrington in Manda. It was a surprise to see gulls after four and a half months in the interior.

Forty carriers had assembled at Florence Bay. Their pay was to be 10 shillings per month, including subsistence.[16] Fortunately, Tanganyikan silver was accepted as legal tender in that district of Nyasaland. Carriers had delivered 50 bags of rice, about three weeks of rations, to the Nyasaland village of Mwakasyunguti, and a policeman and translator had been seconded to the Expedition to oversee the bearers.

The *Guendolen* docked at Manda on September 10. The ship had been armed in 1914 and distinguished itself by sinking the German steamer *Hermann von Wissmann* during the battle for supremacy on the lake. Five hours after departing, the Expedition crew disembarked at Ngara, Nyasaland. Migeod proceeded on foot to Nyungwe Plantation, 9.6 kilometers from Ngara. The plantation would make a convenient base, as it was located midway between Migeod's proposed excavation site, Mwakasyunguti, and Parrington's, at Uraha Hill. It was reported that the former boasted Cretaceous dinosaurs and the latter Pliocene and Pleistocene mammals.

By September 15, the second phase of the 1930 field season was underway as Parrington took 20 men to Uraha. Frank Dixey had originally described the Tertiary and Quaternary beds and fossils. The terrain was much more broken than at Tendaguru, and the nearby settlement had little to offer other than a stream. The southwestern face of a 76-meter-high wooded hill was opened. Quarry T1 was located six meters below the junction of the Chiwondo and Chitimwe Beds. Results were meager: turtle and crocodile scutes, fragments of mastodon teeth, and reptile and cervid teeth.

Migeod made his way to Mwakasyunguti with Boheti and the remaining 20 bearers. The site was two and a half hours from Nyungwe Plantation, and Migeod commuted daily, often in temperatures that ranged from 38 to 44°Celsius. He and Frank Dixey inspected localities on September 16. The geologist had limited time to spend with the BM(NH) crew, as he intended to investigate the Nyika Plateau. Locals were offered a 50-cent reward for reports of bone, but at first the area did not seem promising. It would take time for the trained Tanganyikans to impart their knowledge of prospecting, bone identification, and excavating to their Nyasaland co-workers.

The first dinosaur quarry, Mwakasyunguti No. 2, was located up a steep-sided valley past the settlement of the same name, about 12 kilometers from the plantation. Dixey confirmed that the Dinosaur Beds deposited within the Nyasan rift valley were composed of friable sandstones and marls, which weathered into steep hills overlain by loose debris. They extended over 124 kilometers from north to south, but the richest fossil deposits appeared to be at Mwakasyunguti.

Migeod visited Parrington's camp on September 18 and was impressed, but on the 24th Parrington became ill with severe diarrhea and retreated to Nyungwe Plantation.

Parrington was frustrated. The cliff face at Uraha was disappointing and he shifted the laborers to another exposure. Perhaps the heat and his illness had adversely affected his relationship with his crew, as evidenced by the following sharp note delivered to Migeod: "Herewith Juma (with his detestable musical instrument), a tin of baking powder, and ten prize idiots."[17]

Migeod assigned crews to prospect in specified areas, and paid rewards as an incentive. Site A was 1.6 kilometers north of the tent camp at Mwakasyunguti, and had yielded, in association, two 79-centimeter scapulae, a 58-centimeter humerus, a 72-centimeter radius, a 77-centimeter pubis, and three caudal vertebrae. Migeod believed that at least two individuals were represented at Site A.

Site B, another 1.6 kilometers northeast, had presented a 1.2-meter femur, an 81-centimeter humerus, and several caudal vertebrae. Site C had produced a femur and neural spines of vertebrae.[18] Several vertebrae, a metatarsal, a femur, and a tibia soon appeared. Work was terminated at sites B and C on October 7, after three weeks of excavation.

Site D was opened on a hilltop about 92 meters west of A, and was worked for three and a half weeks. The sauropod at D was developing into an interesting find, easily the best of Migeod's four sites at Mwakasyunguti, with two humeri, two scapulae, one ulna, a femur, a pubis, a metatarsal, two radii, two vertebrae, and an assortment of unidentifiable elements.[19]

A clipping from the *Daily Telegraph* elicited an aggrieved response from Migeod, as it claimed that only in this season had a complete skeleton been found at Tendaguru: "individuals have virtually been told that my finds being only disconnected bones are not worth opening up. . . . Further, regarding the statement in the Press as an invitation to me to resign, I of course hereby tender my resignation."[20] In an indictment of the Museum's inability to deal with the crates from East Africa, he warned that if the bones were not treated soon, even the best would crumble to dust. His outburst appears to have been ignored by the Museum, though much later, Swinton privately admitted dissatisfaction with Migeod's collections:

> I knew Migeod very well and had much sympathy with him but alas, I knew the material he collected even better. . . . The few good bones he collected would not constitute a single limb and but a few feet of backbone. Indeed, much of East Africa was enclosed in plaster with the mistaken impression that bone was contained within.[21]

It is also possible that, with its emphasis on display-quality specimens, the Museum was not aware of the taphonomic information that could be derived from bonebed deposits.

Migeod reported that a shipment of Nyasaland specimens would be carried to Fort Johnston on the *Guendolen* by mid-October. They would be transported to Beira, on the coast of Portuguese East Africa, and onward to London. The process would be repeated in November, when Parrington would leave the Expedition for South Africa. Migeod planned to return to Lindi, sort out M23 at Tendaguru, and leave for England around mid-December.

Rex Parrington trekked to Chiweta, a small village on the shore of Lake Nyasa, 10 kilometers south of Florence Bay, on October 13. He was to investigate Karoo deposits reported by Frank Dixey. Work at Mwakasyunguti ended on October 20, and Migeod, Boheti, and 27 men also left for Chiweta.

Chiweta's setting was picturesque, with Mount Waller to the north and Lake Nyasa to the east. The two bonebeds under investigation lay close to one another and only 1.6 kilometers from camp. They covered an area of about 2.5 square kilometers, but the matrix surrounding the bones was so hard that progress could only be made by using cold chisels. Migeod prospected for several days, visiting coal seams and sulfur hot springs in the Rumpi Valley to the south. He spent nine days around Chiweta. Some bones were found in the difficult Karoo deposits, including part of a skull (designated M.1), a humerus, and rib fragments. The crew was sizable, numbering between 32 and 41 men, so some were sent out to prospect. At month's end Migeod was not feeling well, and he closed the camp on October 31.

There were sufficient specimens and dried plants to fill another five crates, and these were hauled to Florence Bay on November 4. Parrington could claim greater success at Chiweta than at Uraha. Dixey labeled the beds B1 and B2, which became Parrington's Series X and Series Z respectively. A high ridge not far from his camp yielded surface fragments. Eventually four quarries, X1 to X4, were opened. Three of these sites produced, in total, the lower jaw of a dicynodont and the anterior portion of a skull. There were also vertebrae, ribs, and assorted fragments. Three quarries were opened in the Series Z beds. Specimens included a skull of a labyrinthodont amphibian, *Rhineceps,* portions of other skulls, and a lower limb bone. Likely neither Migeod nor Parrington was disappointed to leave Chiweta.

The young Cambridge zoologist had not met Migeod's expectations:

> I am afraid that Mr. Parrington is not gifted with all the qualifications necessary for an African expedition requiring constant personal exertion. . . . I consider the Cambridge Museum of Zoology as under a great debt to the British Museum for giving one of its young men a trip throu⌣ . Africa with the opportunity of learning practical field work.[22]

The other side of the relationship between the two men, which recalls the Cutler-Leakey personality clash, is given by Rex Parrington's obituarist, the late Alan Charig:

> Parrington . . . soon discovered that Migeod's pretentions concealed a profound ignorance of many subjects, and thereafter he delighted in setting deliberate verbal traps for Migeod, into which Migeod invariably fell. Parrington's diaries of the period make amusing reading; Migeod's likewise!

(Parrington's diaries have not been seen and the brief field notebooks that are available at Cambridge mention only zoological specimens and details of local geology.) Charig goes on to dismiss the 1930 field season:

> Unfortunately the expedition (no fault of Parrington's) was ill-conceived and ill-prepared. They did collect the greater part of the skeleton of a huge brachiosaurid

dinosaur; but even that was left for decades to rot in the basements of South Kensington, the only elements that were ever prepared and exhibited being two gigantic vertebrae.[23]

Perhaps the scope of the Expedition had been expanded beyond what was reasonable to expect from two men of differing temperaments, restricted by limited funding. Nevertheless, Migeod's assessment of the potential of the Nyasaland sites he investigated was optimistic.

Migeod boarded the *Guendolen* at Florence Bay on November 4. After 7½ months, the two men parted ways. Rex Parrington carried on to South Africa, where he collected in Karoo beds. On November 15, the truck clattered into Lindi with Migeod aboard. With 21 men on hand, Migeod was able to send eight back to Tendaguru immediately.

Migeod had received a telegram from London that curtly announced the end of the Expedition and ordered him to dispose of its equipment. It was not unexpected: a month earlier he had learned from the chief secretary of Tanganyika Territory that there would be no grant for a 1931–1932 season due to "the present financial position."[24] The Great Depression had taken hold. Still, Migeod closed his final statement to Regan in London with the complaint that there would have been funds enough for another season if Parrington had not been sent. In regard to his relationship with Parrington, Migeod was closer in character to Cutler than he would have cared to admit—the independent expert who regarded the help of European co-workers as competition or interference by the inexperienced.

Boheti was paid, marking the last season he would spend at Tendaguru. Germans and Britons alike owed him a great debt. On November 20, Migeod's safari reached Tendaguru.

Within a few days, a total of 32 men were on site.[25] While excavation was restarted at M23 on November 21, Migeod carefully drafted a plan of the bones in the quarry, presumably working from notes about bones previously removed.

A fresh supply of cement arrived within the first week. The block that contained the two massive vertebrae was overturned and cracked, and was taken out in two sections. Some rib fragments were still appearing as the last of the vertebrae were pulled out.

For some time Migeod had intended to take a closer look at the various strata in the vicinity of his excavations, but the task was in abeyance due to the heavy demands the quarries made on his time. He prospected around the rim of the "plateau" at Kindope, picking up a few invertebrate fossils that he hoped might help resolve local stratigraphy. He again measured elevations with the barometer, this time of the dwelling hut on Tendaguru Hill, M23, and marine beds around Kindope. The excavated and cemented bones were carried into camp by December 2.

Porters struggled along the trail to Matapua on December 9, 10, 11, and 13, hauling out the pieces of M23. One large load, containing the last two vertebrae, could only be moved by a team of eight men. Several more men were sent to

Lindi to pack and prepare the boxes for shipment. Some probing was undertaken below M23 and beyond the immediate area, but the results were negligible.

In mid-December, a scapula, 1.2 meters long, appeared at a higher level than the other elements of M23. A few days later, other isolated elements appeared, such as metacarpals, prompting suspicions of another partial skeleton.

In a subsequent article in the *Times,* Migeod summarized the discovery and excavation of M23. The skeleton had originally been noticed because tree roots had pushed bone fragments to the surface. Light pickaxes had been used to expose eight or nine proximal caudal vertebrae, and then dorsals, which appeared to end in the shoulder region. The quarry had been expanded in all directions, until the cervicals finally appeared. The animal had come to rest on its right side, so the left ribs were scattered, while the right ribs were still in place, 1.15 to 1.2 meters below the level of the vertebrae.[26]

Finally the excavation was halted, on December 24, when M23 was cleaned out and abandoned. Tools were offered to the remaining workers, who claimed the heavy picks. The last crate, number 212, was packed on Christmas Day. For the next three days, Migeod pursued his favorite hobby, collecting plants. It was again raining daily, interrupting the final tasks in camp. Thirty-two men remained, though not for long.

It was on December 29, 1930, that Frederick Migeod departed Tendaguru for the last time. His final effort at the famous locale had lasted 40 days. When that was added to the 123 days he spent there from April to August, the total for 1930 came to almost six months. Only Kibwana and Isa Boli remained at the site. Migeod and a train of porters stepped onto the trail to the coast, reaching Lindi January 2, 1931.

On January 11, 1931, Migeod made his final farewells. The port, with its sandy beaches and swaying palms, was soon lost in the distance. Migeod would never see it again. He had missed Christmas and New Year's with Madeleine, but on February 7 they were reunited in England.

The British Museum East Africa Expedition was over. Altogether, 137 crates had gone north from Africa in the 1930–1931 season: 112 from Tendaguru and 25 from Nyasaland. Migeod's plant collection numbered 1,112 specimens. Several crates of invertebrate fossils had been dispatched, and 20 modern mammal skins had been collected in Tanganyika and Nyasaland. Eleven quarries had been investigated at Kindope, and a partial skeleton of a large sauropod had been excavated. Four quarries had been opened at Mwakasyunguti in Nyasaland, yielding another partial sauropod from a different geological horizon than Tendaguru. One site had been opened at Uraha Hill, yielding some material, and seven Karoo-age sites at Chiweta had produced partial skulls. Other miscellaneous Karoo-age fossils had been found in the Songea region in Tanganyika.

Migeod had measured elevations of quarry sites and geological horizons. He had recorded temperatures and barometric pressures daily. In addition to maintaining his field notes, he had also drafted a scale sketch of the skeletons at M23, at Kindope, and Site D, at Mwakasyunguti. This was not a bad return from an Expedition later labeled "ill-conceived and ill-prepared."

The British legacy at Tendaguru was undoubtedly rich. Between 1924 and 1931, over £11,000 had been expended, a sum comparable to the German outlay. Approximately 589 crates had reached London, representing tens of thousands of kilograms of fossils.[27] The total exceeded 600 crates, if those containing botanical, zoological, and geological specimens are included. More than 700 butterflies and 1,100 botanical specimens had been collected, as well as more than 250 modern birds, mammals, reptiles, lizards, amphibians, and fish. Over 55 quarries had been investigated in detail in Tanganyika, yielding thousands of dinosaur bones and invertebrate and plant fossils. Geological and soil samples had been taken and prehistoric lithic artifacts discovered. Elevations had been recorded with an aneroid barometer and daily temperatures noted. Detailed maps of the distribution of excavations throughout the region had been drafted, as well as individual quarry plans. Numerous geological profiles had been measured and sketched, and hundreds of photographs taken. A fossil preserve had been declared at Tendaguru. Investigations had ranged as far as Kenya and Nyasaland.

One of the most obvious differences between the German and British Expeditions was that the latter did not publish its scientific results. Beyond Parkinson's geological papers, there was no scientific description of the results of seven years of labor. Another was the absence of any skeletons mounted for public display. The contrasting results of the German and British endeavors can be attributed to a variety of factors: political, financial, organizational, and personal.

A fortuitous series of events had linked patriotic Germans who had the influence to develop and sustain a large-scale campaign in the field and museum. Von Branca had excellent connections to powerful and wealthy individuals in German society. Museum representatives like Hennig, von Staff, and Reck were tremendously energetic scientists, whose qualities were complemented by the unparalleled dedication of Janensch. They enjoyed the full confidence of von Branca. The German approach was clearly focused: detailed scientific analysis and popular public displays. At the time of the German involvement at Tendaguru, pro-colonial organizations vigorously promoted foreign ventures. Overseas possessions were brought under strict control. A network of invaluable logistical support was available throughout German East Africa from the government and private firms. Individual representatives like Sattler, Wendt, and Schulze possessed decades of African experience. They functioned as a vital link to the most critical component—the African work force. Skilled African overseers were well trained and, by necessity, were trusted to perform their tasks with minimal supervision.

In contrast, the British program suffered from a chronic lack of support, continuity, and well-defined aims. Initially at least, the goal was to collect enough material to mount a display. There was little scientific emphasis until the stratigraphic controversy developed. BM(NH) directors were not associated with men of means, and were forced to rely mainly on government grants intended for purchasing specimens. Museum representatives changed frequently. Field workers like Cutler and Migeod did not always have the unconditional endorsement of a succession of directors and keepers. Personality conflicts in the field between Museum workers and local officials had a negative impact.

As a mandated territory recovering from the war, Tanganyika was a very different country under British rule. The new administration's objectives differed from those of its German predecessor. Museum and mandate officials varied widely in their commitment to the effort. Poor communication between the many participants hampered the work. Excavation strategy relied heavily on reworking German quarries with smaller crews. The experienced local work force had been scattered by the war, and the relationship between Europeans and Africans at Tendaguru was uneasy. When coupled with a reduced abundance of bone at the site, the outcome was a loss of interest in Tendaguru at the BM(NH).

Yet the sincerity of the undertaking's participants could not be faulted. Britons as well as Africans had faced frustrations similar to those experienced by their predecessors in German times. The African labor force, despite the misunderstandings that inevitably developed, had worked with no less dedication or skill during the British era than they had in the German era, though the work force was considerably smaller. It is to these local men, women, and youngsters to whom the greatest debt is owed, for without their superb efforts, the results achieved by a few Europeans would have been negligible. Sadly, virtually nothing is known of their fate; only the Europeans left a record of their own subsequent activities and achievements.

-18-

1931–1939

Berlin's Museum Triumphs

Tendaguru now lay quiet beneath the brilliant African sun. The storm of activity in the region between 1907 and 1913 and again between 1924 and 1931, though impressive from a human standpoint, was insignificant when measured against the length of time the site had existed geologically.

Aware that there were still funds available, Migeod proposed fieldwork in Nyasaland and the Karoo beds of southwestern Tanganyika. Tendaguru was not in his program, nor did he ask for a salary. When this suggestion was forwarded to Lang, the Museum's keeper of geology, he demurred, proposing instead to apply all the remaining funds, £326, to Louis Leakey's next African project.

This venture was known as the British East African Archaeological Expedition. One of its primary aims was to resolve the question of the antiquity of the human skeleton discovered at Olduvai Gorge by Hans Reck in 1913. Reck, who had half-seriously been invited to return to Africa by Louis Leakey in 1925, was a member of the 1931 team. Past suspicions of a misinterpreted, intrusive burial were upheld, and in the 1970s, with the aid of radiocarbon dating, the skeleton was determined to be about 1,700 years old. Olduvai Gorge would gain world renown in decades to come thanks to the impressive energy of Louis Leakey.

In Germany, the years just prior to the Great Depression witnessed many improvements in the economy and standard of living. Despite political turmoil, unemployment fell to the lowest level since World War I, industrial production exceeded prewar volumes, and in 1930, Germany placed second only to the United States among the world's exporting nations. Bankruptcies and labor disputes declined by half.[1]

Yet the period of prosperity from 1925 to 1929 was in part the consequence of massive American loans. With the 1929 crash, this investment disappeared as foreign bankers wrestled with economic downturn in their own countries and demanded repayment. Trade declined around the world as nations levied protective tariffs. Germany was unable to maintain the exports that would cover foreign debts as well as the rising costs of imported essential raw materials and food. With stunning rapidity, small businesses and plants locked their doors, wages fell, bankruptcies soared, unemployment skyrocketed, and banks collapsed. By January 1930, 3.2 million Germans were out of work.[2] More disturbingly, almost 6.5 million voted National Socialist in the 1930 Reichstag elections.

The global depression gripped Germany as hard as any other nation, yet work continued doggedly at the Museum of Natural History in Berlin. A sad loss had been the death of Wilhelm von Branca on March 12, 1928. The steadfast supporter of the German Tendaguru Expedition had passed away in Munich at age 83. Major inroads had been made into the vast supply of field jackets throughout the 1920s, and an abundance of skeletal elements was available to Werner Janensch, allowing an orderly program of study and publication.

Gastroliths associated with the sauropods *Dicraeosaurus* and *Barosaurus* had been described. Janensch had also published the results of his 1911 trek to search for Karoo sediments. Other papers correlated German quarries and their contents with geological horizons at Tendaguru. The rare mammal *Brancatherulum tendagurense* had been described in 1927. By the late 1920s, sauropods formed the major thrust of Janensch's research.

Sufficient quarry contents had been prepared by 1929 for a detailed study of the vertebral column of two species of mid-sized sauropod. *Dicraeosaurus hansemanni*, a gracile form, had been established in 1914 based on cervical, dorsal, and caudal vertebrae and a femur from Skeleton m. The specific name honored an original Expedition patron, Dr. David von Hansemann. *Dicraeosaurus sattleri*, a robust form, had been based on a dorsal vertebra and pubis from Skeleton M. The species name honored Bernhard Wilhelm Sattler, the European who had recognized the significance of Tendaguru.

Altogether, three partial skeletons of this genus had been collected. The most complete was Skeleton m, from the Middle Saurian Bed at Kindope. Two less well preserved individuals had been excavated from the same horizon at Quarry dd, also near Kindope. A number of elements could be referred to *Dicraeosaurus* in at least seven other quarries. They included Skeleton M (found south of Tendaguru), Skeleton E, Skeleton O, and Localities Ob, GD, s, and La. With the exception of the last two, all were found in the Upper Saurian Bed. Their geographic distribution was broad, from Ubolelo, southwest of Tendaguru, to Kijenjire, north of the hill. There was a range of sizes, from subadult to adult animals, and bone preservation varied from poor to excellent. Some partial skeletons were well articulated, while others were incomplete and scattered. One set of bones had not been discovered by the Expedition. A Defense Force captain had collected numerous vertebrae from somewhere in the Tendaguru region, and they had been forwarded to the Museum of Natural History.

The preservation, articulation, and reasonable completeness of Skeleton m immediately marked it as a prime candidate for display. The dinosaur was found lying on its right side, and had been transported by water prior to burial. The skull, both forelimbs, the right lower hind limb, both hind feet, and the majority of the tail had been detached from the trunk. The skull, atlas, and distal tail section were never found. Janensch surmised that the anterior portion of the neck had separated from the remainder of the carcass. A section from the eighth cervical vertebra to the axis had come to rest in a ventral position, transverse to the remainder of the neck. The left hind leg and left pubis and ischium may have been disarticulated at this time as well. Subsequently, the left side had suffered exposure and abrasion, especially the vertebral processes from the 13th presacral to the sacrum and even a few caudals. The ribs of the left side were scattered, and the left lower hind leg, still in articulation, had been displaced up to the region where the neck joined the trunk. The vertebral column lay in articulation from the 19th caudal through to the ninth cervical vertebra. Pelvic elements, ribs, and the right femur had remained in articulation. The preserved portion of the tail had flexed into the concave bend characteristic of postmortem muscle and ligament desiccation.

Since the two skeletons from dd were so poorly preserved, preparation efforts were concentrated on specimens deemed more important. Thirteen cervicals had been completed by 1929, though the dorsals were untouched. All 63 caudal vertebrae were eventually prepared to supplement the missing elements of Skeleton m.[3]

Gustav Borchert, the senior preparator, had painstakingly freed the majority of the bones from their jackets and surrounding matrix. Some preparation had also been performed by Ewald Siegert, also a senior preparator, and Hans Schober, a preparator. A composite skeleton would be mounted for exhibition. As the elements were laid out, Siegert modeled missing bones and restored partial bones in plaster. Skeleton m contributed a complete articulated series of 24 presacral vertebrae, excluding the atlas and axis, and several cervical ribs, a sacrum, and the first 19 caudal vertebrae, also in articulation. There were also three caudal vertebrae, cervical ribs, dorsal ribs (12 right and 3 left), 17 chevrons, the right ilium, ischium, and pubis, both femora, and the left tibia, fibula, and astragalus.[4]

The two *Dicraeosaurus hansemanni* skeletons from Quarry dd furnished supplementary elements. Another 25 caudal vertebrae were selected from what was called animal A, and Siegert modeled 29 more. A right scapula and coracoid were also chosen from Quarry dd. Fortunately, enough material from a number of sites was similar in size to Skeleton m. A right humerus came from site Q11. The skull was modeled entirely in plaster, but as more cranial elements were prepared corrections were made to the reconstruction. Quarry dd contained numerous useful pieces: two braincases, postorbitals, lacrimals, postfrontals, a pterygoid, premaxillae, maxillae, a squamosal, a dentary, and isolated teeth.

All missing bones for the mount were modeled in plaster. These included the atlas and axis, based on examples from Quarry dd, right cervical ribs 3, 4, 6, 7, 10, and part of 11, and left cervical ribs 3, 4, 6, 7, 8, and 12, as well as parts of 9, 10, and 11. The left coracoid and scapula were built of plaster, the sternal plates were modeled on examples from *Diplodocus,* the left humerus was copied from site Q11, the radius and ulna were modeled from a *Dicraeosaurus sattleri* named Skeleton O, and metacarpals I and II were replicated from originals thought to belong to *Dicraeosaurus sattleri.* The remaining elements of the manus were freely modeled. The left ilium, ischium, and pubis were reconstructed as mirror images of the right side of Skeleton m. The right tibia, fibula, and astragalus were mirror images of the left side of m. All bones of the hind foot were replicated from distorted examples referred to *Dicraeosaurus hansemanni*. The phalangeal formula was that used for *Diplodocus* and *Brachiosaurus*: 2-1-1-1-1.[5] Janensch admitted that he might have included extra vertebrae in the mount. He examined publications on *Camarasaurus* and *Diplodocus* by Gilmore and by Lull and Hay, and other sauropod papers by Matthew and Nopcsa.

In order to distinguish bone from plaster, the surface of reconstructed elements or restored portions was dimpled with shallow pits. The skills of Johannes Schober came into play when the components of the skeleton were assembled. A metal armature was fabricated to support the bones. Limbs and sacrum were held in position by external metalwork that was forged to conform sinuously to the flowing bone shapes. Thirteen upright posts of varying heights supported the mount. The vertebrae were cored and strung onto an internal "backbone" and the ribs were held in place by additional metal straps.

Janensch was not entirely satisfied with the pose, a consequence of postmortem displacement. The articulation of the vertebral processes had been distorted by the pressure of overlying sediments, creating a curvature of the spine that did not appear entirely life-like. It was not possible to correct this distortion, and consequently the mount appeared with the neck curved down and the skull held unnaturally.[6]

The walking stance was based on that of large mammals, since sauropod trackways had not been described in detail at this time. It was decided to place the animal in an upright pose with limbs held under the body. Janensch rejected the sprawling lizard-like posture advocated by Hay, Tornier, and Sternfeld, basing his arguments on the articulations of the bones. He felt that American paleontologists were correct in their recent mounts of *Apatosaurus excelsus*, at the Peabody Museum of Natural History, and *Diplodocus longus*, at the U.S. National Museum. He also rejected Nopcsa's contention that *Dicraeosaurus hansemanni* of the Middle Saurian Bed and *Dicraeosaurus sattleri* of the Upper Saurian Bed were male and female of the same genus. The skeleton was mounted on a plinth that was positioned on the east side of the Museum atrium. It measured 13.2 meters in length and stood 3.16 meters at its highest point. This third Tendaguru dinosaur exhibit was completed in late 1930 or early 1931. It was the latest member of an exclusive club. There were only perhaps half a dozen sauropod dinosaurs mounted anywhere in the world that featured a complete, freestanding skeleton of mainly original bone.

Despair had returned to Germany after only a few prosperous years. By 1932, unemployment stood at 6 million out of a work force of 29 million.[7] In Berlin over 600,000 people, fully a third of the labor force, were jobless.[8]

In 1930 and 1931, the German government cut wages, rents, and commodity prices by 40 to 50 percent.[9] Unemployment benefits were reduced, which only increased dissatisfaction. As public confidence in the government failed, a series of elections placed a succession of leaders in power. The National Socialists promised to create jobs, renounce the hated Treaty of Versailles and its reparation clauses, replace the professional army with a people's army, and build new homes and roads.

The Communist and National Socialist Parties both offered young men, especially, a purpose, in addition to food, uniforms, and discipline. The Nazi Party was particularly popular among university faculties and student bodies. In this community, the proportion of ardent Nazi supporters was twice as high as among the general public.[10] Fears of a Communist uprising contributed untold support to the National Socialist cause among the country's citizens.

Once more violence spilled into the streets as factions like the SA, Ernst Roehm's brown-shirted storm troopers, battled with rival Communists and simultaneously terrorized the public. Hitler was named chancellor in January 1933, supported in his rise to power by industrialists. Less than one month later, the Reichstag building was torched under suspicious circumstances. It was Hitler's excuse to obtain an emergency decree and suspend civil rights. Constitutional government disappeared from the country.

In 1932, in the midst of this turmoil, Hans Reck traveled to South Africa. At the end of a two-and-a-half-month expedition to collect Permian reptiles, he had several hundred isolated bones, 60 to 70 skulls, and roughly a dozen nearly complete skeletons, weighing 2,300 kilograms in all.[11] The finds consisted predominantly of herbivorous parieasaurs, but the skulls of rarer carnivores were also found. He also completed a popular account of his trips to Olduvai Gorge in 1913 and 1931. It was published in 1933, under the title *Oldoway, die Schlucht des Urmenschen: Die Entdeckung des altsteinzeitlichen Menschen in Deutsch-Ostafrika,* or "Oldoway, the gorge of paleolithic man: The discovery of paleolithic man in German East Africa."

The Museum of Natural History in Berlin devoted the years 1930–1932 to study, preparation, and publication of Tendaguru discoveries. The Museum offered German institutes duplicate material in late 1932. The Senckenberg Natural History Museum in Frankfurt received a plaster cast of a complete *Brachiosaurus* forelimb and scapula, possibly the same elements that had been sent to Hamburg in the late 1920s. An original scapula and rib were donated as well, fulfilling von Branca's obligation to his patron Richard von Passavant-Gontard.

A wealth of field jackets containing ornithopod bones was available. In January 1933, Kurt Leuchs of the Senckenberg Natural History Museum informed the Museum of Natural History that, although such specimens were desirable, there was no one to prepare them. The Stuttgart Natural History Collection was sent 32 crates of ornithopod bones in late May 1933 (specimens 7500 to 8399). Dr. Rupert Wild states that the museum also received stegosaur material and other field jackets, some of which bore traces of fire damage. Perhaps these jackets had survived the fire aboard the *Gertrud Woermann* in 1910.

Museums at the Universities of Tübingen, Göttingen, and Munich also received *Dysalotosaurus* material. In January 1933, Edwin Hennig declared the Institute and Museum of Geology and Paleontology at Tübingen University ready to receive the donation, and Berlin reserved 33 crates containing specimens 5000 to 5999 and 8400 to 9999. At the end of May, 18 crates were shipped and then possibly 22 more. Professor Hermann Schmidt of the Institute and

Museum of Geology and Paleontology at the University of Göttingen accepted Berlin's offer in December 1932. By spring 1933, he had been sent 30 crates (specimens 6200 to 6799). The Bavarian State Collection for Paleontology and Historical Geology (Bäyerische Staatssammlung für Paläontologie und Historische Geologie) in Munich was sent 23 crates (specimens 6800 to 7499) in May. In late October, Dr. J. Schröder indicated that preparation of the jackets was underway.

In February 1934, Dr. Roland Brinkman, director of the State Geological Institute in Hamburg, indicated that the German East Africa Line expected a substantial donation for his institute, in recognition of the freight subsidy granted to the Tendaguru Expedition. Berlin offered sauropod material for the Hamburg institute's upcoming exhibit on colonial geology and 27 crates of ornithopod bones (specimens 5700 to 6299). Hamburg accepted, and unpacked the jackets in May 1934. The ornithopod field jackets had all been collected by Hans Reck's crew in a single day.

It does not appear that the Geological Collection in Mannheim ever received Tendaguru fossils in recognition of August Roechling's generous patronage. In 1922, Föhner, director of that institute, indicated that discussions with Pompeckj in Berlin had not been successful. Professor Kalderup of the Mineralogical-Geological Department of the Natural History Museum at the University of Bergen, Norway, was supposedly presented with a sauropod femur from Quarry Sa1 in 1933.

Werner Janensch had spent about 7½ months in South Africa in 1929, studying and collecting Karoo reptiles. Upon his return to Berlin, the Museum donated sauropod bones to the South African Museum in Cape Town: *Tornieria* anterior and mid-caudal vertebrae, *Brachiosaurus* anterior and posterior caudal vertebrae, a *Dicraeosaurus* (?) mid-caudal vertebra, and a *Barosaurus* vertebra.

On March 20, 1934, senior preparator Gustav Borchert passed away in Berlin-Hermsdorf, two days before his 71st birthday. His career had spanned an era ranging from the preparation of *Archaeopteryx* to the preparation of the flood of bones from Tendaguru. He was survived by his wife.

Hitler's assumption of power did not immediately cure Germany's economic and political instability. The new chancellor took swift and brutal steps to consolidate control. By mid-1933, opposition party leaders in the Reichstag had been intimidated, imprisoned, or murdered. The first concentration camp for political prisoners was established that year. Political parties were disbanded and all trade unions were amalgamated into one organization. Non-Aryans were dismissed from public office, teaching positions, and the civil service. Ominously, a massive public burning of proscribed books took place in Berlin and other centers in May 1933. In 1933, only two out of 33 professors at Edwin Hennig's institution, the University of Tübingen, failed to declare themselves enthusiastic about the new regime.[12] Later that year, Germany withdrew from the League of Nations.

All levels of public service came under party control. Dissension was suppressed by a secret police force, the Gestapo. All legal functions came under Nazi control. In early 1934, state parliaments and university constitutions were abolished. Goebbels's Ministry of Popular Enlightenment and Propaganda exerted an irresistible influence on organizations that represented cultural or scientific interests. The media were strictly controlled. Massive annual rallies became known the world over.

With such far-reaching powers that quickly penetrated every level of public and private life, the Nazis could alleviate the effects of the Depression. Prices were frozen to prevent the recurrence of crippling inflation. Unemployment was tackled through massive public works programs. The network of roads was expanded by 3,200 kilometers in the five years after Hitler became chancellor, and included modern four-lane highways.[13] Extensive slum clearance and housing and public works projects fed a construction boom. These programs reduced joblessness by more than 40 percent in a single year.[14] By 1938, a wide-reaching rearmament plan was causing a labor shortage.

Edwin Hennig had been teaching geology at Tübingen University since his appointment in 1917. He had also raised a family, and though many of the hopes of his youth had been fulfilled, unfinished work from his early career nagged at him. British challenges to German interpretations of stratigraphy and dating had been answered in part by subsequent German studies. Researchers like his brother-in-law Wilhelm Otto Dietrich had a great deal of newly prepared material to work with. Yet it was obvious that Hennig felt much had been left undone.

The dating controversy was not settled in Hennig's mind, despite much progress and even a degree of consensus among British and German workers. John Parkinson had offered his views in 1929 and 1930, stressing the difficulty of clearly identifying particular strata when all were deposited almost horizontally. In Parkinson's opinion, there were no distinct events of marine deposition followed by estuarine deposition, but rather an unbroken succession from the *Nerinea* Bed to the Upper Saurian Bed. The Middle and Upper Saurian Beds were really only part of a continuous but localized sequence, where deposition became increasingly estuarine until marine deposits ended altogether. Evidence for this theory was found at Kinjele near Mtapaia, and at Nautope where the *Nerinea* Bed could not be distinguished from the *Trigonia smeei* Bed. Parkinson recognized a disconformity between the *Trigonia schwarzi* Bed and the Upper Saurian Bed, but placed the latter in the Upper Cretaceous.

Parkinson could find no proof of the faults that supposedly bounded Hennig's Niongala Trough to the north and south. It was an area of subsidence through which the Mbemkuru River flowed. He agreed with Kitchin that deposition was rapid in Mesozoic times. In Parkinson's model, a large and powerful river had transported varying amounts of sediment and swept dinosaur bones into a nearby lagoon. Ocean water flowed into the lagoon, carrying shallow marine sediments. Parkinson sent rock samples to Imperial College, London, to have the heavy mineral components analyzed. His article describing the results of this work was his last published commentary on Tendaguru stratigraphy.

In 1933, Dietrich reexamined newly prepared invertebrates. This led to further revisions, in which the German views more closely corresponded with John Parkinson's findings. Dietrich now agreed that the full succession from the

Nerinea Bed to the Upper Saurian Bed was an uninterrupted depositional sequence representing changing environments from lagoonal-estuarine to littoral
and neritic. This gradation throughout the profile was gradual and resulted
from the meandering of the ancient shoreline. Land and sea level oscillation
was the mechanism that had caused the shoreline to wander. The strand was alternately exposed and submerged by shallow seas along the Jurassic coastline.
Rivers meandered from west to east and flowed into coastal lagoons.

Dietrich proposed a new name for the sequence in the interior—the *Smeei*
Stage (*Smeei* Stufe). As Parkinson had argued, the Middle and Upper Saurian
Beds lost their formal status and were considered local, lagoonal-estuarine se-
quences within the *Smeei* Stage. The *Nerinea* Bed was no longer distinguished
from the *Trigonia smeei* Bed. In contrast to Parkinson's view, the entire *Smeei*
Stage was considered Upper Jurassic, ranging from Lower Kimmeridgian to
Portlandian (= Tithonian) times. There were no Lower Cretaceous fossils below
the *Schwarzi* Stage, and those cited by Kitchin were explained as misidentifica-
tions by Krenkel or Lange. The controversial presence of *Trigonia smeei* in the
Schwarzi Stage was viewed as a misidentification, or a specimen transported
out of its original location or accidentally switched with another.

The preservation of the ammonites confirmed they were autochthonous, not
extra-formational as Kitchin believed, though some types may have been carried from greater ocean depths into the littoral deposits at Tendaguru. Dietrich
maintained his assertion that a stratum equivalent to the Middle Saurian Bed
was developed as a marine horizon in the Mahokondo region to the north of
Tendaguru. The fauna in that area had been described in a 1925 paper based on
the abundant ammonites collected by Hans Reck in 1912. Many ammonites
were preserved within septarian concretions, hence the term for the deposits—
the Septarian Marls.

In 1933, Leonard Frank Spath, a renowned ammonite specialist who worked
part-time at the BM(NH), completed his monumental revision of Jurassic invertebrates of India. He challenged several of Dietrich's conclusions about the
Mahokondo ammonite fauna. The British expert did not believe that the Septarian Marls at Mahokondo, which he considered Kimmeridgian, could be
equivalent to the younger Middle Saurian Bed at Tendaguru, which he considered Portlandian (= Tithonian).

Spath also rejected Kitchin's argument for Cretaceous Saurian and marine
beds at Tendaguru. He placed the five main divisions into the Upper Jurassic,
Portlandian (=Tithonian) times. In 1935, he pointed out that Kitchin had erred
in using an evolutionary or morphological stage of a fossil, *Trigonia smeei,* to
precisely date sediments at Tendaguru as Cretaceous. Kitchin died suddenly in
January 1934, and after Spath's comments, the controversy over stratigraphy at
Tendaguru subsided briefly.

American paleontologists, including Charles Schuchert, William Diller Matthew, and George Gaylord Simpson, supported a Jurassic age for all the Saurian
Beds. They also agreed in seeing a disconformity between the Upper Saurian
Bed and the *Trigonia schwarzi* Bed. Here the matter rested until Edwin Hennig's surprising return to the field at the height of the Great Depression.

Hennig had long been disappointed by the inability of the German Tenda-guru Expedition to carry out geological fieldwork. In fact, 23 years had passed before he was able to satisfy his craving for scientific closure. In 1934, he was middle-aged at 50, with four teen-age daughters. Somehow, he convinced the Foreign Office, his university, and scientific agencies, as well as private donors, to sponsor a research trip to southern Tanganyika Territory in 1934.

The Maggi Corporation again provided tinned foodstuffs, while Leitz (Wet-zlar) donated a Leica camera. The Trossingen firm of Hohner supplied harmon-icas as gifts for Africans who assisted the expedition.

Hennig was accompanied by two younger colleagues. Dr. Ernst Nowack, an Austrian from Salzburg, was a veteran of a number of lengthy voyages to re-mote locations. His interests focused on Africans and their languages. Dr. Al-fred Mayer-Gürr of Württemberg had completed his doctorate under Hennig at Tübingen.

Chief among their research goals was a more detailed understanding of the stratigraphical succession, geological age, facies relationships, structure, and tectonic development of the sediments of the southern coastal strip. The region under scrutiny would eventually include the Matumbi Hills inland from Kilwa, the Mahokondo-Mandawa area, the Pindiro Valley, and the environs of Tenda-guru and Lindi. No excavations were planned at Tendaguru, since Hennig had interpreted the fossil preserve established by the British to be off limits.

The final major goal was the examination of Karoo deposits in the Ruhuhu Valley. They had been discovered by Stockley in 1931, and briefly visited by Mi-geod and Parrington that year. Their geology had been described in 1932 by Stockley, the same year Haughton of South Africa had described the fossils. Rex Parrington had returned in 1933, to undertake a program of excavation for Cambridge University.

It was a difficult farewell for both Hennig and Nowack. Throughout his many dozens of letters home between March and November 1934, Hennig was troubled by leaving his family while he realized a quarter-century-old dream. Nowack's wife was experiencing financial difficulties, and Nowack had no prospects of employment when he returned to Germany.

Hennig and his co-workers boarded ship in Hamburg on March 24, which caused Hennig to miss the 16th birthday of his twin daughters. Landfall in Tanga occurred on April 30, 1934. Unsettlingly, the rainy season was late and could wreak havoc with travel plans. A new Ford 1½-ton truck was purchased from the dealership in Tanga in early May for 3,400 shillings. Rain began fall-ing in torrents and the truck was frequently mired; the wearisome sequence of unloading it, raising the chassis with a jack, and placing stones under the tires was repeated for weeks on end. Black laborers at one German-run plantation gave Hennig's group the Hitler salute. Tensions existed as some Germans en-thusiastically embraced National Socialism while others shunned the party. Hennig regretted missing his daughter Ingeborg's 18th birthday.

The exact relationship between the German Foreign Office and Hennig's ex-pedition is not clear, though he was obliged to report on German progress in the country. He also wrote popular articles about the trip.

On July 4, a quarter of a year since their departure, Hennig's team reached the Ruhuhu Valley. Porters were engaged to safari further afield into Karoo deposits, but heavy rains stymied all efforts.

By the time they drove into Songea, the Ford had carried them four thousand kilometers.[15] Heading east, they passed between the Rondo and Makonde Plateaus. Rumors of political upheaval at home swirled about them. Indeed, Hitler had ordered the infamous Night of the Long Knives that purged the ranks of the SA.

It was July 17 when they finally halted at their long-sought goal of Lindi. It had been two and a half months since they had departed from Tanga, nearly four months since Hennig had left Germany. Two former supervisors from Tendaguru had followed Hennig's truck into Lindi on foot. Hennig could recognize only one of the men—Seliman Nyororo, a Wangoni who had been a supervisor 25 years earlier. The other old crew leader was Saidi bin Ali, whose good humor was intact after 25 years.

Hennig discovered that his former servant Ali had died in December 1930, but that his assistant Wilhelm was still alive. Janensch's servant Saidi now lived near the Mbemkuru. Hennig and Boheti had corresponded during the 1930s, when Boheti mentioned Cutler and listed the quarries operated by Migeod. Unfortunately, Boheti was too ill to travel to Tendaguru. His eyesight was poor and he had asked Hennig to send a pair of glasses. The able overseer and the former cook Bernardo, who had also gone with Reck to Olduvai Gorge, now wrote to Hennig.

Tertiary sediments along the coast would be examined first. About four months were available for their investigations. Motorized transport allowed distances to be covered more quickly than by the safaris of 1911. Base camps could be reached with minimal delay, specimens could be transported more easily, and food supplies and mail were delivered more rapidly. Rampant growth of vegetation due to the belated rains, however, restricted freedom of movement and the visibility of exposures. The traditional safari would still form a large component of fieldwork.

The passage of years added to their difficulties:

> the maps were obviously incomplete enough [in 1909–1911] and totally out of date today. The names of over 25 years ago are already largely forgotten, localities abandoned, new ones established, names transplanted to other places. . . . landscape, settlements, etc., have different names today than then, different then than in Bornhardt's time.[16]

Niongala Village, where wartime excavation had been undertaken by the British, could not be re-located. Wherever it was possible to identify sites that had yielded significant fossils or exposures during the German Tendaguru Expedition, Hennig would insist that Nowack or Mayer measure the geological sections to compare with his original profiles.

With few landmarks to guide them, they failed to notice that they had driven inland as far as the Pindiro Valley. Hennig hoped to investigate more thoroughly than in 1911. A favorite locale from decades past, Mto Nyangi, was reached, still populated by hippos that had pleased Hennig, Janensch, and the Recks a quarter century earlier. Caves in the Matumbi Hills northwest of Kilwa were examined for human and animal fossils during the third week of August.

Locusts had stripped native crops to the stalk, and unless the grass could be burned to prepare fresh fields, famine was a real threat. Hennig could not even hunt game for his own crew in the dense bush. They passed through Miteja, the site of Bernhard Sattler's untimely death in 1915, though Hennig made no mention of this or of Sattler's gravesite in Kilwa.

The Germans drove southwest along the road to Liwale in early September. Near Makangaga, they found trenches that dated to the war. Three locals were hired to excavate a bone discovery made by Seliman. For six hours half a dozen men dug and scraped at the ground. Two limb bones were collected but other fragments were abandoned.

Hennig commented that the British had seemed unaware that there were fossils at Makangaga, and also accused the administrators of limited knowledge of their districts:

> The British officials get out into the countryside astonishingly rarely, are not even seen beyond the motor roads. (. . . ca. 30 km. from the coast, we were the first whites seen by the generations born during and after the war, despite the multi-year proximity of the British Tendaguru excavations!)[17]

The Germans and their porters trekked along the Mavuji River south of Makangaga. Ammonites of Albian age had eroded out of the banks where the river emptied into a bay at Namasatu. While Hennig and Nowack continued to Tendaguru on foot, Mayer returned for the truck. Hennig already wished he could cover some of the ground again: "Now I should see the entire story again, and even some more with all this new knowledge. Well, 1959! Who will join me?"[18]

They investigated the region around Mandawa and Mahokondo, where Hennig had recognized an anticline in 1911. Ammonites, gastropods, bivalves, and other fossil invertebrates were collected to correlate and date strata.

Five days after arriving at Mandawa, Edwin Hennig again stood at Tendaguru. It was September 27, 1934. The hill had been burned before his arrival and as Hennig trudged through black ash to the summit, he was overwhelmed with gratitude. A grass hut in which meals were taken was built at the base of the hill. Another shelter served as kitchen and sleeping quarters for the Africans.

Hennig's joy at returning to Tendaguru was unabated after four days, despite the trying living conditions: "profound feelings of gratitude . . . that I was permitted once more to set foot on a very modest little patch of ground, upon which a little scrap of youthful love was indeed caught." Rain and vegetation had obscured the work done over seven field seasons by BM(NH) crews. Yet Hennig detected few changes even since 1911:

Map G.
Hennig's Investigation in Makangaga Area, 1911; Kilwa Area 1934
Redrawn from Hennig, 1937, Der Sedimentstreifen des Lindi-Kilwa-Hinterlandes, Palaeontographica, Suppl 7.

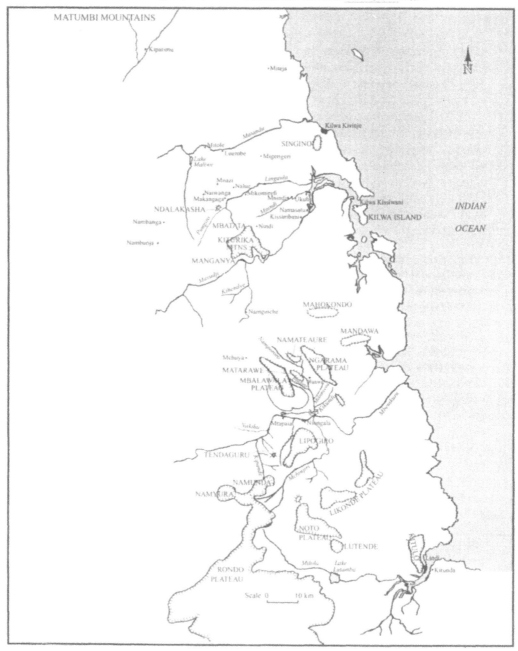

nothing at all has been altered. The quarries, even those of the British, are naturally long since essentially worked into the relief, rain and erosion have actually not prevailed at all on the hill. The vegetation is just as meadow-like . . . as it once was, the Wamwera just as dull and indolent. The paths, with minute exceptions, run exactly as they did in those days; only the remains of an automobile bridge have been added by the English.[19]

Almost magically, familiar faces from the German Tendaguru Expedition gathered in a few days: "in numerous cases the joy of reunion was truly great on both sides. Hardly one of them placed themselves at the service of the British; to us they came with 1000 joys."[20] Hennig may be forgiven for making the unlikely last statement. It is probable that many men had applied to the British for work, but far fewer had been hired. Excited reunions took place as another of Janensch's former servants, Abdallah, arrived from Mtapaia. Another old supervisor, Sefu, visited Hennig often. Abdallah Kimbana walked all the way from Niongala when he heard of the arrival of the Germans. Hennig passed greetings from all to Janensch in a letter.

Hennig walked through streambeds, down to the Mbemkuru, measuring sections and collecting invertebrates. "The purpose this time was not dinosaur bone. In the First World War, English troops who marched through insultingly misinterpreted our quarries as defensive trenches. Now the last vestiges of them could hardly be discerned in the terrain."[21]

By October 6, ten days after their arrival, Hennig had examined every accessible local exposure. His group returned to Kindope, where the gaping excavations at the ornithopod and stegosaur quarries were overgrown with grass. He composed the last letter he would ever write from Tendaguru on the evening of October 10. Three men and an overseer would carry geological samples to Lindi. Early the next morning Edwin Hennig left the memory-rich site of his youth and middle age. He and his companions had spent just over two weeks around Tendaguru.

Hennig would return repeatedly to the theme of how difficult safaris had become since the advent of the automobile, as it represented a marked change in the region since German occupation:

The natives have drawn closer to the scarce settlements . . . [and] have completely abandoned even larger areas to nature than before [in 1911]. Above all, the sparse native paths have also largely disappeared because the few and poor stretches opened for motor vehicles relieve the blacks of the effort of keeping their own trails open, even when they must make detours of 20 kilometers and more as a result. . . . Thus far larger areas than before the war are not only empty but rather have actually become inaccessible.[22]

The familiar streets of Lindi appeared a week later. Unlike the others, Nowack was not going home, despite having been absent already seven months from his wife. He was returning to the Songea area to collect Karoo reptiles for the University of Tübingen. Hennig was able to transfer between 2,000 and 2,500 shillings that remained from the 1934 funding, along with 1,000 reichsmarks that had been raised by Friedrich von Huene, specifically for excavation in Karoo deposits.

The trio had traveled 7,000 kilometers through Tanganyika Territory. About 2,000 had been covered on foot, 1,500 of them during the safaris of the last three months. This was an average of 15 kilometers per day through difficult terrain.[23] The final cost of the expedition was 7,000 reichsmarks.[24] Hennig recognized that without the assistance of the British administration from the outset, little would have been accomplished.

Many memories had been reawakened. Those that stirred Hennig most were recollections of the many Africans whose lives had once been so closely linked with his own. Those times inevitably seemed more peaceful than the present:

> in the mind's eye the days, months, years rose again vividly. Where scanty tree and wild grass growth shot up, our little camp village had stood. . . . Comfortable lack of worry extended over all. . . . On Sundays, even the lords of creation sat in the shade . . . gossiped, laughed, smoked. . . . Laughter too, the best of all skills, is common to all people internationally. Only they exercise it too rarely, indeed forget it entirely, because in the earth-omnibus's circuit through space everyone demands a window seat, and because of the strife over it, the pleasure from it is lost to all.[25]

Hennig was a different person than he had been in 1911, and longing for family drew him home. A spartan steamer pulled away from the port on October 23. About a hundred days had been spent in the southeast of the African mandate. Like Migeod, Hennig would never return. In Dar es Salaam, Clement Gillman generously placed the drafting department of Tanganyika Railways at Hennig's disposal. Maps and profiles could be sketched while the material was still fresh in the geologist's mind. Hennig left Dar es Salaam on November 7, 1934. He was home in time for Christmas, 10 months after he had last seen his wife and daughters.

Hennig produced a monograph that appeared in *Palaeontographica* in 1937. He described ammonites, bivalves, corals, gastropods, brachiopods, and arthropods from Tanganyika, and presented his conclusions on stratigraphy and tectonics. Samples from the 1934 expedition were forwarded to other specialists. The results of these studies were monographs on fossil foraminifera by Helmut Fahrion in 1937, and crinoids by Hertha Sieverts-Doreck in 1939.

Edwin Hennig retained much of the stratigraphical framework he had developed in 1914. He rejected the assertions of Parkinson, and thus also of Dietrich, that the Saurian Beds were only local occurrences of estuarine facies in a much longer sequence. On his marches around and far beyond Tendaguru Hill, he had observed evidence of a repeated series of dinosaur-bearing estuarine sediments, alternating with marine horizons. There were no major faunal changes among the estuarine or marine layers. Hennig still insisted that the deposition was uninterrupted, that there was no recognizable disconformity between the Upper Saurian Bed and the *Trigonia schwarzi* Bed. A new sequence was established: the Pre-*Smeei* Layer (Vor-*Smeei* Schicht). It represented a marine transgression extending southward from the Mandawa-Mahokondo area to the north of Tendaguru. The original *Trigonia smeei* Bed was now correlated with the Main *Smeei* Layer (Haupt *Smeei* Schicht) to the north. Another subdivision at Mandawa-Mahokondo was recognized, the Late *Smeei* Layer (Spät *Smeei* Schicht). The latter was placed in the Lower Cretaceous. The Jurassic-Cretaceous boundary was once more shifted back into the Upper Saurian Bed. Hennig also confirmed the Niongala Trough along the Mbemkuru.

Germany's turmoil worsened in late 1934. During the first five years of Hitler's regime, 30 percent of the country's university professors had left, a third of them having been dismissed outright. Student enrollment at centers of higher learning had dropped by half.[26] Much of Germany's international reputation in physics, chemistry, and natural sciences had been lost. The same held true in the arts, as thousands of writers, musicians, and actors fled the repressive state.

General conscription was introduced in the spring of 1935. The army (Reichswehr) expanded dramatically, and was 14 times larger in 1937 than in 1933.[27] Enrollment in party youth organizations such as the Hitler Youth (Hitlerjugend) and the League of German Girls (Bund deutscher Mädchen) mushroomed. Hennig encouraged his eldest daughters to become involved in the latter. Nearly three million children joined these groups in 1934.[28] Two years later membership was compulsory.

Jews had been forbidden to hold public office and business positions since 1933. They were stripped of citizenship in 1935. The Treaty of Versailles was repudiated in 1935, removing even the pretence of reparations. Hitler grew ever bolder, occupying the Rhineland in the early spring of 1936.

No one could fail to feel this threatening degree of control and coercion, yet the standard of living had undoubtedly improved. Crime and juvenile delinquency rates were lower, thanks to fuller employment and the availability of youth organizations. Civil service pay was frozen at pre-Depression levels to encourage economic recovery, yet in 1936 industrial workers earned on average 10 times what they had received in unemployment insurance four years earlier.[29]

A comprehensive program of social benefits was instituted, one of the original planks of the National Socialist platform. The elderly, disabled, and impoverished received financial support and women's pay began closing the gap with men's. Slum clearance, especially in Leipzig and Cologne, provided better living conditions, but it was housing construction that greatly aided overall recovery.

A much-transformed Berlin was the focus of world attention in August 1936, when the city hosted the Olympic games. Over one million visitors were drawn as spectators. By now, Berlin was home to 4.2 million people, making it the fourth largest city in the world by population.

August 1937 saw the celebration of the seven hundredth anniversary of the city's founding, and also the death of Hans Reck. This collector of Arab art, skilled pianist, and experienced vulcanologist was in Lourenço Marques when his heart failed on August 4. He was just 51 years old and was survived by his wife Ina, the lifelong partner who had shared equally in his demanding career. Reck had undertaken another lengthy foray around Africa in late September 1936, announcing it would be his last. Over the course of a year, he investigated volcanoes on the island of Palma and in the Cameroons, and Karoo sites in South Africa. He was en route from Natal, South Africa, to Lake Eyasi in Tanganyika to investigate Ludwig Kohl-Larsen's excavation of prehistoric human remains when a congenital heart weakness finally took its toll.

Over 130 publications bore his name, including the results of excavations at Olduvai Gorge, which appeared between 1914 and 1937. Reck had been co-editor of the journal *Zeitschrift für Vulkanologie* since 1923. Arthur Tindell Hopwood, who was a member of Leakey's 1931 British Museum East African Archaeological Expedition to Olduvai Gorge, wrote fondly of the German geologist: "Tall, fair, and blue-eyed, Reck was a Bavarian of the finest type. His sunny disposition and upright character endeared him to all who had the privilege of his friendship. Himself incapable of an ungenerous thought or act, he was unable to understand them in others."[30]

The same year, 1937, also featured a triumph that honored the epic efforts at Tendaguru. As the culmination of 27 years of painstaking labor, another display was opened at the Museum of Natural History. It had been a long road from Tendaguru to Berlin, fraught with delays due to a world war, hyperinflation, mass unemployment, civil unrest, and a host of other difficulties. Yet what a display! The magnificent sauropod, *Brachiosaurus brancai,* now towered over enthusiastic visitors in the glass-ceilinged atrium of the Museum.

Work on this large sauropod had begun while the Berlin Expedition was still in the field. Werner Janensch had initially erected two species in 1914. *Brachiosaurus brancai* was based on one of the two enormous skeletons dubbed SI and SII, from the Middle Saurian Bed, three kilometers south of Tendaguru Hill. The species name *brancai* was assigned to the type animal, SII, using a cervical vertebra, a humerus, an ulna, and a radius as the diagnostic elements. Wilhelm von Branca, who at the time was the director of the Geological-Paleontological Institute and Museum that had launched the Expedition, was thus honored. It was the larger skeleton, SII, which formed the foundation of a museum display.

Skeleton Y had been found in the Middle Saurian Bed as well, northwest of Tendaguru Hill near Kindope. *Brachiosaurus fraasi* had been erected based on a scapula and humerus from this skeleton. The species name honored Eberhard Fraas, who had urged that a large-scale expedition to Tendaguru be organized.

There were no complete skeletons but rather articulated sections of vertebrae, disarticulated but associated accumulations in bonebeds, and isolated individual elements. Remains that were eventually referred to *Brachiosaurus* came from about 40 sites. They originated in all three Saurian Beds as well as the transitional sediments at the base of the Middle and Upper Beds. Janensch listed skeletal elements at the following quarries: S, W, D, Q, U, I or J, p, Aa, Gl, Bo, Ma, Lw, Sa, R, T, X, no, XX, GD, TL, F, IR, XV, G, Ng, II, XII, IX, XIV, Nr. 4, Nr. 37, Nr. 12, and AR. *Brachiosaurus* was also identified at St (EH), dd, t, Y, and IG (WJ), all near Kindope. To this list must be added Skeleton be from Makangaga, the poorly preserved vertebrae Hennig had struggled with in 1911. The horizon in which specimens from these sites were found is not always clear from Janensch's later monographs. Individuals varied greatly in size, representing a range from juveniles to adults, and, likely, both male and female animals.

Preservation ranged from poor to excellent, depending on visible postmortem effects such as scavenging by carnivores, water abrasion, and duration of surface exposure.

Several monographs had been published by 1937. The volume of jackets shelved in the wooden storehouse behind the Museum was daunting to the most energetic preparator. Smaller dinosaurs were to be prepared, described, and mounted first. They included *Kentrurosaurus, Elaphrosaurus,* and *Dicraeosaurus.* The First World War had slowed the pace, as manpower and resources were diverted. By the beginning of the war, little had been done with the sauropod jackets, though the disarticulated skull elements had been prepared. By war's end and shortly thereafter, assorted limb bones and other elements had been assembled and placed on temporary display.

Postwar inflation and social and political unrest had caused further delays. In 1929, the journal *Palaeontographica* published Janensch's tabulation of sauropod material. With a few exceptions for display purposes, the intricate vertebrae had still not been prepared by the end of the decade. There was a great selection of other skeletal elements available for examination at this time. It had proved impractical to prepare all the bones from a single genus, and then describe and mount a skeleton. Janensch was the only paleontologist at the Museum who was attempting to cope with tonnes of material representing thousands of individual bones. As an alternative, he planned to compare and describe skeletal complexes of several genera in theme monographs. For instance, there was to be a study of the vertebral column of Tendaguru sauropods, a comparison of the cranial elements, the appendicular skeleton, and so on.

It was not until the turbulent 1930s, almost 20 years since the inception of the Expedition, that a significant number of in-depth papers appeared on Tendaguru sauropod anatomy. Portions of the *Brachiosaurus* hyoid apparatus were described in 1932. The new director of the Geological-Paleontological Institute and Museum, Dr. Walter Gross, analyzed bone anatomy in 1934. Humeri, tibiae, and femora of adult and subadult *Brachiosaurus* were thin-sectioned. It was determined that limb bones were composed of spongiosa of secondarily remodeled osteons surrounded by a thin layer of periosteal bone that contained fibrolamellar zones. The dense outer compacta was composed of secondary osteons. Juvenile bone and bone in zones of growth was laminar.[31]

Palaeontographica again published major monographs by Janensch in 1935 and 1936 in which the skulls of the Tendaguru sauropods were featured in detail. In 1935, the preparator Hans Schober assembled the *Brachiosaurus brancai* skull from Quarry t into a mount, which was displayed in a glass case. The skull was complete, including lower jaws, hyoid elements, and sclerotic ring plates. The field jacket had been opened so that preparation could proceed from the ventral side. As they were exposed, elements were removed individually. There was fracturing and distortion from overlying sediments, making it difficult to articulate some of the palatal bones.

Skeleton S also yielded two partial skulls. The skull of the smaller animal was found disarticulated, and included two hyoid elements as well as a cast of the left labyrinth, or inner ear. The disarticulated skull of the larger animal included hyoid elements and about twenty plates from the sclerotic ring of the eye. A braincase was collected with Skeleton Y. Cranial elements had also been collected at St (EH) and at Ig (WJ) by Hans Reck. There were several dozen isolated teeth, including some from juvenile animals. By this time, Janensch was doubtful that *Brachiosaurus fraasi* was a legitimate species. Baron Nopcsa thought *Brachiosaurus fraasi* and *brancai* might be male and female of the same species, as they both came from the Middle Saurian Bed. Janensch had an endocast poured, which went on display with the skull. To measure the volume of the brain cavity, nerve and blood vessel openings were blocked with plasticine wrapped in tissue paper. Fine seed was poured into the cranial cavity five times and the volume measurements averaged: approximately three hundred cubic centimeters for Skull t1.[32]

Janensch puzzled over illustrations of skull elements that had appeared in a paper by O. C. Marsh. A dentary labeled *Brontosaurus excelsus* resembled that of *Brachiosaurus*. When Janensch wrote to Walter Granger at the American Museum of Natural History, he was assured that Marsh's figures illustrated *Brontosaurus,* and that the dentaries of both genera were likely to appear similar. Lull at Yale was more cautious, stating that the skull illustrated by Marsh had been restored and mounted at the Smithsonian Institution in Washington, and that Marsh had used a skull that had not been found in association with the *Brontosaurus* skeleton. He admitted that both the Smithsonian and Yale skulls were assumed to be from *Brontosaurus,* but this could only be confirmed when further examples were uncovered. *Brontosaurus* is more correctly known as *Apatosaurus.*

The long-awaited publications garnered acclaim from paleontologists worldwide. Janensch received letters of congratulations from peers like William K. Gregory, Walter Granger, Richard S. Lull, George Gaylord Simpson, Charles W. Gilmore, Barnum Brown, Charles Schuchert, William Swinton, Arthur Smith Woodward, Friedrich von Huene, and Franz Nopcsa, and corresponded with others.

Around the difficult year of 1930, as the worldwide economic slump sparked a dreadful spiral of unemployment across Germany, plans were formulated for the mounting of a *Brachiosaurus* skeleton. Skeleton SII was the candidate that would supply the majority of bones. At least three preparators had labored diligently since the crates poured into Berlin in 1909: Gustav Borchert, Ewald Siegert, and Johannes Schober. Borchert would pass away before he could see the ambitious result to which he contributed so much.

The remains of SII consisted of anterior cranial elements, 11 cervical and 11 dorsal vertebrae, some partial cervical ribs and most of the dorsal ribs (all but 4 ribs out of 11 pairs were recovered), the left scapula, both coracoids, both sternal plates, a complete right foreleg including foot, the left humerus, ulna, and radius, a partial right femur along with tibia and fibula, both pubes, and some poorly preserved metatarsals and phalanges from the hind foot. All the cervical vertebrae but C8 were found in articulation, with cervical rib heads still attached, though the majority of the dorsals were disarticulated and lay deeper in the ground. Many of the vertebrae exhibited signs of crushing, but C9 to C15 had suffered the most damage. Only the centra remained. Janensch surmised that postmortem maceration and disturbance by moving water had destroyed the neural processes. More recently, distal ends of dorsal ribs had been scattered when water flowed in the streambed. Still, a good number of fragments could be traced and reassembled. Janensch maintained that the vertebral formula might have been 13 cervicals, 11 dorsals, 5 sacrals, and at least 50 caudals. He surmised that SII might have been an adult male, SI a subadult male, and t a female.

Skeleton SI was much less complete, consisting of a disarticulated partial skull, five anterior cervical vertebrae, and assorted other elements. Other quarries that contained articulated or associated *Brachiosaurus* bones included D (29 vertebrae, 23 of which were articulated with the sacrum), Aa (18 vertebrae articulated with a sacrum), no (50 caudal vertebrae), Gl (16 caudal vertebrae and some limb bones), dd (14 caudal vertebrae and some dorsals from a subadult individual), and p (11 anterior caudal vertebrae from a subadult animal).[33]

The design of a display was modified over time:

> The original plan to mount the entire skeleton out of plaster casts and models of the individual bones was abandoned, although its execution would have offered the advantage of significantly easier technical mounting work, and such a reconstruction, consisting of only cast or modeled bones, could have attained a high degree of scientific accuracy with careful correction of all defects of preservation. However, another path was adopted . . . since it seemed more proper to show the museum visitor real remains of the giant dinosaur.[34]

A complete right forelimb had been cast or modeled in plaster and erected as a trial in the 1920s. A copy had been presented to the Tübingen museum prior to 1923, and was illustrated in a museum guide. The State Geological Institute in Hamburg and the Senckenberg Natural History Museum in Frankfurt had also received copies.

Using real bone demanded compromises. Since SII was incomplete, a composite mount was unavoidable. Some missing elements would be modeled in plaster as mirror images of corresponding elements from the opposite site of the body. They could also be modeled freehand if the bone was not known for *Brachiosaurus*. Finally, bones referred to *Brachiosaurus* that were of the appropriate dimensions could be substituted from other quarries. Janensch admitted that this would allow inaccuracies to creep in, but it was a standard museum practice.

The vertebrae posed the most challenging problems. It had been technically taxing and time-consuming to prepare them, especially the cervicals, which reached 1.6 meters in length. Their structure, fragility, preservation, and great weight forced the inevitable conclusion that only plaster reproductions could be mounted. It was feared that if originals were used, so much external metal bracing would be necessary that the support structure "would distract the eye of the observer so much that the details of the highly complex external architecture would no longer be very visible."[35] Such an outcome was anathema to Johannes

Schober, who set himself the demanding goal of ensuring that the metalwork was as unobtrusive as possible throughout the mount. Consequently, the entire presacral vertebral column was to be modeled as accurately as possible in plaster, using original bones as a guide. The neural processes of many could be only approximations at best. Modeling was the task of Ewald Siegert.

The skull was too scientifically valuable to be suspended 12 meters above the floor. It would be modeled in plaster and the original would be available for further study. The skull of Skeleton SII was used as a guide, and missing elements were copied from the more complete skull t1.

The right scapula, right dorsal ribs 1, 7, and 11, left dorsal rib 5, the entire left forefoot and hind foot, the left ischium, and the sacrum were also constructed entirely from plaster. The feet were based on bones from Bo and SII; the remaining elements were mirror images of counterparts from SII. The sacrum was modeled after those found in Aa and T. The left ischium was also reconstructed in plaster. Cervical ribs were based on several reasonably complete examples from SII. The first caudal vertebra and various chevrons were made of plaster. Chevrons of the anterior third of the tail were based on those found at Aa; the rest were modeled freehand. The last 26 centimeters of the tail, four terminal caudal vertebrae, were a freehand reconstruction.

To complete the skeleton, the stockpile of bones in the basement of the Museum was searched. Elements of similar size were used as replacements: a left femur from Ni or Ng, a left tibia and fibula from Bo, a right ilium from Ma, a right ischium from L, and a tail, 50 articulated caudal vertebrae, from site no. Walter Gross later reported that preliminary work on the mount had begun in 1934 under Werner Janensch.

Siegert built a scale model of the proposed skeletal mount. It was about a meter long and a meter high, and judging from archival photographs, it appears to have been made of plaster bones supported by a wire armature. With the aid of this model, a pose was established. Siegert sculpted full-size replicas of dorsal vertebrae. With Schober's assistance, the dorsals, resting on square bases, were set atop display cabinets. Adjustments were made to ensure correct articulation while the vertebrae were held in alignment by a metal rod that was passed through a hole drilled through the centra. It is not known how long this reconstruction stage lasted, but certainly it must have taken months.

Schober next took over, with Marquardt, to produce a full-scale mockup. A large space was cleared in the attic of the Museum, as there was not enough room elsewhere to begin a life-sized trial assembly. A variety of equipment, such as lumber, wire cables, ropes, and pulleys, was waiting. The modeled vertebrae were sawn in half along the median axis. The left halves of the dorsal and cervical vertebrae were laid out on the floor of the attic and articulated in a curve they would later assume permanently. The caudal vertebrae were added. Skeletal elements from the left half of the body were transported from basement workshops to the attic in a large freight elevator.

A network of stout wooden beams supported the roof, and these were supplemented with additional horizontal members. The left scapula, forelimb, ilium, pubis, and hind limb were suspended from these horizontal beams in a life-like position, with cables and ropes. Metal straps were heated and twisted to conform to the shape of the giant ribs. Sections of real bone as well as plaster reconstructions were pinned to the metal straps. Now the ribs of the left side could be positioned. The skeleton lay on its side as though the entire animal had been sliced in half.

Separate support mechanisms were devised for the dorsal and cervical vertebrae. Two metal girders were bolted together to form an I-beam in cross-section. They were heated and bent in a forge in the basement, until they matched the desired curvature of the dorsal vertebral column.

When measurements and adjustments were finalized, the mockup was disassembled. A wooden framework was built to support the curved metal I-beam in an upright position off the floor. Channels were cut into the inner faces of both halves of the centra of the dorsal vertebrae. These grooves corresponded to the flanges of the I-beam that now extended outward laterally. The two halves of each vertebra were slid onto the I-beam and pinned together through the centrum with two metal rods. The flanges of the I-beam and the channels in the vertebrae locked the vertebrae in position. The pins held the two halves of the vertebrae together.

Metal crosses were affixed to the upper edge of the I-beam, to which each of the enormous and heavy ribs would be attached. The arms of each cross were concealed by the transverse processes of the vertebrae. Each plaster component of the backbone was painstakingly fitted and fixed to the rigid central girder that carried the great weight securely. Ribs could now be attached and the angles adjusted to Janensch's satisfaction. When the backbone and ribs were set to an acceptable position, the assembly was dismantled.

A 10-millimeter-thick flat metal bar had been heated and shaped into a gentle S-curve to support the articulated cervical vertebrae. The plaster reconstructions were sawn in half through their long axis. Once affixed to the metal bar, they would form the articulated neck.

A wooden cradle was constructed in the attic, and pelvic elements were suspended from it by cables, heavy chains, and thick ropes. Metal brackets were built into the pubes. In turn, these brackets could be bolted to metal rods that extended outward from the plaster sacrum. The ilia and ischia were also bolted to the sacrum by thick metal straps. Once the best articulation was achieved, the trial assembly was taken apart.

Mounting the tail was difficult, as real bone was used. Anterior or proximal caudals were fitted onto a metal T-bar. Mid-caudals were strung onto a round metal rod that passed through the neural canal. The final posterior or distal caudals rested on a thin metal wire.

The armature that supported the skeleton required a solid platform. A massive, well-braced rectangular metal base was bolted or riveted together as a foundation to which two upright posts would be anchored. Two metal I-beams, each perhaps 15 centimeters across the flange, were obtained from an ironworks supplier. Another short metal bar was bolted across the front upright and riveted in place with heavy metal plates, forming a cross. This pillar would serve as the forward upright load-bearing unit of the final skeletal assembly. It would support the neck, forelimbs, pectoral girdle, and dorsal vertebral column. The rear upright girder would support the massive pelvis, hind limbs, and tail.

Major forelimb, hind limb, and pelvic elements possessed a multitude of natural breaks. According to Janensch, the larger fragments were split or sawn lengthwise, though the method is not obvious either in contemporary descriptions or in examination of the mount. Channels were then chiseled into both halves of a section of bone, and a metal rod laid into them. The sections were held together with metal pins. A fragmentary bone was securely reassembled around its metal support in this manner. There is no photographic record of this method, though a rotating drill bit is shown poised above a section of bone. Perhaps solid lengths of bone were cored or drilled and the rod inserted.

The labor-intensive modeling and cramped mockup stages may have consumed years, but the final assembly proceeded remarkably quickly. All was ready by the summer of 1937. The rectangular metal base was placed in the center of the main-floor atrium, where light from the glass ceiling would supply natural illumination. The two heavy uprights were bolted or riveted to the metal base, like the masts of a ship. Wooden scaffolding was built up around all sides of the metal armature. As the scaffolding rose higher, ladders were attached. Wooden uprights and crossbars were joined by bolts and some wooden beams were slotted, making them adjustable. Block and tackle was hung from the horizontal wooden beams, in order to lift individual bones or bone assemblies into place. Planks spanned the crossbeams to allow Schober and Marquardt to walk around the skeleton as they assembled it. The surrounding scaffolding had to be built higher than the skeletal mount, so one can imagine how carefully the preparators avoided dropping any tools, to avert damage to the skeleton and constant ladder-climbing. By the time the metal bar that was designed to hold the cervical vertebrae was attached, the "snakes and ladders" scaffolding towered 12 meters above the floor.

The massive I-beam designed to support the dorsal vertebrae was bolted to the front and rear upright posts of the rectangular base. One end of the 10-millimeter-thick flat metal bar for the cervicals was bolted to the front upright post and the other end was secured to the ceiling with a metal cable. The armature for the skeleton's neck and trunk was now in place.

One of the first components to be attached to the metal armature was the 1.07-meter-long plaster sacrum. It too must have been sectioned and pinned to the backbone girder and the rear upright post. Ilia, ischia, and pubes were bolted on. Femora were hoisted on pulleys, articulated with the acetabulum, and attached to ironwork emerging from the sacrum. Metal rods embedded in the femora were connected to others that held the lower hindlimb assembly. The metalwork emerging from the distal ends of the tibiae or metatarsals was anchored to the podium.

Cartloads of sectioned dorsal vertebrae were hauled down from the attic and pinned to the I-beam that spanned the upright posts. The hind limbs, pelvis, and dorsal vertebrae were now in place. Flat metal straps were affixed to the front upright post, and curved back on both sides of the armature to the ilia. The straps braced the big double-headed dorsal ribs. The longest of these weighty and fragile elements measured 2.63 meters. Scapulocoracoids, cradled in thick ropes, were hauled up. Long chains and ropes snaked downward, as the shoulder blades had to travel six meters up from the floor. Humeri followed, and then the lower forelimbs.

One by one, the sectioned cervical vertebrae were bolted onto the flat metal bar to form a gracefully articulated neck. At last, the plaster replica of the 77-centimeter-long skull made its ascent. It was articulated with the atlas and axis so that the lateral semicircular canal of the labyrinth was horizontal. Carefully working their way back down, Schober and Marquardt attached cervical ribs. From the start of assembly to this point had taken a scant two months, a testament to the care exercised in the attic preassembly.

The last major effort involved the caudal vertebrae. One by one they were strung onto the T-bar, then the rod, and finally the wire. A thin metal upright supported the tail midway along its length. Janensch noted that the vertebrae had suffered from postdepositional chemical processes. Calcium carbonate had crystallized, causing the centra to appear smaller than normal. It is likely that the metapodials and phalanges were the last bones to be attached.

Minor repairs and touchups were performed before the plaster reconstructions were painted to match the color of the bone. After a thorough inspection, the scaffolding was disassembled, level by level. A wooden platform covered the rectangular metal base. A display case containing the original skull was set in front of the skeleton. For some years, a human skeleton was placed on the wooden platform for scale.

Janensch had searched the scientific literature for interpretations of sauropod poses. He studied illustrations of the mounts of *Diplodocus longus* and *Apatosaurus louisae* at the National Museum in Washington and photographs of Yale's *Camarasaurus lentus* skeleton. His final decision was conservative, but as accurate as the knowledge of the era allowed. He had never traveled to the United States to view mounted sauropods. In light of his examination of the articular surfaces, he gave the forelimbs an elbows-out position. The right forefoot was just lifting slightly off the ground to take a step forward. Similarly, the right hind foot was about to be raised. It was placed closer behind the right forefoot, in a normal two-meter stride. The left forefoot had just been set down.

A titan had been resurrected after 150 million years. It stood 11.87 meters tall and measured 22.65 meters along the neural spines from the snout to the tip of the tail. It was 5.8 meters tall at the first dorsal vertebra, and its 8.78-meter-long neck stretched upward nearly to the ceiling. Compared to those of other sauropods, its tail was short at 7.62 meters. It was the largest skeleton of a land animal mounted anywhere in the world. Yet it was by no means the largest example of its kind. Another *Brachiosaurus brancai* femur in the collections was 13.5 percent longer than that of the mounted specimen.[36]

With swastika banners hanging from the walls as a backdrop, the exciting new exhibit opened in August 1937. A curious public, especially schoolchildren, formed long lines, waiting to see Berlin's latest attraction.

Response from the international paleontological community was gratifying, considering the many years between the excavations and the final display. Barnum Brown, one of the premier fossil collectors in the world, remarked that it was "a perfectly magnificent skeleton of which I am frank to tell you we are envious. Please accept our congratulations on this grand achievement."[37] Similarly, Charles Schuchert of Yale was deeply impressed: "Surely you have the largest mounted dinosaur in all the world. I congratulate you on this wonder of life."[38]

Exciting as this achievement was, it was soon overshadowed. In the latter half of the 1930s, the ominous mixture of an improved standard of living and disturbing coercion and repression continued in Germany. An extensive social welfare program, the Winter Assistance Program (Winter Hilsfwerk), benefited 10 million less-fortunate Germans in 1937. Tax relief, rent rebates, and subsidies for women were so successful that it was difficult to persuade women to enter the work force when a labor shortage began to develop in 1937. In the summer of 1938, the situation worsened as 400,000 workers from all corners of the country began an elaborate series of fortifications known as the Westwall.[39]

A totalitarian state had been created. Denunciations were encouraged in all levels of society. The number of political prisoners increased dramatically. Capital punishment was extended to embrace dozens of violations, compared with a handful before the ascent of Hitler. Already excluded from public life and expelled from schools and public recreation facilities, Jews were next dispossessed and physically harassed. Kristallnacht, in November 1938, when thousands of shops were destroyed and hundreds of synagogues were burned, was but a foretaste of the nightmare to come.

Another sinister development occurred in late 1937, with Hitler's announcement of a plan to incorporate all Germans into a single state. Austria was annexed in the spring of 1938, and in the fall, the Munich Conference ceded the Sudetenland in Czechoslovakia. The rest of the country was invaded in 1939. When Germany invaded Poland in September 1939, war was declared by Britain, France, and the Commonwealth states.

-19-

1939–1976
Destruction and Renewal

A wind had been sown, and the whirlwind that would be reaped would bring misery to the lives of millions around the globe. On January 16, 1940, a participant in the British Expedition to Tendaguru died. Thomas Deacon passed away at age 75 in Waterloo, Liverpool. In a few months, the Battle of Britain would darken the skies over his homeland, and London, Coventry, and Plymouth would sustain serious bomb damage.

At the behest of the trustees, keepers at the BM(NH) prepared plans for the evacuation and protection of the collections in 1938. In January 1939, specimens were moved to Tring, basement rooms at the Kensington facility were strengthened as air raid shelters, and the staff was trained in first aid.

London was bombed by the Luftwaffe starting in August 1940. Although the Museum received no direct hits that first month, windows were shattered by concussion and rain entered the building. Serious damage occurred in the early morning hours of September 9, when two incendiary bombs were dropped through the roof of the east wing. Fire destroyed the roof, some rooms, and part of the herbarium of the Department of Botany. In the intense bombing of the ensuing months, the Museum suffered additional hits and near misses. September 10–11 saw hits on the east colonnade, courtyard, and forecourt. Incendiary bombs started a blaze in the shell and ornithological galleries on October 16. When more bombs struck the courtyard on November 16, Museum authorities were convinced to move more specimens to country houses. After being closed temporarily, the Museum was reopened between 1942 and 1944, and remained a popular wartime attraction.

Another incendiary device struck the roof over a fossil gallery on February 24, 1944. The resulting fire was quickly brought under control. Fortunately, the collections suffered no major damage and Tendaguru material was unharmed. The building was exposed to the elements, however, with serious consequences.

Civil defense measures were organized across Germany by late 1939. In addition to air raid drills, blackouts were enforced and decoys and camouflage nets were strung in cities, including Berlin. No single government agency had authority to divert supplies and manpower to protect sites or objects of cultural significance. It was not until the fourth year of the conflict that preparations were made to safeguard historical treasures. Statues and monuments that could not be moved were shielded with sandbags or brick walls. Most museum, library, and archive directors were left to take whatever measures they could on their own.

The German Tendaguru Expedition lost another member on January 30, 1942. Ina Reck passed away in Berlin at age 69. On May 10, Georg Wetzel, the preparator at the University of Tübingen, passed away at age 58. Born August 25, 1883, he was a carpenter by trade. Joining the Institute and Museum for Geology and Paleontology shortly after the First World War, he learned fossil preparation from Christian Strunz of the Senckenberg Natural History Museum in Frankfurt. Georg Wetzel taught his nephew Wilhelm, who had been trained as a machinist-locksmith. The plaster cast of a *Brachiosaurus* forelimb donated by Berlin had been mounted by the Wetzels some time before 1923.

By 1943, British and American bombers were targeting not only cities supporting the war industry but many German population centers. As homes were destroyed, millions of people were evacuated. By the end of the year, 700,000 had been moved out of Berlin.[1] Those who remained in the cities began lining up outside bomb shelters around dusk to ensure a spot during the nightly raids.

The Württemberg Natural History Collection in Stuttgart had been shielded with additional masonry, and vulnerable collections were relocated to fortified basement rooms. Beginning in 1940, and continuing as often as air raids and available manpower allowed, museum specimens and library materials were packed and crated. They were transported to three other sites within the city and to another 30 castles, salt mines, and schools outside Stuttgart.[2] The main displays were left intact until the fall of 1942, so that the public and injured servicemen could visit. However, by late September 1943, even the displays had been taken down.

Max Böck, a preparator who had mounted sauropod limbs from Tendaguru and Wyoming, was forced to retire in the summer of 1943 at age 66. He suffered from silicosis developed during his career as a stonemason and sculptor. Böck had been drafted in the First World War and served in France. Wounded and partly disabled, he rejoined the Museum in 1918. He was classed as unfit for military duty in the Second World War. Böck was married and had four sons. He came out of retirement to help supervise the packing of thousands of museum specimens. When air raid sirens wailed, he led crews to the basement, where they continued wrapping ammonites and other invertebrates.

On the evening of February 20–21, 1944, 598 RAF bombers heavily damaged the center of Stuttgart. The main structure of the Württemberg Natural History Collection was struck by several bombs, and crumbled.

Another raid by 557 RAF aircraft on March 1–2 again damaged the Museum. A series of three RAF attacks started on July 24–25, 1944, shortly after the Allied landings at Normandy. Between five and six hundred RAF aircraft appeared each night, and the city center was obliterated. Museum buildings suffered minor damage, but collections transported to other facilities within Stuttgart were destroyed.

On September 12–13, 1944, the old center suffered severely under the payloads dropped by over 217 RAF bombers. Buildings on both sides of the Museum were enveloped in flame and the main Museum building was also hit. Staff risked their lives to pull specimens from the basement as wooden ceiling beams collapsed around them. When the fire trucks arrived on the 13th it was too late. Concrete floors and iron doors saved some specimens, but other sections of the building collapsed. Historically valuable mammoth tusks and Holzmaden plesiosaur skeletons were destroyed.

The ruins were sifted until mid-December by Eberhard Fraas's son-in-law, the curator Fritz Berckhemer, along with Max Böck and prisoners of war. It is not known whether Tendaguru bones had been evacuated, but they were not damaged in the war. Eberhard Fraas's house was flattened during these raids. All photographs and notes of his 1907 trek to Tendaguru were lost.

As Germany's second largest city, the continent's largest port, and a center for U-boat construction, Hamburg was the target of heavy bombing in the summer of 1943. In the space of 10 days between July 24–25 and August 2–3, about 9,100 tonnes of bombs were dropped by an average of over seven hundred aircraft per raid. An intense firestorm developed in residential areas, prompting two-thirds of the population of 1.75 million to temporarily flee the city. Portions of the collections at the State Geological Institute were relocated to the basement of a nearby school. On July 27, the institute burned to the ground, and its entire contents were lost. The adjoining school and its contents were destroyed as well. None of the Tendaguru sauropod, stegosaur, or ornithopod material survived the devastation.

RAF raids were launched on Frankfurt in October and December 1943 and again in March and September 1944. Cultural buildings in the city center suffered irreparable damage, but the *Brachiosaurus* forelimb, scapula, and rib at the Senckenberg Natural History Museum survived unscathed. Tübingen escaped aerial bombardment, which was fortunate, as there was no facility to which the collections could be relocated.

On October 14–15, 1944, 240 RAF bombers struck Braunschweig, destroying 150 hectares of the historic city center. It is not known whether the centrally located Polytechnikum had received Tendaguru material, as Expedition patron Johann Albrecht had requested, but the institute was 70 percent destroyed. A large unidentified dinosaur humerus still exists in the collections of the State Natural History Museum.

A bomber attack on Munich on April 24–25, 1944 reduced up to 80 percent of the collections and correspondence of the Bavarian State Collection for Paleontology and Historical Geology to ashes. Fortunately, many type specimens had been relocated. A *Dysalotosaurus lettow-vorbecki* skull (catalogue number AS 834) survived, though it is not known how many other African specimens out of the 23 crates received in 1933 disappeared.

Sometime in the first two weeks of April 1945, the United States Army was operating in the Göttingen area. Artillery fire struck the Geological-Paleontological Institute and Museum of the University of Göttingen, with catastrophic results. An independent museum established at Berliner Strasse in 1877, it housed valuable collections dating back to 1773. Fire broke out in the building and spread rapidly, gutting much of the institute. As floors and ceilings collapsed, a jumbled layer of rubble and fossils formed that was one meter deep. So widespread was the destruction that no one could determine whether Tendaguru material sent from Berlin in 1933 had survived.

Departmental curators at the Museum of Natural History in Berlin were responsible for safeguarding their respective collections. In January 1941, some specimens were moved into the basement and a bank vault was leased to house 40,000 bird skins, zoological type specimens, minerals, and the contents of the Willdenow Herbarium.[3] Fritz Marquardt, formerly a preparator at the Paleontological Institute, was serving in the army.

Invertebrate fossils were moved into tunnels in the Alvensleben limestone quarry at Rüdersdorf outside Berlin in the summer of 1943. Only a few vertebrate fossils were safeguarded there. The paleontology library was relocated to the exhibit halls on the main floor of the Museum in July.

Berlin was a priority target. Not only was it the seat of the Nazi government, it also housed wartime industries that produced aircraft engines, submarine motors, and torpedoes. As the nation's capital, Berlin was invested with the greatest array of anti-aircraft weapons and the most night fighter squadrons of any German city.

On the night of September 3–4, 1943, 315 RAF Lancaster bombers dropped their loads over the city. Some damage was sustained at the Museum of Natural History, possibly to the glass skylight over the main atrium. This incident presumably galvanized the staff into action. The preparators Siegert and Schober partly dismantled the skeleton of Brachiosaurus and carried all elements but the replica skull and plaster vertebral column into the basement. The remaining Tendaguru dinosaurs were also pulled down at this time.

From November 1943, British Bomber Command was given free rein over Germany for 4½ months, and Berlin alone would receive sixteen major attacks in what became known as Main Battle—Berlin. The Museum's paleontology section suffered damage in attacks by RAF Mosquito bombers on November 11–12 and 17–18.

On the night of November 22–23, 690 Allied aircraft dropped 2,270 tonnes of high explosive and incendiary bombs in about 40 minutes. An estimated 175,000 people lost their homes, and about 2,000 were killed. Smoke from widespread fires towered six thousand meters into the sky.[4] The Museum of Natural History suffered a direct hit on the west wing. Fire broke out on the second floor and destroyed the zoological teaching collection. A large lecture hall, a lecture preparation room, and the reptile hall were lost. Only iron doors prevented the fire from spreading to the east wing. Members of the Zoological Institute participated in the desperate fire-fighting efforts.

Allied bombers had destroyed transportation facilities, industrial plants, and as many as 11 million dwellings across Germany by the beginning of 1944.[5] Hoarding, black marketeering, and smuggling became common despite draconian penalties.

A graphic description of the conditions at the Museum of Natural History during this period is provided by Jurij "George" Jeletzky, a specialist in fossil invertebrates:

> many individual panes of the glass roof were shattered by concussion from nearby explosions of bombs. Puddles of water and patches of snow covered parts of the floor, fossil cabinets and some of the vertebrate skeletons, including that of <u>Brachiosaurus</u>. I vividly remember how forlornly its towering rumpf [body] and long neck looked with patches of snow on them and against the background of shattered window panes in the glass roof above. . . . Everybody, including myself, worked in heavy winter clothing and gloves because the temperature inside the domed hall of the building was the same as outside (i.e. around zero Celsius). It was not much warmer elsewhere in the building, including the personal offices even though their window panes were either intact or replaced by cardboard or plywood.[6]

On February 3, 1945, 937 B-17 bombers of the United States Army Air Force's Eighth Air Force crossed the city in a daylight raid. An estimated 2,060 tonnes of mixed ordnance was dropped. About 3,000 Berliners perished and 120,000 lost their homes.[7] The east wing of the Museum of Natural History was hit, and blast and fire totally wrecked the whale hall. These enormous skeletons had filled the glass-ceilinged atrium for decades before Tendaguru dinosaurs replaced them. The zoological library, the entomological collection, the anatomical collection, and the preparation rooms were consumed by fire.

RAF Mosquitoes attacked again on March 17–18 and 18–19, 1945. On March 18, the U.S. Eighth Air Force sent 1,221 planes to Berlin and dropped 2,810 tonnes of high explosives and incendiaries.[8] It is likely that this raid demolished mineralogical and paleontological collections housed in the top floor of the western wing of the Museum of Natural History. These actions decimated the type specimens of the Tendaguru dinosaur *Dysalotosaurus lettow-vorbecki.*

Germany's transportation system was in chaos, and essential services in cities such as Berlin were operating at low levels. Mounds of uncollected garbage and rubble littered streets and the supply of gas, water, and electricity was intermittent.

From the middle of February 1945, the bombing of Berlin continued almost without letup for 30 days and nights. The night of April 20, 1945, marked the end of the aerial bombardment by British and American forces. The RAF and USAAF had dropped approximately 64,545 tonnes of bombs on Berlin between 1944 and 1945.[9] An estimated one-third of the city's built-up region was devastated and perhaps 50,000 civilians died.[10] Now the Soviet Air Force pounded the city as the final land assault brought the capital to its knees.

On April 16, the final Russian assault on Berlin began. The city was encircled by the 25th and many of the defending forces were trapped or badly mauled. Shelling of Berlin by artillery and rockets was relentless after April 24. The civilian population, now numbering about half of the prewar total of 4.2 million, was forced into hiding in bomb shelters and the cellars of ruined buildings.

Fires burned unattended throughout the city, as it was too dangerous to go out to fight them.

A massive bombardment opened for an hour each morning and continued at a lesser rate throughout the day and night. Soviet general Berzarin would later report that 32,700 to 36,300 tonnes of shells were fired in a two-week period.[11] Fighting moved from house to house, and buildings were demolished by rockets and flame-throwers to clear out defenders and snipers.

On the afternoon of May 2, 1945, Berlin surrendered. Artillery shells struck the top floor of the south and east wings of the Museum of Natural History on May 2 and May 3. The roofs burned through, and the contents of the rooms beneath were consumed by the flames. Mineralogical and paleontological collections were lost. Galleries on the second floor were badly damaged in the south wing. The situation was dire, as water flowed down to the main floor after every rain.

Frankfurt am Main was captured on March 21, 1945. Stuttgart was occupied by French forces some time after April 19, and Tübingen fell around the same time. The Americans took Munich on April 30.

Within a week, hostilities ceased across most of Europe. The tally of dead reached 30 to 40 million, with many millions more wounded and millions more missing. Cities were reduced to rubble. The scale of human dislocation was incomprehensible. Streaming across Western Europe in every direction were about seven million displaced persons—forced laborers and prisoners searching for their homes and relatives. A similar number were moving around Central and Eastern Europe. Adding to this mayhem in the following weeks were about eight million German soldiers released by the Western Allies and another two million who had been Soviet prisoners. By late summer 1945, as the waves of displaced persons abated, they were replaced by a tide of German refugees. Some 12 million ethnic Germans who had lived in areas now under Soviet control were expelled as a result of the Potsdam Agreement of July 1945.[12] The Allies occupied the entire country, split it into four zones, and established a divided Berlin as the seat of government. The Soviet Union occupied Berlin until the arrival of the Western Allies in July 1945. In this two-month period, approximately 80 percent of the intact industrial equipment was dismantled and transported back to the Soviet Union as part of the $10 billion in reparation demanded by the Russians.[13]

Over 55 million cubic meters of rubble filled the streets of Berlin.[14] The food supply was so poor that the average daily intake stood at eight hundred calories for the first months.[15] Thousands of women cleared debris from the streets, but there was little paid employment, since the economy was at a standstill. As winter approached, there were still severe shortages of housing, food, fuel, and medicine. A quarter million expellees began to arrive each month from Eastern Europe.

The Museum of Natural History, the university, and the Prussian Academy of Sciences were all located in the Russian zone, and the Soviets appointed two curators of zoology to organize the cleanup and repair work at the Museum. Volunteers stepped forward on May 11 and 12, barely a week after Berlin's surrender. By mid-May, a horse-drawn wagon had been obtained to retrieve the zoological and botanical specimens from the bank vault where they had been deposited four years earlier. The Museum of Natural History was the first museum to reopen in the city, on September 10, 1945.

The Paleontological Institute was slower in reviving its activities, due to lack of staff. Some employees were prisoners of war, others were located in distant parts of the country as a result of evacuation or military service. Those who had been members of the National Socialist Party were dismissed by the Soviets. As of June, the institute consisted of its director, Stille, Janensch and Dietrich as curators, Schoenenberg as an assistant, and the preparators Siegert and Schober. Marquardt was a prisoner of war, and upon his return to Berlin was prohibited from rejoining the Museum.

The condition of the building was poor. Rain and melting snow ran into a dishearteningly large number of rooms. The central steam heating was long out of service so stoves were obtained, with great difficulty. Little wood or coal was available. The collection of invertebrate fossils that had been stored in the limestone quarries at Rüdersdorf was discovered by the Russians and taken to the Soviet Union. It would not be returned until 1958. The first difficult postwar year came to a tragic close for the Paleontological Institute when Hans (Johannes) Schober, the preparator who had mounted four Tendaguru dinosaurs, succumbed to pneumonia in December at the age of 59.

Stuttgart preparator Max Böck passed away on August 6, 1945, at age 68. Of his four sons, two had died in bomb attacks on Augsburg in 1945, and a third was missing at the end of the war. His health had improved in the summer of 1945 until he felt strong enough to plan a return to limited duties in Stuttgart, when double pneumonia abruptly ended his 40-year career at the Museum.

In Tübingen, Edwin Hennig was dismissed from his post at the university in October 1945. The French occupation authorities took this step as part of a countrywide campaign to remove former Nazi Party members from public office. Though Hennig had not joined the party, his pro-regime stance had cost him his position.

In the last months of the war, French occupation forces had commandeered several buildings in the Stuttgart Museum complex that were still habitable. The most important goal for Museum officials was to reassemble the collections that had been evacuated between 1941 and 1944. Portions stored in both American and French zones had suffered to varying degrees. The contents of the biology gallery were smashed by French soldiers, a bird collection disappeared, tusks were removed from elephant skulls, minerals and fossils stored in a tunnel were smashed, and a Steinheim mammoth skeleton was broken into fragments.[16]

In September 1945, a French commission led by paleontologist Camille Arambourg selected all but a few pieces of the surviving collection of Holzmaden fossil slabs. Carpenters crated them for transport to Paris. Alerted at the last moment, Berckhemer, the senior conservator, intervened, and all the priceless fossils were returned to Stuttgart in the fall of 1947.

Some Stuttgart specimens were stored in a salt mine at Kochendorf, including the Steinheim Neandertal skull and anthropoid fossils collected by Fraas and

Markgraf in the Fayum of Egypt. During the fighting, the mine works were badly damaged and the water pumps were now barely functioning. American forces were keenly aware of the value of the three hundred crates threatened by rising waters and assisted in the rescue. By 1947, those crates containing museum specimens had been returned to Stuttgart.

In the face of great odds, professors at the Paleontological Institute of the Museum of Natural History resumed limited teaching duties with the reopening of Berlin University in January 1946. It was renamed Humboldt University.

The black market became the sole form of economic activity as the reichsmark lost virtually all value. Barter became the exchange mechanism, with cigarettes, coffee, and butter as currency.

In London, the BM(NH) was in poor condition. Broken windows and a hastily repaired roof allowed heavy rain to enter. Exhibits had been damaged, and staff levels were low. As workers returned from war service, the evacuated collections were moved back during 1946.[17]

Early in the spring of 1947, the promoter of Tendaguru, Charles William Hobley, died. He passed away at his home in Oxted, Surrey, on March 31, at age 79. In honor of his contributions to Kenya, one of the three main valleys on the eastern slope of Mount Kenya was named Hobley Valley. He was predeceased by his wife, and was survived by a son and a daughter.

Barely four months later, another individual connected with the British effort at Tendaguru died. Geologist John Parkinson's long and productive career ended at his London home on July 19, 1947. He was 75 years old. Parkinson had continued his wandering for many years after his tenure at Tendaguru. At the start of the Second World War, he had taught geology in Kenya. When he was nearly 70, he undertook a geological survey of the Tsavo area for the Mining and Geology Department. This rigorous fieldwork broke his health and he returned to Great Britain. He was survived by his wife and son. John Parkinson's career spanned 40 years and produced 66 scientific papers. W. E. Swinton offered a personal view of Parkinson:

> he was constantly travelling, always finding pleasure in it.
>   . . . all his life he was intensely attracted to authorship.
>   . . . There are many at home and abroad who will miss him, whether in the field or at the Athenaeum [Club], for he was a good companion, knowledgeable, and with an enthusiasm that would have well become many half his age.[18]

In Berlin, a new preparator was hired in 1947 to replace Hans Schober. Günter Neubauer was born in Berlin and trained first as a machinist, then as a preparator in geology and paleontology. His father had prepared fossils for the State Geological Institute, next to the Museum of Natural History.

There had been no scientific publications on Tendaguru since 1939. In 1947, *Palaeontographica* printed a study by Janensch on pneumaticity in sauropod bones. It had been submitted in 1943, but in early 1944, the publisher announced that the plates and figures had been destroyed when a phototype firm in Leipzig was bombed. By the summer of 1944, Janensch had resubmitted this first paper, and another paper on the vertebral column of *Brachiosaurus bran-*

cai. All the materials survived the destruction of one of the publisher's buildings in Stuttgart. Then both papers were lost when the printing firm was bombed. Replacements were again produced, and the pneumaticity paper finally appeared in 1947.

After the war, the Stuttgart publisher recalled the plates and figures for the *Brachiosaurus* paper from the wrecked Leipzig phototype firm. The division of Germany made it impossible to send these documents from Leipzig in the Soviet-occupied zone to Stuttgart in the Allied-occupied zone. Another company in the American Zone was commissioned to produce the plates, and the *Brachiosaurus* paper was finally published in 1950.

In the fall of 1947, steps were underway to mount another stegosaur skeleton from Tendaguru. Duplicate bones had been shipped from Berlin to Tübingen many years previously and now missing elements were modeled in plaster. Friedrich von Huene directed the efforts of the preparator Wilhelm Wetzel. By December 1947, the vertebral column was nearly complete, a mix of original and modeled elements. Von Huene used Hennig's illustrations as an example, but had difficulty restoring the caudal vertebrae from the 26th caudal to the final vertebra. Janensch sent sketches, von Huene referred to the Gilmore *Stegosaurus* monograph, and the tail was completed by mid-January 1948. A shortage of plaster caused the next delay. One hand, one foot, several ribs, a pubis, hemapophyses, and some spines were among the unfinished elements. Von Huene predicted that several more months would pass before the mount was complete, so one may assume the skeleton was on display later in 1948.

Following approval of the American Marshall Plan in March 1948, billions of dollars in machinery, food, and capital began to flow into Europe. Factories were rebuilt and workers were hired. A new currency, the deutschemark, was introduced into western Germany by the Western Allies around mid-June 1948. These measures were perceived by the Soviet Union as flagrant breaches of the Four Power Agreement undertaken at Potsdam. The Russians responded by introducing their own currency in the east zone of Berlin, the ostmark. When the Americans distributed deutschemarks in the western zones of Berlin, the Soviets attempted to force the Western powers out of Berlin by blockading the city.

Beginning June 24, 1948, all traffic into and out of the city was halted. Electricity, supplied predominantly from power plants in East Berlin, was no longer transmitted to the western zones. Coal shipments from the east to the west halted. The 2½ million people in the western sectors required roughly 11,000 tonnes of supplies daily, three-quarters of which were imported.[19] The Allies embarked on an ambitious plan to supply their sectors by air, which became the famous Berlin Airlift. In January 1949, it was obvious that the airlift was succeeding, and the Soviets lifted the blockade on May 12.

Currency reform, so instrumental in economic recovery, brought difficulties in divided Berlin. The Soviet sector refused to recognize the deutschemark and paid all salaries and wages for workers in the east in ostmarks. Anyone residing in the west but working in the east found it difficult to live on ostmarks, as this currency rapidly assumed a far lower value than the deutschemark. Werner Janensch and other Museum employees were caught in this onerous situation. In the spring of 1949, the Western governments declared the deutschemark the sole legal tender in their sectors.

The Western powers unified their zones and the Federal Republic of Germany came into being on May 23, 1949. The Soviet Union felt equally strongly that Germany must never present a threat to its security again, and to frustrate any reconstitution, created the German Democratic Republic on October 7, 1949. East Berlin was declared the capital city of the GDR and Bonn that of the FRG.

All repair work on the Paleontological Institute of the Museum of Natural History was organized by members of the Institute. Rebuilding proceeded slowly, due to the scarcity of money, labor, and building materials. Once the roof over the south wing was fixed in 1949, the rooms on the third floor could be rebuilt. A new roof was also installed over the east wing, replacing the emergency covering.

By 1950, all damaged roofs were finally replaced at the Museum, and work could proceed on the interior. The third floor of the east wing was a priority. The glass skylight in the large display atrium that formerly housed the Tendaguru dinosaurs was replaced. Burned-out rooms on the second and third floors of the east wing were finally repaired in 1951. Brighter workspaces were created on the third floor of the south wing, allowing the geological and paleontological preparators to emerge from their basement quarters.

Fritz Marquardt was reinstated in 1949, possibly at his previous rank of institutional aide. After 49 years of service, Werner Janensch retired in 1950 at the age of 72. Dr. Walter Gross became the head of the Paleontological Institute and Museum a year later.

The Stuttgart museum was renamed the State Museum of Natural History of Stuttgart (Staatliches Museum für Naturkunde Stuttgart) in 1950. It was so severely damaged that its operations were relocated to Schloß Rosenstein. This building too was badly hit, but would be repaired as a historical site.

In Göttingen, it was not until 1950 that rooms in the Geological-Paleontological Institute of the university could be reopened. The devastated ruins remained without a roof until 1948, by which time plant life sprouted from them. Student volunteers undertook the immense task of sifting through the rubble to retrieve any of the valuable fossil collections. In a letter to Janensch dated March 1950, Professor Hermann Schmidt sent the heartening news that 30 crates of *Dysalotosaurus* material had survived, and that he was anxious to begin preparation.

In the summer of 1952, the life of another loyal participant at Tendaguru ended. On July 8, Frederick William Hugh Migeod died at home in Worthing in his 80th year. Active in politics, he had been elected to local seats in Worthing in the 1930s and 1940s. Migeod's final years were not happy. His wife, Madeleine, had predeceased him in 1933. His health was poor, which forced an uncharacteristic inactivity. "Fortune seldom shone on him. His married life was happy but brief, and his wife died many years ago. For the last year or two he had been ill and much alone." His obituarist, William E. Swinton, also hinted at the unsatisfying legacy of his Tendaguru years:

On the whole he had little success in this, and it is paradoxical that through the nature of the work he came into the public eye much more than he ever did through his able and successful studies in folk-lore and language. It is only fair to say that Migeod in these later years did not claim much for his East African adventure.[20]

It was a sad end, as Migeod's life had been a productive one. The eleven books he had written were evidence of his desire for knowledge about the world.

The Museum of Natural History in Berlin witnessed a significant success in 1952. Weather-damaged walls and ceilings in the collection rooms of the Geological-Paleontological Institute were given fresh coats of plaster and paint. Hundreds of new specimen labels were handwritten, a tedious but essential job that took years. In the atrium ceiling, beams and elaborate moldings were refurbished and painted. The stage was set for the triumphant return of the Tendaguru dinosaurs.

In the fall of 1952, Günter Neubauer pulled hundreds of bones from basement storerooms where they had lain for a decade. The precious *Brachiosaurus* skull and *Archaeopteryx* slab had already been recovered from their places of concealment in the basement. They had suffered little harm in all that time. However, the plaster skull and cervical and dorsal vertebrae of *Brachiosaurus* had stood beneath practically open skies and the effects of rain and snow were marked. Neubauer repaired and reassembled *Brachiosaurus*. Fritz Marquardt lent an experienced hand with the heaviest bones. *Dicraeosaurus, Elaphrosaurus, Kentrurosaurus,* and *Brachiosaurus* were soon standing again. The Carnegie *Diplodocus* was cleaned and damaged pieces were restored. Holzmaden fossil slabs were repositioned. With immense pride, the gallery was opened to the public in the spring of 1953, after 10 of the blackest years in German history.

Werner Janensch doggedly dealt with the tension and uncertainty of the times to continue his life's work. In 1955, he published more monographs in *Palaeontographica,* including detailed descriptions of the small ornithopod *Dysalotosaurus lettow-vorbecki.* He had begun formally describing the dinosaur in 1952, but the figures were delayed because the scientific illustrator, who lived outside of Berlin, was hampered by newly imposed travel restrictions.

Janensch had assumed responsibility for these specimens upon the death of Pompeckj, the Museum's director, in 1930. Pompeckj had initiated work on the blocks during the First World War and had continued until his unexpected early death. At that point, Janensch was fully engaged with the overwhelming mass of sauropod remains. The type material of the small bipedal dinosaur had been destroyed in the Second World War, so Janensch had Siegert and Neubauer work up more jackets that were stored in the basement. It took surprisingly little time for the skilled technicians to ready enough skeletal elements for description, including a skull. Neubauer was promoted to senior preparator in 1955.

In 1958, amidst growing political tension in Berlin, the fifth Tendaguru dinosaur skeleton was mounted for display at the Museum of Natural History. The ornithopod *Dysalotosaurus lettow-vorbecki* had been formally established in 1920 by Josef F. Pompeckj, though Virchow had published the name a year earlier. Much of the material originated from the Middle Saurian Bed at Kindope, about three kilometers northwest of Tendaguru. Most of the bones were disarticulated, although some short sections of articulated vertebrae were uncovered. dyI comprised 20 presacral vertebrae, six sacrals, elements of the pectoral girdle, the sacrum, and forelimbs. dyII consisted of nine presacrals, six sacrals, and eight caudals, as well as chevrons and the left ilium. dyIII was composed of two presacral and six sacral vertebrae. dyIV was an assemblage of caudal vertebrae, sacral elements, and hindlimb bones. dyV included three sacral and six anterior caudal vertebrae, chevrons, right and left ilia, and bones of both hind limbs. dyVI was a right hind limb.[21]

Skulls A and B were newly prepared after the Second World War, and Janensch featured them in his monograph. Skulls C, D, E, and F had been prepared and illustrated for Pompeckj's papers, but his work had never been published. Numerous isolated *Dysalotosaurus* teeth had been found in the deposit, which also included a few stegosaur and subadult sauropod bones.

Sufficient skeletal elements had been prepared by Ewald Siegert and Günter Neubauer prior to Siegert's retirement in 1958. Neubauer assembled similarly sized bones to form a composite mount. Specimen dyI supplied the greatest articulated segment, from the second to the 21st presacral vertebra. The sacrum came from dyIII, and the first five caudals came from an articulated series. It was estimated that there were 48 caudals in total, and the remaining 43 were assembled from a variety of sources, including dyII. In all, 27 vertebrae were chosen from original isolated bones, and the remainder were modeled in plaster and inserted among the real bones. The last 12 tail vertebrae were also modeled as a unit. Hemapophyses were modeled after those of dyII. Only the chevrons of the first caudal vertebrae were original and were articulated with the vertebrae. Janensch concluded that the vertebral formula was 8 cervicals, 15 dorsals, 6 sacrals, including a dorso-and caudo-sacral, and an unknown number of caudals, possibly 48.[22]

Posterior skull elements from dyC were used in the mount. The skull was completed by employing plaster replicas of bones from other skulls. Only one right cervical rib was original bone, but eight left and eight right dorsal ribs were. Skeleton dyI also provided scapulae, coracoids, sternal bones, humeri, the left ulna, and the left radius. The right ulna and radius were modeled. Hand elements were very rare and ultimately were sculpted according to those of *Camptosaurus*. The same individual, dyI, supplied both ilia, the pubes, and the left ischium, along with both femora. Isolated original tibiae were selected from the collections, while fibulae were modeled after other original isolated bones. The four-toed hind feet were assembled from bones belonging to dyVI and dyV, and from isolated real bones. Tarsals and metatarsals were built up from a combination of isolated originals and plaster models.

Janensch made a number of assumptions, such as the number of tail vertebrae and the final length of the vertebral column. The latter problem arose from uncertainty over the thickness of intervertebral discs. Neubauer assembled the mount to portray a gracile animal in rapid motion. Measured in a line from head to tail, it was 2.28 meters long and stood 1.13 meters tall.[23]

A wide range of sizes of *Dysalotosaurus* was present in Ig (WJ). Some individuals were one-third larger than the mounted skeleton, while others were only two-fifths as large, according to femur length. Janensch's last paper, which described this mount, appeared in September 1961, so this display was among the last Tendaguru assignments completed by Günter Neubauer.

Ewald Siegert, one of the long-lived original technicians, who had prepared and mounted four Tendaguru dinosaurs, retired on December 31, 1958 at the age of 75. He had been associated with the museum for 47 years.

Edwin Hennig's fortunes were improving. The censure against him was lifted by the French on December 31, 1950, and he was named Emeritus Professor of Geology and Paleontology at the University of Tübingen. A Swiss publisher, Albert Müller Verlag, printed Hennig's second popular book about Tendaguru, *Gewesene Welten: Auf Saurierjagd im ostafrikanischen Busch,* or "Past worlds: Hunting dinosaurs in the East African bush." This 1955 volume updated the story of the German Tendaguru Expedition of 1909–1913 and recounted Hennig's 1934 odyssey through Tanganyika. It was an engaging story told from the perspective of an older man.

In the late 1950s, the painstaking efforts at Göttingen had advanced to the point where *Dysalotosaurus* remains from Tendaguru could be displayed. Professor Schmidt had the bones "reconstructed and displayed" in the newly repaired facility at Berliner Strasse 28.[24] It is not clear whether isolated skeletal elements were laid out or whether a skeleton was articulated in some form.

Dissatisfaction and disillusionment in the GDR had grown to the point where 130,000 to 330,000 people had fled the republic every year since 1949.[25] West Berlin was an escape route, as one had only to walk or ride the subway across a boundary zone to leave the Communist world behind. By 1961, an estimated 2.7 million had fled the German Democratic Republic.[26] The GDR was rapidly losing the youngest and most highly skilled members of its population, people desperately needed to rebuild the state. In July 1961, 30,000 escaped via West Berlin. Signs became ominous—60,000 East Berliners working in the West were told to find alternate employment. It became a crime to flee the GDR, and border patrols were increased.

The crisis came to a forceful head. On August 13, 1961, soldiers and armored cars of the German Democratic Republic, supported by Soviet troops, moved swiftly to seal the entire length of the boundary dividing East and West Berlin with barbed wire. In the coming days, subway, bus, road, rail, and postal links were cut. Barriers were placed across the 12 remaining border-crossing points. Houses facing the boundary were emptied of their residents and the windows bricked up. Soon a reinforced concrete wall was built along the entire length of the border between the western and eastern zones. Over the years, the full extent of the border between the German Democratic Republic and the Federal Republic of Germany was fortified.

The Berlin Wall isolated the Museum of Natural History in East Berlin. The lives of Werner Janensch, Günter Neubauer, and Fritz Marquardt were affected. Though Janensch was long retired, he could no longer conveniently visit the site of a lifetime of dedication. Both preparators were compelled to leave

Humboldt University on September 30, 1961. Neubauer joined the Technical University of Berlin (Technische Universität Berlin) in the field of electron microscopy. Marquardt transferred to a position in Wiesbaden, West Germany, which involved micropaleontology, in 1962. On October 10, 1963, he died in Wiesbaden in an industrial accident.

One of the preparators of Tendaguru fossils in London, Louis Parsons, passed away on June 10, 1964, at age 75. Parsons had been chosen to further prepare the London *Archaeopteryx* in the early 1930s. One year prior to his retirement in 1954, he had been named a member of the Order of the British Empire. Parsons had continued working part-time at London's Natural History Museum from 1954 to 1964, for a total of 46 years of service in the Department of Geology. He was survived by his wife and at least one child.

Silence had reigned at Tendaguru for decades. In 1951, the Geological Survey of Tanganyika undertook geological mapping of Mesozoic and Tertiary outcrops along the coastline of the country. William George Aitken, a geologist with the Survey, worked northeast of Tendaguru from 1951 to 1954. Aitken's first visit was to the Mandawa-Mahokondo anticline in Kilwa District, a geological feature that Hennig had described in 1914 and again in 1935 and 1937. He wished to determine whether gypsum deposits were of economic value.

Aerial photographs were now available. After studying them, Aitken returned to the region, traversing the area around the Makangaga-Ruawa anticline and collecting fossils for a refined stratigraphic correlation of the geological horizons that had been vigorously disputed.

Aitken's review of stratigraphical questions culminated in a major monograph published in 1961. Several of Hennig's assertions were revised. The opinions of Schuchert, Spath, Parkinson, and Dietrich were supported, namely that the Upper Saurian Bed was entirely of Jurassic age. Hennig had maintained that the Jurassic-Cretaceous boundary was to be found within it. Aitken also contended that there was a faunal break between two *Trigonia* layers—Hennig's *Trigonia smeei* Stage of the Upper Saurian Bed and the overlying *Trigonia schwarzi* Stage. Hennig's view was that there was an unbroken faunal sequence.

The British geologist went further, collapsing Hennig's Upper and Lower Septarian Marls into one, and recommending that the Pre-*Smeei* Layer be abandoned. He also revised the Dietrich-Hennig correlation of the Septarian Marl. The German authors had linked its presence in both anticlinal structures with the Middle Saurian Bed at Tendaguru. According to Aitken, the Mandawa-Mahokondo *Nerinella* Bed, being Kimmeridgian, was older than the *Nerinea* Bed at Tendaguru, which was Tithonian. In 1956, paleontologist W. J. Arkell considered the Septarian Marls to be Lower Kimmeridgian, and the *Trigonia smeei* Stage Upper Kimmeridgian, but the implication was unchanged: the two sites could not be correlated as coeval. Finally, Aitken saw no evidence of faulting to the south of the Niongala Trough. A fault trough did not exist, but some tectonic activity had taken place. There was no response from the German researchers, most of whom were now of advanced years.

The potential for hydrocarbons along the coast prompted British Petroleum and Shell International Oil to launch a joint exploration project. Aeromagnetic surveys were flown, a gravity survey was completed, and seismic reflection and refraction surveys were performed from the mid-1950s to the mid-1960s. Fifty-two exploratory wellbores were drilled to establish stratigraphy or for seismic purposes. Among them were several in the Mandawa Basin (Mandawa No. 7) and others at Pindiro, Keranjiranji, Lindi, and Tendaguru. Dinosaur bone was identified in the Narunyu borehole south of Lindi.

In the 16 years after the erection of the Berlin Wall, little or no research was undertaken on Tendaguru. The death and isolation of the Expedition partici-pants and the death or dismissal of the last preparators linked to the Museum meant that there was no one left at the Berlin institution to continue the work. Perhaps there was a perception that there was little else to accomplish with the collections. The British Museum (Natural History) had prepared material, but no one ever described what was collected.

Great changes had taken place in the country where the finds were made. On May 1, 1961, Julius Nyerere was elected head of state of Tanganyika, and on December 9 of that year, the country celebrated its independence as the Repub-lic of Tanzania, with Nyerere as president. A revolution took place on the island of Zanzibar on January 12, 1964, and in April, Zanzibar and Tanzania united.

In 1965, a collection of ostracods made by the Tanganyika Geological Survey prompted a return to the field. Raymond Bate, Mike Howarth, and Noel Morris of the Natural History Museum in London joined Survey members in an expe-dition to three areas, including the Mandawa Anticline. They spent two months working in Jurassic and Cretaceous sediments along the coast, and made a one-day trip to Tendaguru. Morris investigated brachiopods and bivalves of the *Trigonia* Beds, while ostracods were collected at the anticline to the north. Re-sults were published in 1975.

The British Museum East Africa Expedition had been directed westward to Nyasaland in 1930–1931. Sauropod remains had been excavated, but the speci-mens were never described. A joint group of North American and Malawian paleontologists would return in the 1980s, but in the interim, and quite by acci-dent, a Tendaguru-like dinosaur assemblage was discovered in Southern Rhode-sia (now Zimbabwe) about 1,100 kilometers to the southwest of Tendaguru.

In 1965, Mike Bingham, a glossinologist in the Rhodesian Ministry of Agri-culture, participated in the survey of a fence used to control the movement of game. He was working in the Zambezi River valley, 230 kilometers north of Sal-isbury (now Harare). The Kadzi River runs north to south, intersecting the Musengezi Tsetse Control Fence, which runs east to west. Bingham was operat-ing near the fence, between a road and the Kadzi River, when he found bone fragments, including a vertebral centrum measuring 30 centimeters in diameter. Professor Geoffrey Bond of the Department of Geology, University College of Rhodesia in Salisbury, examined the find and published a brief note.

On June 1, 1967, 13 students from the Mennell Society of the University Col-lege of Rhodesia produced a geological map of the area. They were joined by Mike Raath, the acting curator of the Queen Victoria Museum, Salisbury, and an assistant. Over six days, the students traversed an area of 78 square kilome-ters in the summer heat. Raath and his crew excavated two major and a few minor bone-producing localities. The badly weathered remains of sauropod dinosaurs amounted to several articulated caudal vertebrae, a nearly complete humerus, and other fragments. A rough date of Upper Jurassic to Lower Cretaceous could be assigned to the sediments.

Additional specimens were collected between 1967 and 1970, and the site was revisited by experts in 1970 and 1972. Large bones were found 24 kilometers west of the Kadzi River, near the Ambi River, in December 1970, again by members of the Mennell Society.

Rhodesia unilaterally declared its independence from Great Britain in 1965, as the outcome of the dispute over the introduction of majority rule. Britain placed an embargo on trade with Ian Smith's government in 1966, and in that same year black African nationalist groups attacked. This was but a harbinger of years of bitter guerilla fighting between 1972 and 1979, which became known as the Bush War. Fieldwork in the isolated Kadzi area was too dangerous during the campaign. A cease-fire and disarmament were negotiated, and in March 1980, elections were held. The outcome was an independent nation called Zimbabwe under Robert Mugabe as prime minister.

Raath, now working in South Africa, was able to visit the Kadzi River site again in 1985. By the late 1980s, he and American sauropod expert John S. McIntosh had identified Kadzi fossils as those of *Barosaurus, Brachiosaurus, Dicraeosaurus,* and *Tornieria,* all genera present at Tendaguru. Another dinosaur likely represented at the site was *Camarasaurus.* Two theropod femora, possibly allosaurid, were found as well. Elements included vertebrae, femora, humeri, fibulae, sacral bones, a scapula, a coracoid, and rib fragments. The finds are stored at the National Museum of Natural History in Bulawayo and the geology department of the University of Zimbabwe in Harare. The coeval presence of *Tornieria* and *Brachiosaurus* implied an Upper Jurassic age for the Kadzi site.

A less-well-known dinosaur fauna had been uncovered even before Kadzi, about three hundred kilometers to the southwest, at Gokwe, Rhodesia. In 1961, K. Bromley found fragmentary dinosaur remains. Geological mapping parties were active in the area almost every year between 1966 and 1970. Sauropod femora and vertebrae, theropod teeth, and turtle bones were collected from sediments that were dated as Upper Jurassic or Lower Cretaceous. The environment of deposition was different from Kadzi, being further inland, but the time period represented was roughly equivalent to Tendaguru's. Here too the Bush War disrupted further studies.

Another link to past successes in Berlin was broken on April 26, 1968. Ewald Siegert, the sculptor and preparator who was involved with all of the Tendaguru dinosaurs, passed away at the age of 84. He was survived by a daughter.

The long and distinguished career of Dr. Werner Janensch ended in West Berlin on October 29, 1969. He would have celebrated his 91st birthday two weeks later, but succumbed after a brief illness. He was survived by his wife, who moved to a seniors' home in Lübeck. In recognition of a lifetime devoted to understanding the finds made in his youth, the Geological Society of the German Democratic Republic (Geologische Gesellschaft der Deutschen Demokrat-ischen Republik) had named him an honorary member in 1964. West Ger-many's Paleontological Society (Paläontologische Gesellschaft) had elected him an honorary member in 1958. His body of work, done over a span of 59 years, defined an entire fauna of African dinosaurs.

Janensch had lived through stunning changes in his lifetime. He had wit-nessed the dissolution of imperial Germany after the Great War and the frantic but short-lived blossoming of the Weimar era. He had experienced the despera-tion of hyperinflation and then the Depression. At an advanced age, he had sur-vived the destruction of the city he called home for decades and the harsh years of occupation and recovery following the Second World War. Finally, he had suffered the disillusionment of life in a divided city where he was no longer able to easily visit his life's work. In his time, Werner Janensch had lived to see Hal-ley's Comet and man walking on the moon. Obituarists all recalled the charac-ter of the man whose extraordinary dedication overcame immense obstacles, with similar respect and affection. According to Hermann Jaeger,

> Of mid-sized slim build and unbowed to the end, Janensch gave the impression, to the young students and assistants, of an utterly modest yet almost timid and very reserved person. Therein a mantle of dignified distinction and strength surrounded him. Carefully weighing words, he spoke little and quietly, but with warmth.
>
> . . . Even when Janensch stood at the pinnacle of his creative work, he was—according to reports—hardly any different. He was an introverted person and an extremely careful scientist. It has been passed down that he examined, measured, and compared the skeletal elements of the dinosaurs again and again, and that he exhaustively studied the anatomy and the gait of modern large mammals before he decided on a particular skeletal pose. One even speaks of an occasional indeci-siveness in him. Yet once he struggled to a conclusion he remained steadfast in his viewpoint.[27]

Werner Janensch was buried at the Berlin-Frohnau Cemetery. Colleagues in East Berlin were not permitted to attend the funeral.

The youngest British participant at Tendaguru went on to a career in East African prehistory that brought him and his family worldwide acclaim over a period of 45 years. Louis Leakey became famous for the research he and his wife Mary undertook at Olduvai Gorge in Tanzania, the site first excavated by Hans Reck in 1913. Leakey's productive life involved him in wide-ranging pro-jects, from discoveries of humanity's earliest ancestors, government service dur-ing the Mau Mau Emergency in Kenya, and decades of fund-raising efforts, to the support of famous primatologists like Jane Goodall and Dian Fossey. He authored roughly a dozen books and numerous scientific papers, and his first autobiography, *White African,* described his experiences at Tendaguru. Honors were bestowed upon him by the universities and academic societies of several countries. Leakey passed away at age 69 on October 1, 1972, in London. He was buried in Limuru outside Nairobi, Kenya. His son Richard devoted many years to the study of African prehistory.

Another preparator of the Tendaguru fossils housed in the Natural History Museum in London passed away on February 25, 1972. Cyril Castell was 64 years old when he died, having spent 40 years in the Department of Geology. Castell had earned a B.Sc. in geology in 1933, and spent much of his time at the Museum identifying and organizing Mesozoic and Tertiary molluscs. Castell was a Fellow of the Geological Society of London and a Member of the Geologists' Association and the Paleontological Association.

The Institute and Museum for Geology and Paleontology of the University of Göttingen moved to new quarters at Goldschmidtstrasse 3 in 1974. Dr. Hans Jahnke assumed responsibility for the displays of *Dysalotosaurus*. Michael Sosnitza, a preparator, reworked the 1950s-vintage exhibit into an attractive panel mount of an articulated skeleton.

Curiosity brought the next European to Tendaguru around 1976. Urs Oberli was a private fossil dealer based in St. Gallen, Switzerland. Fascinated by fossils since youth, Oberli had trained under a Swiss sculptor. He learned the skills of fossil preparation under Dr. E. Kuhn-Schnyder at the Paleontological Institute and Museum of the University of Zürich.

While in his late twenties in 1976, Oberli made an impulsive decision to visit Tendaguru. Having obtained a copy of Hennig's *Am Tendaguru* from an antiquarian book dealer, the Swiss preparator was so fascinated by the story that he resolved to journey to the site in the fall, during his three-week vacation. He flew to Nairobi and met with Richard Leakey, whose father had worked at the site. Undeterred by warnings of a border dispute between Tanzania and Mozambique, Oberli traveled to Dar es Salaam. His intention to catch another bus and complete the four-hundred-kilometer drive to Lindi was frustrated, so Oberli boarded a ship to Mtwara. Before long he was in Lindi.

Lindi was a city of about 17,000 inhabitants in 1977.[28] Despite its location in the isolated and relatively impoverished south, and the destructive effects of a tropical storm in the 1950s, it was a market center for sisal, rice, and cashews.

Oberli left his minimal luggage at the Beach Hotel, rented a motorbike from an Indian auto repair shop, and bounced over a rough track 30 kilometers inland to the village of Rutambo, where he was interviewed by the local policeman. Up to this point, no one he met knew of Tendaguru or the work done there. A local German missionary, Pater Berno, was also skeptical of Oberli's story until two locals insisted that there was a place in the bush, a day's march away, where there were "large holes." Older family members spoke of a site where groups of Africans had worked with whites year after year.

Oberli was advised to return to Lindi and obtain permission from the regional police authorities. Instead, he visited the regional development director, who was responsible for cultural affairs. His quest was greeted with suspicion and he was prohibited from returning to the area. Still determined, Oberli presented himself to the police and according to his account, received permission in exchange for a ballpoint pen and a necktie.

A day later, he was back in Rutambo. The missionary drove him and two young men who were familiar with the site 15 kilometers inland to Nahino. Oberli spent the night in the village, too excited to sleep.

The next morning, five men led the resolute Swiss into the bush. After a fast paced three-hour march, a hill came into view, which Oberli took to be Tendaguru. Two hours later the party reached their goal.

> Only just arrived, my companions drew me to a roughly 15-to-20-meter-long and three-to-four-meter-deep trench, which for the most part was overgrown with smaller trees. . . . Amongst tall grass and thorn bush I saw the first larger bone, which obviously must have belonged to a dinosaur. My helpers also dragged over the first finds.[29]

A sauropod foot bone was pulled out of a second nearby ditch. These few excited moments were the total duration of his time at the site. Oberli took photographs, and by evening was back in Nahino, and in Lindi again the next day. He never returned to Tendaguru. But, around the time of his visit, a serious effort was underway to renew scientific research at the famous site.

1971–2001

Russell to Africa,
Brachiosaurus to Tokyo, Berlin to Tendaguru

In the early 1970s, the National Museum of Natural Sciences in Ottawa, Ontario, was planning new paleontology galleries. Director Louis Lemieux, a wildlife biologist, had worked at the Ngorongoro Crater in Tanzania for two years around 1970, and was an enthusiastic promoter of international projects. In November 1971, he suggested to the chief of the Palaeontology Division, Dale A. Russell, that the Ottawa museum might share its expertise in gallery development and scientific research with Tanzanians involved in similar undertakings. Russell replied that this collaborative project could include a fresh look at Tendaguru.

The director's memorandum was sent to a Tanzanian colleague, Solomon ole Saibul, minister for natural resources and tourism. Saibul forwarded the offer to A. A. Mturi, conservator in the Antiquities Division of the Ministry of National Education. Mturi welcomed the suggestion of joint development, as a study was underway in Tanzania to chart the course of the country's museums. One plan was to create a natural history museum. Mturi offered his department's resources to organize and coordinate a visit to Tendaguru, but lacked finances and paleontologists to study any finds.

Dale A. Russell's enthusiasm and intense interest in a wide range of research projects was essential to advancing the scientific understanding of Tendaguru. Born in San Francisco in 1937, he received a B.A. at the University of Oregon and an M.S. at the University of California, Berkeley. Pursuing a career in vertebrate paleontology, he continued his education at Columbia University in New York in 1964 under Dr. Edwin Colbert. Following a postdoctoral fellowship at Yale in 1965, Russell accepted a position as curator of fossil vertebrates in the Palaeontology Division of the National Museum of Natural Sciences in Ottawa.

In October 1976, Mturi, now the director of antiquities in the Ministry of Natural Culture and Youth, again opened discussions with Lemieux. Tanzania would provide field equipment for Tendaguru if the Canadians supplied scientific experts and two or three vehicles. In the longer-term plan for a natural history museum, Canada was to train Tanzanian technical and scientific staff.[1] Lemieux suggested a meeting with Mturi in Dar es Salaam later in 1977, followed by a preliminary orientation trip to Tendaguru. An excavation program would commence in 1978.

Dale Russell began an exhaustive review of the German publications with the aid of a government translator. He collated data on the lithology and contents of the German quarries, for each of the Saurian Beds and intervening horizons. His approach was biostratigraphic; he analyzed the fossil elements of each quarry according to genera, and added sedimentological details where possible. Quarries were classified on this basis, to generate insight into the distribution of the various Tendaguru dinosaurs temporally and geographically, and to understand the relative abundance of the genera. Russell also contacted Dr. Hermann Jaeger, a graptolite specialist at the Museum of Natural History in Berlin, in an attempt to locate field notebooks that might supply data on the sedimentology of the German quarries.

Russell's progress report of mid-June 1977 correlated the contents of quarries with stratigraphic levels. The elements and genera of the Tendaguru bones in British hands were identified by sauropod expert John S. McIntosh and included in the study. Russell recommended concentrating on the Upper Saurian Bed and Quarry dd of the Middle Saurian Bed near Kindope, if it could be rediscovered.

On July 3, 1977, Dale Russell, who was 40 years old, and Louis Lemieux, 52, flew to Africa. They met with Saibul and Mturi in Dar es Salaam nine days later. Topographic maps and aerial photographs were purchased. On July 17, the Canadian party, consisting of Louis Lemieux, Dale Russell, and Robert McLaren Jr., the son of the Canadian high commissioner, left for Lindi. Their four-wheel-drive vehicle and its driver were provided by the Tanzanian Department of Antiquities. They reached Kilwa late that evening, and Lindi the next day. There they spoke with government officials, including the regional commissioner, an officer of the Regional Cultural Office of Lindi Region, and a member of the regional office of the ruling Revolutionary Party. After purchasing provisions, the Canadians left for Mpingo, where they camped.

On July 20, 1977, they were joined by five men from the nearby village, who acted as guides. After a 3½-hour drive over a thickly overgrown path, they stood at Tendaguru Hill. It was around 12:30 in the afternoon, and a demanding day was about to begin: "Tse-tse flies were extremely abundant en route. . . . it was staggeringly hot—the land was very dry. . . . the quarries (at least one) are probably thoroughly overgrown—the larger trees are perhaps 35′ high and about 6″ in diameter. What a desolate place!"[2]

They collected matrix samples for pollen analysis from the southwest face of the hill and from a site less than a kilometer away. One of the African workers explained that his father had been employed by the British, and his grandfather had been with the Germans at Tendaguru. After a brief reconnaissance, the Canadians drove back to Mpingo. It was a calm evening after the intense heat and dust of the day: "night was soft and clear with the Milky Way brightly visible in the sky and a waxing crescent moon. The evening was very peaceful, and we could hear the drummers far into the night in the nearby village."[3] Camp was struck the next morning, and the crew drove north, having spent about half a day at Tendaguru.

With firsthand experience of the area's limited infrastructure, Dale Russell soon decided that heavy earth-moving equipment would be the most efficient method of stripping overburden at Tendaguru.

At the Canadian high commission, senior representative Robert McLaren offered to transmit a statement of support to Canadian prime minister Pierre Trudeau, who would be meeting with Tanzanian president Julius Nyerere in Ottawa shortly. A telegram drafted by Dale Russell was sent to the Department of External Affairs in Ottawa.

In Canada, a meeting was held with a representative from the international sales division of Caterpillar, a heavy equipment manufacturer, in September, who recommended a suitable piece of equipment and provided information on its fuel consumption. The Canadian International Development Agency, CIDA, was approached later that month, to discuss possible agency financing of the Tendaguru project. A budget was developed by November 2, 1977, and estimated the cost of a four-year excavation program at over C$800,000.[4]

Major expenses included a diesel-powered excavator, a diesel D4 Caterpillar tractor with attachments and spare parts, the operators, fuel, and four-wheel-drive vehicles. The heavy equipment would be hauled to Mpingo by transport truck, once the trail had been improved.

It was estimated that five months of excavation with mechanical equipment would suffice, and that under the local conditions, fuel consumption would amount to 720 liters per day. About 25 and 50 African workers would be on site for the first and second years respectively, along with two heavy equipment operators and three museum staff members. Food, shelter, and wages would be paid by the Tanzanians. The excavated fossils would be trucked to Dar es Salaam, where 10 preparators would be employed for six months to ready the material for display. For a project of this magnitude, external sponsorship was essential, and the search for funding was undertaken by Lemieux.

Dr. David Jarzen of the National Museum of Natural Sciences found no fossil pollen in the matrix samples from Tendaguru. They were also tested for magnetic polarity. In Tanzania, the project was announced in the Dar es Salaam *Daily News* of August 17, 1977.

Unknown to anyone on the Canadian team at the time was the death of Edwin Hennig in Tübingen, West Germany, on November 12, 1977. He had attained the advanced age of 95. His wife predeceased him the same year, after 64 years of marriage. For 28 years, he had been active at Tübingen, lecturing, researching the geology of the surrounding countryside, and, as director, developing the Museum with Friedrich von Huene into a renowned institute. He had served as co-editor of the journals *Neues Jahrbuch für Geologie, Paläontologie und Mineralogie* and *Palaeontographica*. A festschrift dedicated to Hennig by his students on his 70th birthday listed 198 scientific publications as of 1952, on a remarkably broad range of topics. With Hennig's passing the final participant in the fieldwork of the German Tendaguru Expedition was gone. Hennig's eldest daughter eventually donated an invaluable collection of field notes, correspondence, and photographs to the Tübingen University Archives.

Lemieux's efforts to build the required support led him to the board of trustees of the National Museums of Canada. The president of Massey-Ferguson Corporation was invited to participate in the venture. By spring and summer 1978, a year after returning from Tendaguru, Lemieux had been forced to temper his optimism: "Because of governmental measures of economy and a general atmosphere of economic intrenchment in this country, we have thus far been unable to generate adequate support for a full scale Tendaguru program of excavations from either the public or private sector."[5]

Despite this gloomy report, the effort to find assistance did not end. The National Film Board of Canada was producing a documentary on dinosaur ecology, which would include footage from Tendaguru. It was hoped that broadcasting this program would encourage interest in and support of the Tendaguru project.

Lemieux suggested another visit to Tendaguru in late 1978, to enable the director of the proposed dinosaur film to scout the location. Additional matrix samples would be collected and analyzed for fossil pollen, an important indicator of paleoenvironmental conditions. It was hoped that sites of Cretaceous age could be inspected in Malawi, to determine the type of dinosaur fauna that existed there. The team would include Dale Russell, Pierre Béland, and technician Gilles Danis, all of the Canadian museum, and Roger Tétrault of the National Film Board. Modest travel plans proposed that the team take a bus from Lindi to the road leading to Mpingo, then walk to Tendaguru over the course of a day.

Tendaguru, as a four-year program of excavation, research, and public interpretation, formed only part of the overall plan that the Tanzanians were devising. Tanzanian museum resources would be developed nationally, with the expectation of substantial Canadian commitment to train technical and scientific staff. Canadian specialists would establish research and conservation facilities in the natural history museum, to be located in Arusha.[6]

Mturi and Dr. Fidelis Masao, director of museums in Dar es Salaam, supported another site visit. Abel T. Nkini of the Antiquities Division was assigned to the Canadian team. Mturi stipulated that Russell should spend about two weeks in other parts of Tanzania examining important sites, before proceeding to Tendaguru.

Russell and Béland drafted a paper on the paleoecology of Tendaguru, summarizing their exhaustive literature study and McIntosh's museum investigations. It was read at an international vertebrate paleontology colloquium in Paris in early September.

Dale Russell's next African visit began in Ottawa on October 17, 1978. The first stop on the tour with Abel Nkini was at Moshi, to the northwest of the capital, followed by Arusha. From here, they drove to the Ngorongoro Crater and Olduvai Gorge, where Mary Leakey was their host. Russell and Nkini returned to Dar es Salaam eight days after arriving on the continent. Mturi informed Russell that an effort was underway to obtain UNESCO World Heritage Site status for Tendaguru.

Russell and Nkini reached Lindi on October 28. The Beach Hotel was closed due to cholera, so lodgings were found elsewhere. The remaining members of the Canadian party, Béland and Tétrault, flew into Lindi a day later.

On October 30 they departed for Tendaguru, via Mpingo and Nyangara. The trip lasted about 3½ hours. Guides were collected and the party reached Tendaguru in the afternoon. Three additional matrix samples were gathered for fossil pollen analysis.

Russell and Béland visited the hill again on the 31st. Russell reflected that there had been great changes over the decades: "It was amazing to think that it took us 3½ hours to reach T[endaguru] when it took 5 days for Fraas. . . . much of the road to the village had been improved—graded."[7] Reaching the site by mid-afternoon, they split into two groups and prospected for an hour and a half. Two dinosaur caudal vertebrae were found and brought back to Lindi.

A newspaper interview was arranged in Dar es Salaam on November 2 and a photograph of Dale Russell and the dinosaur vertebrae appeared in the Tanzanian *Sunday News* of November 12. However, the significance of these proceedings was overshadowed by the events of October 30, 1978, the very day the Canadian team visited Tendaguru. President Idi Amin Dada of Uganda had ordered his army to invade Tanzania.

Border provocations in northwest Tanzania had occurred sporadically since 1971. Thousands of refugees had fled to Tanzania following Amin's 1971 coup, including President Milton Obote. On October 28 and 29, 1978, two towns in Tanzania were bombed. On the 30th, three thousand Ugandan troops crossed the border and pushed south as far as the Kagera River, to the west of Lake Victoria. They occupied 1,839 square kilometers of Tanzanian territory, stripping the region of food and cattle, looting valuables, destroying schools, sugar mills, sawmills, churches, and medical dispensaries, and taking civilians to Uganda as prisoners.[8]

As the Canadians prepared to depart, Tanzania mobilized for war. Permission to visit Malawi had not been received, so Dale Russell returned via Berlin to visit the Museum of Natural History and the Tendaguru collection. When Abel Nkini was informed of these plans, he contacted the embassy of the German Democratic Republic to ask it to repatriate Tendaguru material.[9] This sudden and unexpected move left Dale Russell in an awkward position, as the issue was fraught with political complications. Repatriation of fossils was in no way the intent of the Canadian team.

Russell passed through Checkpoint Charlie in Berlin on November 6 and was gratified to see the mounted Tendaguru dinosaur skeletons on display, having left the site of their origin only days earlier.

In Canada, Dale Russell advised Abel Nkini to reexcavate Tendaguru with German technical and financial support, rather than pursue the transfer of collections from Berlin. Russell also planned to co-author a semipopular booklet about Tendaguru in both Swahili and English, to be distributed in Tanzania.

Vertebrae collected from the Upper Saurian Bed at Tendaguru were identified as *Brachiosaurus* caudals from midway down the tail.

At the end of March 1979, a three-kilogram package was sent to A. A. Mturi. It contained Dale Russell's typewritten analysis of Tendaguru quarries and specimens, along with translations from the German literature.

David Jarzen's preliminary analysis of the latest sediment samples from Tendaguru again demonstrated a disappointingly sparse pollen record. Only one of the samples collected in 1978 contained any recoverable examples, a total of 15 taxa.[10] However, there was a possibility that angiosperm pollen was present. If this could be confirmed, it would indicate the presence of flowering plants in Upper Jurassic times.

Additional Tendaguru matrix was requested from Dr. Jaeger in Berlin and Dr. Alan Charig at the Natural History Museum in London. Three samples from Berlin's collections had surrounded bones from the type locality of *Barosaurus africanus* and *Brachiosaurus brancai*. London sent three more samples, but none of the Berlin or London matrix yielded any pollen. Dale Russell wrote to Mturi in June 1979, hoping to return to Tendaguru in October of that year and cut a deep trench from the base to the top of the hill. It was hoped that pollen from this excavation would resolve the question of the angiosperm taxa. The palynology materials were sent to G. Herngreen in Holland for further analysis.

Monetary support for the planned National Film Board project never materialized. At the end of July 1979, Dale Russell expressed his frustration to Dr. Jaeger and Mturi, doubting that a return to Tanzania would be feasible that year. Though no final assessment has been seen, it appears that the inability to raise the considerable sums required for a large-scale project ended any hope of returning to the site.

On April 17, 1981, the last participant in the fieldwork of the British Museum East Africa Expedition died. Francis Rex Parrington passed away at the age of 76. After his service with the Expedition had ended, he returned to the Karoo of Tanganyika and collected fossils in the Ruhuhu Valley for six months, unaccompanied by Europeans. In 1938, he had become the director of the University Museum of Zoology in Cambridge. After the war, he had supervised students working toward doctoral degrees at Cambridge, among them Alan Charig and A. W. Crompton. In the late 1950s and 1960s, his significant contributions to vertebrate paleontology had been recognized with his election as a Fellow of the Royal Society and the bestowal of a D.Sc. Parrington had been offered the position of director of the Natural History Museum in London, but declined. He had married in 1946, and had a son and daughter who survived him.

Curators at the Museum of Natural History in Berlin were mobilized in 1984, in a project with seemingly impossible deadlines. Japan's largest-circulation daily newspaper, the *Yomiuri Shimbun,* coordinated a series of traveling exhibits of fossils to satisfy growing Japanese enthusiasm for dinosaurs and other prehistoric life. Throughout the early 1980s, the newspaper had sponsored shows that included *Homo erectus* from Java and China, and Chinese dinosaurs in 1981.

The scientific advisor for the early programs was Dr. Yoshikazu Hasegawa, a paleontologist at Yokohama University.

The success of previous displays was such that Mr. Ota of the *Yomiuri Shimbun* was searching for new ideas. Hasegawa mentioned the Tendaguru dinosaurs in East Berlin, which he had visited in 1973. Ota felt that the gigantic *Brachiosaurus* would make a stunning impression. The first venue suggested was Japan's National Science Museum in Tokyo, but museum planners rejected the idea due to insufficient exhibit space. In Berlin, Dr. Hermann Jaeger vigorously resisted dismantling any of the skeletons, especially the enormous and fragile *Brachiosaurus*. Hasegawa was aware that the skeletons had been disassembled for safekeeping during the Second World War, and insisted the project was feasible.

Opposition from German museum curators was disregarded, since approval had presumably come from the highest political level. German Democratic Republic state secretary Erich Honecker had made an official visit to Japan and as part of the agenda agreed to closer cultural ties. Formally, the project was promoted as a celebration of the 175th anniversary of the founding of Humboldt University and the 110th anniversary of the founding of the *Yomiuri Shimbun*. The show was titled "175 Years of the Humboldt University of Berlin—Capital of the GDR. The Greatest Dinosaur Exhibition on Earth."

Deadlines were extremely tight, which added to the concerns of German scientists and museum specialists who opposed the idea. Only 10 weeks were available to select specimens, disassemble mounts, and pack everything for transport to Japan. Once in Japan, everything had to be unpacked, reassembled, and arranged for display in a scant few weeks.[11] The rarity, fragility, and size of the objects caused grave technical concerns. The head of the Association for the Promotion of the Natural Science Museums of Berlin, the directors of the Institute for Paleontology and the Institute of Zoology of the Free University of Berlin, and the chief of the German Museum Alliance in Karlsruhe all went on record stating that the risk of damaging irreplaceable specimens was alarmingly real. They were outraged that protests by researchers had been overridden: "Every dismantling and remounting of paleontological objects often incurs a partial destruction of the object. This applies particularly to older specimens." "The decision to lend originals to Japan has been reached by politicians. All objections of the scientists were ignored."[12]

A West Berlin newspaper speculated that an even more cynical reason lay behind the agreement. It suggested that the government of the GDR was knowingly accepting the possibility of damage or even partial destruction of the specimens in order to claim a portion of the purported $43 million insurance money as hard currency. These and other comments had no effect on the subsequent course of events.

The exhibit would have two components. One celebrated the achievements of the university's researchers over 175 years, as that institution had produced 27 Nobel Prize laureates. Artifacts, photographs, and text that illustrated the history of the university would be displayed.

The second component showcased outstanding objects of natural history that had been purchased or collected by the university's affiliate, the Museum of Natural History. A Japanese team of four negotiators included Hasegawa, who worked alongside German curators. A translator was provided by the Japan International Culture Association, and the sponsoring newspaper sent two or three representatives. Dr. Gottfried Böhme, acting head of the Paleontological Institute and Museum, mobilized German curators and technicians. A total of 350 specimens were selected that would depict the history of life on earth.[13] There would be Devonian invertebrate fossils, fish from the Mesozoic, Permian reptiles from South Africa, ichthyosaur slabs from Holzmaden, pterosaurs from Solenhofen, early land plants, an ancient fossil turtle, Mesozoic dinosaurs such as *Plateosaurus* from Halberstadt, and all five original Tendaguru dinosaur skeletons. Adding to the controversy was a specimen of incalculable value, known the world over—*Archaeopteryx lithographica*. These fossils had never been displayed outside Berlin.

Dicraeosaurus, Elaphrosaurus, Kentrosaurus, and *Dryosaurus* were pulled from their armatures. Chief preparator Hans-Hartmut Krüger played a major role, coordinating a combination of mechanical lifting devices and human hands. Metal scaffolding was erected around the skeleton of *Brachiosaurus* as the limbs, caudal vertebrae, and pectoral and pelvic girdles were removed. The intricacy and method of attachment of Ewald Siegert's modeled vertebrae were such that they would sustain considerable damage if pulled off the metal armature. It was decided to create a mold of the presacral vertebral column. Four Japanese artist-technicians, each of whom was a display specialist, model maker, or sculptor, climbed the scaffolding to apply silicon rubber to the vertebrae and ribs.

As quickly as the individual bones were detached from the armature, they were packed for transport by the West German firm of Hasenkamp. Wooden cases were custom-built to the dimensions of the various fossils, which were wrapped in protective air-bubble plastic or cushioned in sponge foam. The disassembly and mold making were underway by early May 1984. The Japanese Yamato Transport Company sent four men, who assisted in packing and arranging the shipment over the course of about two weeks on site. The entire Japanese team of 14 lived in East Berlin and occasionally crossed into West Berlin.

The heaviest *Brachiosaurus* skeletal parts, along with the other Tendaguru dinosaurs, were transported by truck to Hamburg, though newspaper accounts in the GDR disingenuously stated that the displays were driven to Rostock. In Hamburg, the crates were transferred to two containers that were loaded aboard the *Thames Maru* of the Mitsui O.S.K. Lines. Some time in May, the African dinosaurs and other treasures from Berlin began the long sea voyage to Japan. Molds of the vertebrae and ribs of *Brachiosaurus* were crated and flown to Japan by Japan Air Lines, the national air carrier. *Archaeopteryx,* accompanied by Dr. Manfred Barthel, the director of the Museum of Natural History, "flew" to Tokyo via Copenhagen, also on JAL. According to Dr. Hasegawa, this route was chosen when an "international criminologist" strongly suggested that a flight through Moscow be avoided. The great dinosaur hall in East Berlin stood nearly empty. It was a scene like that of 1943. Museum staff did not know exactly when the fossils would return, presumably sometime in December 1984.

Freight containers were unloaded in Tokyo on June 2. The second major urban center in Tokyo, the district of Shinjuku, would host the show. The Japan Rail station at Shinjuku had the space and offered the public easy access to the display. A dome consisting of a plastic-clad metal framework was erected in the switching yard. It would provide two to three thousand square meters of display space.[14] A promotional campaign was launched by Japanese media, with notices appearing on television and radio, and in newspapers and magazines. Tokyo subway stations featured color posters of the upcoming exhibit, and a color guidebook was printed.

When the molds of the *Brachiosaurus* vertebrae and ribs arrived in Tokyo, technicians at the Maruo studio in Sanuki produced polyester resin casts in two weeks. Mrs. Uematsu, a specialist on the conservation team, translated for Krüger and Japanese technicians. Work was feverish because the exhibit was scheduled to open in four weeks. Steel bases were welded for the two large sauropods around June 22. Two days later the reassembly of *Dicraeosaurus* began. It was completed on July 1, in nine days. *Brachiosaurus* installation commenced on June 23 and was finished about 11 days later, on July 2. The priceless *Archaeopteryx* slab lay on red velvet under a case of bulletproof glass.

The exhibit opened on July 7, 1984, three months after leaving Berlin. Prince Mikasa, the youngest brother of the emperor of Japan, opened the display along with the rector of Humboldt University, Prof. Dr. Helmut Klein, and Dr. Manfred Barthel.

Estimates of 6,000 to 7,000 visitors daily proved to be optimistic, though a respectable 40,000 crowded into the windowless dome in the first two weeks.[15] The busiest day was August 26, when 22,600 visitors passed through the pavilion. During the show, articles appeared in countless magazines and newspapers, among them the *Daily Yomiuri*. The sponsoring newspaper gave away 150,000 free admission passes. A contest for children called for drawings of the dinosaurs, and more than four thousand adorned the walls of the pavilion. A selection was sent to Berlin when the specimens were returned. When the exhibit closed after 14 weeks, on October 14, 1984, more than 600,000 people had walked through the dome, including Crown Prince Akihito and his younger brother Prince Hitachi. Despite this success, which closely matched projections, the *Yomiuri Shimbun*, which covered all expenses, lost about 20 million yen.[16] Total costs ran into the hundreds of millions of yen.

By Christmas 1984, the dinosaurs were back in Berlin. All the skeletons except *Dryosaurus* were remounted by January 1985. The ornithopod did not reappear for several more months. *Archaeopteryx* was stored in a vault at Humboldt University, and returned to the Museum when a fire-resistant storage case was purchased. Molds of *Brachiosaurus* vertebrae were destroyed and the casts sent back to Berlin. The original bones suffered minor damage. In ensuing years traveling exhibits that featured dinosaur skeletons became common, and employed a mix of original skeletons and casts.

The next European known to reach Tendaguru was Dr. Gundolf Ernst, a paleontologist at the Free University of Berlin. He completed the difficult trip in 1985, by four-wheel-drive vehicle and on foot, following a route which ran south of that taken by Dale Russell and his crew.

In the late 1980s, Soviet president Mikhail Gorbachev's policies signaled a period of liberalization and turmoil that soon spread to neighboring countries. Gorbachev attended the German Democratic Republic's 40th anniversary celebrations in East Berlin, and reminded Germans that historic events were inexorable. Following mass protests, Erich Honecker was removed from office on October 18, 1989. The premier and his cabinet resigned on November 7. The Politburo of the Communist Party (SED) resigned the following day, to be replaced by a reform group. Mass protests continued in the GDR, and on November 9, the East German government allowed citizens to travel or emigrate. The Berlin Wall fell.

By mid-February 1990, a plan for the eventual reunification of Germany had been endorsed by the Soviet Union, the United States, Great Britain, and France. Free elections were held in the GDR in mid-March 1990, bringing a conservative coalition into power. The deutschemark became the official currency on July 1. It was agreed that the unification would take place on October 3, 1990, with nationwide elections to follow in December. A vote taken in 1991 established Berlin as the capital of a united Germany. What had been inconceivable a few months before was now reality.

Despite initial euphoria over the unification, its economic effects included high unemployment, and required massive investment that increased taxes. Four decades of life in radically different social and political systems could not be overcome instantly, and the early years of the 1990s were particularly difficult. Attitudes of both westerners and easterners regrettably led to acrimony.

In 1990, the director of the Museum of Natural History completed an agreement to loan original bones of *Brachiosaurus* to a Dutch museum. Although the complete skeleton was not being offered, the bitter struggle against the loan to Japan in 1984 was fresh in Dr. Jaeger's mind. He fought the latest arrangement.

A new director was named at the Museum on May 14, 1990. A general policy was instituted under which no original dinosaur bones would be loaned, with the exception of unprepared jackets, and no molding and casting of the skeletons would be allowed. In North American museums, casts were often sold to generate revenue. Jaeger believed strongly that the unique and hard-won displays were devalued if duplicates appeared elsewhere; they were the proud historical legacy of the Museum. The risk of damaging the originals in the molding process was another concern. Jaeger felt that the Tendaguru collection had been entrusted into his care by Werner Janensch. Janensch had expressed the wish to Jaeger that a composite skeleton of *Dicraeosaurus sattleri* be mounted.

A cast of a *Brachiosaurus* skeleton is on display at the Dinosaur Museum Aathal (Sauriermuseum Aathal) outside Zürich, Switzerland. H. J. "Kirby"

Siber obtained the replica from a dinosaur exhibit held in 1990 or 1991 in Denekamp, The Netherlands. The Gunma Museum of Natural History in Japan also features a *Brachiosaurus* mount in its galleries. In 1993, Gilles Danis's firm Prehistoric Animal Structures erected a *Brachiosaurus* cast for the Field Museum in Chicago, and in 1999 another, which is currently on display at O'Hare Airport. There are plans to produce a very different dinosaur gallery at the Museum of Natural History, which would entail remounting the Tendaguru dinosaurs.

Dr. Jaeger and his associates breathed fresh life into the programs at the Paleontological Institute of the Museum. Dr. Wolf-Dieter Heinrich, a specialist in Cenozoic mammals, curated the Tendaguru collections. By the summer of 1990, Dr. Heinrich and a new assistant had brought much order into the basement storage rooms. Heinrich treated matrix from field jackets with hydrogen peroxide to gently break up the sediments. The resulting concentrate was examined for microvertebrate remains and interesting dinosaur teeth were found. Heinrich and Jaeger made plans for further steps in a revived program of Tendaguru research. An administrative change took place on December 10, 1990, when the director of the Geological-Paleontological Institute and Museum was replaced by Dr. Hermann Jaeger. Dr. Heinrich was his deputy.

In 1989, Heinrich had the lower jaw of the Tendaguru mammal *Brancatherulum* X-rayed at the Charité, a hospital associated with the Humboldt University of Berlin. It was more fully prepared by chief preparator Hans-Hartmut Krüger during the tumultuous autumn of that year. Heinrich published a revised description in 1991, in which he was able to more precisely classify this important find. About five hundred kilograms of Tendaguru matrix had been dissolved with acetic acid, yielding a wealth of interesting specimens.

Sadly, the man who had so steadfastly promoted and fought to protect the Museum's Tendaguru legacy, who had unstintingly offered his cooperation and openhearted hospitality to a myriad of international researchers, was suddenly gone. Dr. Hermann Jaeger passed away on September 22, 1992, at age 63. He would not know how fully his hopes for a rejuvenated scientific investigation of Tendaguru would be realized.

Paleontologists worldwide independently reexamined the Tendaguru collections at several German museums. Peter Galton of Connecticut authored a series of papers between 1977 and 1989, following his visits to Europe. He published detailed descriptions of the cranial and postcranial anatomy of *Dryosaurus* in 1983 and 1989, *Elaphrosaurus* in 1982, and *Kentrosaurus* in 1982 and 1988, as well as a description of a pterosaur tibiotarsus in 1980.

Greg Paul proposed a new subgenus name for *Brachiosaurus brancai: Brachiosaurus (Giraffatitan) brancai,* in 1988. Rupert Wild of Stuttgart resolved the tangled systematic position of *Tornieria* by assigning the genus to *Janenschia,* in a paper that appeared in 1991. This sauropod was studied by Wild, Jose Bonaparte of Argentina, and Heinrich in Berlin, starting in 1994. A comprehensive redescription appeared in 2000, along with the erection of a new Tendaguru sauropod—*Tenduguria tanzaniensis.*

David Weishampel and Ron Heinrich of Johns Hopkins sectioned a number of *Dryosaurus* limb bones from the Tübingen collection for a 1994 study of the dinosaur's ontogeny and biomechanics. Anusuya Chinsamy of the University of Pennsylvania in Philadelphia prepared a study of *Dryosaurus* bone histology, also based on the Tübingen collection, in 1995.

The invertebrates found in the areas traversed by the German and British Tendaguru Expeditions also received attention. While revising the Jurassic and Cretaceous ammonites of Ethiopia, Arnold Zeiss of Erlangen, Germany, assigned two specimens identified by Dietrich as *Haploceras* to the genus *Sutneria*. The fossils had been collected by Hans Reck at Mahokondo, and Zeiss published his paper in 1979. In 1995, Saidi Kapilima and Manfred Gröschke described ammonites that Kapilima had collected at the Mandawa-Mahokondo anticline.

Oil and gas exploration continued, and in 1979, Agip (Africa) Ltd. drilled the Kizimbani exploration well southwest of Kilwa. The results of palynological and micropaleontological analyses of wellbore cuttings revised earlier dating for the Pindiro Shales. The outcrop, first noticed by Hennig, was now considered Upper Jurassic.

The Karoo-age basins south of the Uluguru Mountains that Janensch had traversed in 1911 were revisited on a number of occasions. Several geologists had returned to Hatambulo Hill, the Sumbadzi River, and the Luhombero Hill area and published revised geological profiles, beginning with G. M. Stockley in 1936, and J. Spence in 1957, both of the Geological Survey of Tanganyika. More recently, there were T. Kreuser in 1983, O. Hankel in 1987, and H. Wopfner and C. Kaaya in 1991. Only a few fossil plant remains were recovered, but some fossil pollen was analyzed. The most recent study placed the Hatambulo Formation of this area as roughly equivalent in age to some of the Karoo deposits of the Ruhuhu Basin, where Parrington had collected in the 1930s.

In the mid-1980s, geologists from the mining company Uranerzbergbau in Bonn prospected for uranium near the Madaba River, a hundred kilometers south of the Rufiji River. In the Lower Jurassic Madaba Formation, they found sauropod remains, including a scapula, a femur, a fibula, and caudal vertebrae. Philippe Taquet identified the bones as camarasaurid, extending the history of Gondwana sauropods in Tanzania.

Dr. Louis Jacobs of Southern Methodist University in Dallas had worked in Nairobi with Richard Leakey in the early 1980s. He hoped to collect fossil mammals at Tendaguru by means of washing and screening techniques. In 1981 and 1982, he had applied to the Tanzanian authorities to work at the site, but political relations with Kenya were tense, the border between the two countries was closed, and Jacobs was never granted permission. Fortunately, he persisted upon his return to the United States, but in another direction. He obtained a National Geographic grant to excavate fossils in Malawi. From 1984 to 1992, his highly successful joint venture with the Malawi Department of Antiquities worked the Dinosaur Beds, last visited by Migeod and Parrington 53 years earlier.

In January 1984, Jacobs's team made their first trip to Malawi. In June, excavations commenced around Mwakasyunguti. Camp was set up at Ngara, about ten kilometers north of Nyungwe, another locality well known to Migeod and Parrington. Fieldwork continued in 1987, 1989, 1990, and 1992. Over 150 specimens were collected at nine localities, representing 13 taxa of vertebrates.[17] About five species of dinosaurs were identified, including a titanosaurid and diplodocid sauropod, two theropods, and a stegosaur. Remains of three crocodiles were recovered, including a notosuchid crocodile, as well as evidence of frogs, fish, crustaceans, and snails. Washing and screening of sediments produced many of the microvertebrates. Fossil plant diaspores and ostracods were recovered from samples of matrix brought back to the United States.

The sauropod bones included caudal, dorsal, and cervical vertebrae, sternal plates, cervical ribs, an ischium, humeri, and skull elements such as a dentary, maxilla, and teeth. Jacobs was able to resolve some of the taxonomic confusion around the name *Gigantosaurus dixeyi,* which had been established by Sidney H. Haughton in 1928. Tendaguru material in Berlin and Migeod's collections from Mwakasyunguti at the Natural History Museum in London were compared to the recent Malawi finds. Jacobs felt that the Malawi sauropod was sufficiently different from Tendaguru's *Janenschia* (which was originally *Gigantosaurus,* and then *Tornieria*) to warrant a new genus, and erected *Malawisaurus dixeyi.* It was collected in the same rock units and the same area as the specimens found by E. C. Holt, Migeod, and Parrington, and was considered a Lower Cretaceous titanosaurid. The majority of the specimens are in the custody of the Malawi Department of Antiquities in Lilongwe.

Exciting developments were taking place outside of the Berlin Museum during the early and mid-1990s. In 1993, a group of three geoscientists in Berlin were preparing for a brief return to Tendaguru. Their goal was to determine the feasibility of collecting more microvertebrate remains of the type the Museum of Natural History in Berlin was discovering. Such material could lead to a more detailed understanding of the paleoecology and faunal composition of the area.

Dr. Christa Werner, of the Technical University Berlin, Dr. Wolfgang Zils, of the International Association for Promoting Geoscience, and Andrea Moritz established contact with the Department of Geology at the University of Dar es Salaam. The dean, Dr. Mruma, supported the venture. Funding was obtained from private sources, and the Tanzanian Commission for Science and Technology, COSTECH, granted a research permit.

The team flew to Dar es Salaam in January 1994, where they were joined by Charles Saanane, a student at the university. Dr. Sugo of the Department of Geology also joined the group. A four-wheel-drive vehicle took them to Lindi, where the regional district director cleared their papers and provided them with an introduction to the local authorities around Tendaguru.

Two possible routes existed to the site, the southern route taken by Urs Oberli and Gundolf Ernst, and the northern route taken by Dale Russell. The 1994 team decided on the northern approach. In two hours, they drove 70 kilometers north of Lindi to the village of Nkwajuni. From here, they turned inland for 10 kilometers over a rough track to Mpingo, the village at which the Canadians had camped 17 years earlier. As they went further along an even narrower path, Tendaguru Hill came into sight to the south. Weathered old quarries were visible along either side of the trail. The rainy season foiled their careful plans, and they were forced back to Lindi. Undaunted, they made preparations to return. Their vehicle was loaded aboard a ship at Mtwara that took them back to Dar es Salaam.

In November 1994, Christa Werner, Wolfgang Zils, and Andrea Moritz returned to Dar es Salaam. Their trip was again privately funded. Along with Charles Saanane, they chose a longer route through the interior of Tanzania, covering a distance of 1,600 kilometers in the process. From Dar es Salaam, they drove west to Morogoro, Iringa, and Makamsako, about 700 kilometers along paved roads. The next 340 kilometers, south to Songea, were also on well-maintained pavement. From here, they were retracing the journey made so painfully slowly by Edwin Hennig in 1934. To cover the 275 kilometers between Songea and Tunduru took almost 10 hours. Tunduru to Masasi, 200 kilometers, was little better, but the road along the final 150-kilometer leg to Lindi was in good condition. On this occasion, the weather held and they were able to spend three days at Tendaguru. They pitched their tents at Namapuia and hired several villagers as guides for the surveys they undertook in every compass direction.

In the course of the fieldwork, the team made numerous stratigraphical observations at a variety of points. The Saurian Beds and transitional horizons were identified, profiles were recorded, and geological specimens were collected. Petrographic thin sections were later produced. Marl samples were brought back to be processed for microvertebrates and pollen. Geodes or septarian nodules were split in the search for fossil inclusions. Ammonites were later identified to date the sediments. An abundance of old quarries was evident to the north and south of Tendaguru Hill. The team inspected a number of regions worked by the German and British investigators, including Tingutinguti and Maimbwi Creeks, where they hoped to find the Lower Saurian Bed that had not been recognized by Parkinson. When two tires on the vehicle went flat, Mohamed Mpamba and Ahmed Bakari Kiwamu strapped the heavy wheels to their bicycles and pedaled 80 kilometers to have them repaired.

The Germans made a third visit to the site in December 1995. They experienced serious difficulties, including vehicle breakdowns and a severe water shortage. The team left Namapuia after two days. Copies of their papers were distributed in the village, and the significance of the nearby site was explained to the inhabitants. On their return to Dar es Salaam in mid-January 1996, the Germans presented their findings at the university's department of geology. Their lecture was titled "Tendaguru, the most famous dinosaur site in Africa." They returned to Berlin at the end of January 1996.

Between July and September 1996, a group of British forestry experts was in southeast Tanzania as part of the Cambridge Mpingo Project. Their program was designed to investigate the ecology of *Dalbergia melanoxylon*, locally known as the mpingo tree. Team member Justin Ormand made the difficult journey to Tendaguru, confirming the conditions experienced by virtually all visitors:

> Even with a 4WD vehicle and hired "road clearers" (men with pangas) the journey from Mnyangala to Namapuia, a distance of 47 km took 9 hours! Although rumours that there were man-eating lions terrorising the area proved to be unfounded, lions and other large mammals are common (at least in the dry season).[18]

Wolfgang Zils, now of the International Association for Promoting Geoscience in Berlin, Christa Werner, and another German completed a fourth trip to Tendaguru. This time Zils's organization provided travel funds, and in December 1996 they were back in Dar es Salaam. Two Tanzanians, Charles Saanane among them, joined the team.

Rough trails along the northern route to Tendaguru caused a leak in the fuel tank and split a fuel line. Dutch development experts in Lindi recommended the southern route. After reaching the Lukuledi River, they discovered that two trucks had broken down and blocked the bridge to all traffic. As an alternative, local villagers were employed as guides and led the way to Namapuia via a path that did not appear on any map.

There was a severe shortage of drinking water, forcing the group to leave after three days. Nevertheless, a careful survey was made of the Middle and Upper Saurian Beds and the intervals between them. Old quarry sites were visited, which contained poorly preserved bone fragments discarded by earlier excavators. Discussions for future projects at Tendaguru were held with the Tanzania Petroleum Development Corporation, TPDC, and the National Museum of Tanzania.

The nature of research at Tendaguru had changed, even from Dale Russell's concepts of paleoenvironmental study, to embrace a highly detailed examination of the surrounding area using a variety of modern analytical techniques in a multidisciplinary approach. Dinosaurs were no longer the focus, but rather precise geological investigations, accompanied by microvertebrate sampling in cooperation with Tanzanian specialists.

> The samples could be studied from modern sedimentary-geological viewpoints, for example magneto-, bio-, litho-, and chemostratigraphy, organic/inorganic petrography, soil mineralogy, isotope and geochemistry, and could yield fundamental data for the facies development, stratigraphy, paleogeography, and paleoecology of the Tendaguru sequence.[19]

This approach was taken at the Museum of Natural History throughout the 1990s, following the groundwork encouraged by Hermann Jaeger and spearheaded by Wolf-Dieter Heinrich. The new director of the Institute of Paleontology, Dr. Hans-Peter Schultze, strongly supported this research. Additional sources of funding had become available following the unification of Germany. The German Research Foundation (Deutsche Forschungsgemeinschaft), in both

Bonn and Berlin, awarded grants to the project "Dinosaurier-Lagerstätte Tendaguru." The result was a Tendaguru renaissance.

A multidisciplinary group of scholars from across Germany met at the Institute of Paleontology of the Museum of Natural History on January 22, 1998, for a project colloquium. Participants from Berlin, Bonn, and Bochum presented preliminary results of studies that in some cases had been conducted over a number of years. Significantly, the processed Tendaguru matrix had yielded two new mammals: the triconodont *Tendagurudon janenschi* and the eupantothere *Tendagurutherium dietrichi*. Several papers were published.

Berlin was the site of another meeting in November 1998. The investigators finalized the contents of manuscripts to be published in *Mitteilungen aus dem Museum für Naturkunde in Berlin, Geowissenschaftliche Reihe.* The entire volume, which appeared in December 1999, contained the findings of the team. Not since the original scientific results had been printed in *Archiv für Biontologie* and *Palaeontographica* had such an accumulation of data been released simultaneously on the topic of Tendaguru. Papers included an analysis of quarry taphonomy, a biomechanical study of the forelimbs of *Brachiosaurus,* a reconstruction of *Brachiosaurus* lung structure, an estimation of the mass of *Brachiosaurus* and *Dicraeosaurus* by means of photogrammetry, an examination of sauropod bone histology, and descriptions of a new pterosaur, *Tendaguripterus recki,* a new fish, *Lepidotes tendaguruensis,* a lizard not found at the site previously, *Paramacellodus,* and a new haramiyid mammal, *Staffia aenigmatica.* Vertebrate remains were not the exclusive focus of the volume. There were also papers reviewing the palynology of Tendaguru, and preliminary descriptions of the fossil macroflora and charophytes.

Apart from this Berlin Tendaguru research group, a number of paleontologists in the late 1990s were making use of the materials collected so long ago in Africa. Per Christiansen in Copenhagen examined sauropods as part of anatomical and biomechanical studies. Oliver Rauhut at the University of Bristol worked with *Elaphrosaurus.* Brian Curtice, a sauropod specialist in the Southwest Mesa Museum in Arizona, was studying *Brachiosaurus* and returned to the Kadzi Formation in northern Zimbabwe in the summer of 1997. He was part of an expedition of the State University of New York, and was looking for sauropod material similar to that collected by Mike Raath in the 1970s.

Tendaguru had attracted researchers for some time, but the political climate and funding problems intervened. Eldon S. West had obtained approval from the Tanzanian government to study the geology of Tendaguru in 1977, but was not able to realize his project. More recently, in 1991 C. B. Wood, Mark Goodwin, and Charles Schaff had received a National Geographic grant to work on the mammals of Tendaguru, but could not proceed for various reasons.

Jack Horner, of the Museum of the Rockies, Bozeman, Montana, and Mark Goodwin, of the University of California (Berkeley), led a crew of ten to Tanzania between March 14 and 28, 1997. The group hoped to find hypsilophodontid remains and eggshell at Tendaguru, but were caught in unexpected monsoon rains, which flooded the landscape and frustrated their attempt. Unable to get closer than 24 kilometers to the site, the team examined the local geology before heavy rain ruined roads and flooded rivers.

A most welcome outcome of this concentration of effort was the popularization of Tendaguru for the people of Tanzania. Christian Magori and Charles Saanane co-authored an illustrated book for children in Swahili, *Dinosaria wa Tendaguru*. It was printed in Dar es Salaam in 1998 with German financial support, and there are plans for an English translation. Thomas Thiemeyer illustrated the story after visiting Tanzania with Manfred Ewel of the Goethe Institute in Dar es Salaam. Others have reached the isolated site. Elke Frey, author of a guidebook to Tanzania, visited Tendaguru in 1996. A team from the Arusha museum also made the trip.

The Tanzanian government has also recognized the importance of Tendaguru. In 1999, it was announced that a permanent bridge would be constructed over the Rufiji River, to enable easier access to southern Tanzania's attractions, like the Selous Game Reserve, the archaeological site at Kilwa, and Tendaguru. The US$25 million structure was to be completed in two years.

One of the most historically significant developments was still to come. In late 1999, it became apparent to the research group at the Museum of Natural History that a return to the field was necessary. A multidisciplinary project, Tendaguru 2000, was drafted and submitted to the German Research Foundation for funding. Its primary aims included further clarification of stratigraphy, dating, and paleoecology at Tendaguru. With careful stratigraphic control, sediments would be collected and analyzed for microflora and microfauna. Pollen, spores, and charophytes were sought, as well as macrofloral remains. Microvertebrates, such as mammals, and microfaunas, such as dinoflagellates and ostracods, were important components. The latter are small bivalve crustaceans that are geographically widespread and geologically short-lived. This makes them excellent index fossils for biostratigraphic dating. The location and lithology of old quarries would be recorded in support of taphonomic studies. By describing numerous geological profiles in detail and recording locations with global positioning system (GPS) devices, a geographic database would be constructed to aid in geological mapping. Tendaguru 2000 was envisioned as a pilot project, and its historical implications were profound. The same Berlin museum that had operated on such an extensive scale in the first decades of the twentieth century was returning to Tendaguru after 87 years.

Wolf-Dieter Heinrich, head curator of the Paleontological Institute of the Museum and an expert on fossil mammals, was responsible for the research program. Vertebrate paleontologist Oliver Hampe organized logistics. Other members from the Museum of Natural History included Martin Aberhan, who specialized in Mesozoic marine faunas, and Stephan Schultka, the head curator of paleobotany. The team was complemented by specialists from the Technical University Berlin: Eckart Schrank, palynologist, and Robert Bussert, sedimentologist. Benjamin Sames, a graduate student at the Free University of Berlin, was searching for ostracods and foraminifera to complete his master's degree.

Oliver Hampe drafted dozens of letters to potential sponsors. Several Berlin firms donated valuable equipment or supplied gear at reduced prices. Biwak and Toolco offered camping supplies and equipment to clear away undergrowth.

Applications for research permits were approved by Tanzanian authorities at the Commission for Science and Technology (COSTECH). The Department of Geology at the University of Dar es Salaam was the research partner in the venture, and was represented by Saidi Kapilima. The Tanzania Petroleum Development Corporation was an invaluable partner, and seconded Emma Msaky, a palynologist in the micropaleontology department. Archaeologist Remigius Chami of the Department of Antiquities completed the scientific team. The author, too, had the good fortune to join the project.

Hampe, Bussert, and Schrank flew to Dar es Salaam on August 22, 2000. They cleared crates of equipment through customs and purchased supplementary food and camping items. A water purification system was used to filter several hundred liters of water, always a scarce commodity at Tendaguru. On August 29, Heinrich, Schultka, Aberhan, and Sames arrived, and the Tanzanian team members assembled a few days later.

Final administrative requirements were completed, and in the early morning of August 31, three vehicles left the hotel compound. Toyota Land Cruisers from the university and the Tanzania Petroleum Development Corporation soon joined the procession. At noon, a ferry carried the trucks across the Rufiji River. After a 12-hour drive, two flat tires, and a demolished shock absorber, the convoy reached Lindi, about 400 kilometers south of the capital. This was a stunning contrast with the two-day ship journey of colonial times, or Louis Leakey's epic walk.

Buildings along Lindi's shore once served as headquarters and warehouses of the German East Africa Company and the imperial government. Only collapsed ruins remained, or walls supported by trees growing through them. William Edmund Cutler rested peacefully in the old European cemetery on a hill overlooking the Indian Ocean. It was a neatly maintained graveyard, and only the coral core of his headstone, exposed by flaking cement, hinted that 75 years had passed since his death. The Lindi beach seemed unchanged, kilometers of golden sand lined by palms and dotted with outrigger canoes. It was still as beautiful as described by the Tendaguru veterans who had marched along it at the vanguard of legions of bearers.

Formalities were completed with Tanzanian authorities, and on September 2, the four-wheel-drive trucks pulled out of Lindi. After a four-hour journey along a rough path, through dry creek beds, and past a few tidy, isolated villages, Tendaguru Hill appeared. Gear was unpacked and camp was set up at the base of the hill. A chameleon peered down at the cook tent the first day, a good omen.

Permits were presented to the local authorities at Mipingo, and a daily routine was established. During the day, the group split into teams and hiked or drove to localities of interest. The Middle and Upper Saurian Beds and the *Trigonia smeei* Bed were soon located. The walls of streambeds were cleaned to expose a profile, and each profile was pinpointed using GPS devices. The lithology of the profiles was recorded in minute detail. This information would be used to produce geological sections, which could be correlated over a broader area.

Sediment samples were collected for processing in Berlin. Eckart Schrank supplemented the one hundred matrix samples he had already processed in Germany—sediment that adhered to dinosaur bones in the Museum of Natural History. Emma Msaky also brought back microflora probes. Robert Bussert and Martin Aberhan ranged widely over the countryside on foot, recording geological profiles. Ben Sames filled his sample bags, hoping for ostracods and forams. Wolf-Dieter Heinrich was searching for additional mammal finds, and was compelled to return with the greatest volume of matrix. Saidi Kapilima accompanied different teams, recording local geology. Remigius Chami found a variety of lithic artifacts. Unfortunately, there seemed to be few plant macrofossils for Stephan Schultka.

Villagers from Namapuia were hired to cook for the Tanzanian drivers, who quickly made themselves comfortable by building tables, benches, and a gazebo with the plentiful bamboo. There were 16 people in camp, and water was a constant concern. Two trips were made into Lindi to purchase several hundred liters of bottled water, as well as fresh fruit, vegetables, and additional hardware. A one-way journey from Lindi to Tendaguru could be completed in three hours. It was remarkable to bounce along in air-conditioned comfort, impervious to tsetse flies, heat, and dust, and reflect on the exhausting three- to five-day walk that Expedition pioneers had faced.

A grassy flat south of the hill was strewn with pottery shards, fragments of china, broken bottles, and countless bits of rusted tins. This formed the debris of field seasons of decades past, the rubbish of the Europeans and African laborers. Schultka discovered stone foundations on the slope of the hill. Difficult to identify precisely, they marked former German or British dwellings or storage huts.

Old quarry sites were impressive. Historical excavations at Kindope, like Ig (WJ) and St (EH), formed deep, broad gullies meandering along for dozens of meters, as much as five meters deep. They were fully overgrown with grass and trees, but the volume of earth that had been shifted was humbling when one realized it had all been removed by hand. Walls were eroded to soft outlines, and gently rounded piles of earth were heaped high around their perimeters. Other quarries south of Tendaguru Hill consisted of parallel or interconnected trenches of startling length. Some gaping circular holes resembled bomb craters. A profusion of pits existed, difficult to reconcile precisely with archival field notes due to the extensive deepening and extension undertaken by British crews. At a few localities, rejected bones, usually the weighty articular ends of sauropod limbs, were piled in heaps. Quarry coordinates were determined by using GPS equipment.

Observations and descriptions, recorded during past years of fieldwork, evoke a rich picture of Tendaguru. To the modern visitor familiar with these diaries and letters, very little seemed to have changed in almost 90 years, and countless details were immediately recognizable. Fist-sized *Achatina* shells still littered the landscape, bleached white in the tropical sun. Whitened exoskeletons of millipedes were strewn about underfoot. Segments up to two centimeters in diameter belonged to individuals up to 15 centimeters long. Lithic artifacts abounded—cores and reworked flakes were commonplace.

Locals still burned the undergrowth, and smoky fires rose in the distance. Ash quickly blackened clothing during hikes through these areas. Brilliant butterflies floated through creek beds, and it was easy to understand why Cutler and Migeod had been captivated by them. The days of summer were hot and dry. Temperatures peaked at 33°C, and rarely fell below 21°C. Tsetse flies were sometimes a plague, and ticks were noticed, but thick grass and thorn vines were the bane of the region. In contrast, little flowers that grew on vines along paths, possibly a type of morning glory, added rare spots of color to the sun-bleached landscape.

Evenings were, as Ina Reck had so lyrically described them, mild and delightful. Parrots screeched overhead, doves called, and the air felt clear and cooler. A breeze inevitably arose. While sitting on Tendaguru Hill, next to the foundations of a decades-old structure, it was easy to imagine the days of Janensch, Hennig, Cutler, and Migeod. As the wind blew from the plain, one could recreate an image of the hundreds of Africans who had cooked and laughed, argued and danced, and passed weeks or even months in the campsite below. Locals still retrieved water from deep pools in nearby streams. One such pool was found in 2000, undoubtedly a water source for the German Tendaguru Expedition. As the sun set, the red walls of plateaus to the east glowed, exactly as Hennig had portrayed them in words and paintings. Venus and the moon rose over Tendaguru Hill as the crew discussed the next day's plans. The African sky was a wonder at night, bright with sparkling stars and the Milky Way.

The last evening at the site was September 17. Shortly after 9:00, one or two lions called. Their deep, hoarse roars galvanized the camp into wakefulness. A blaze was lit and watch was kept in shifts, as the chorus continued well past midnight. The next morning heavily laden trucks left Tendaguru after a successful two-week pilot project. A stop was made at Kilwa Kivinje to search for Bernhard Wilhelm Sattler's grave, but the effort was unsuccessful. The European cemetery suffered from neglect and vandalism.

In Dar es Salaam, approximately eight hundred kilograms of sediment were cleared for export to Germany. Despite tremendous changes in the capital, including a proliferation of Internet cafés, many colonial-era buildings survived. Janensch and Hennig would still have recognized the Roman Catholic and Lutheran churches. The Ocean Road Hospital of the German era was now a cancer institute, not far from the rebuilt governor's palace. The German Club now served as the headquarters for the department of Hotel Tourism and Training. The National Museum displayed a porthole of the ill-fated *Königsberg* next to the propeller of the crashed Vickers-Vimy aircraft that had carried Peter Chalmers Mitchell to Africa.

Back in Berlin, the samples underwent analysis and field notes formed the basis of papers and plans for future work at Tendaguru. The University of Dar es

Salaam and the Tanzania Petroleum Development Corporation have expressed their desire to continue the partnership, yet the difficult question of funding remains. An attractive certificate is available for purchase from the Museum, in varying denominations. Funds generated through its sale will support future research in the field and at the Museum. Fittingly, the certificate reproduces an oil painting by Edwin Hennig. No doubt he would be pleased to know that the importance of Tendaguru to paleontology is not forgotten.

-21-

A Significant Contribution

When the results of the Tendaguru expeditions are examined, the sheer tonnage of Jurassic dinosaur remains is immediately evident. Not only is a sense of wonder evoked by the magnificent skeletons mounted in Berlin, it is evident that the discoveries made by the expeditions at such immense cost of labor, and investigated with such determination in the face of such daunting circumstances, have lasting scientific significance.

A fundamental characteristic of scientific research is the persistent reexamination and revision of concepts developed by previous investigators. Just as the original interpretations of findings from Tendaguru have been modified over time, so too will current views about the environment and the inhabitants of that intriguing site. Many issues of dating, stratigraphic succession, and paleo-environmental reconstruction that proved controversial in the past remain unresolved. Nevertheless, the volume of recent work confirms that Tendaguru remains an important and rewarding area for research. Current studies have greatly improved our understanding and, inevitably, raised new questions.

According to the current model, the protocontinent of Gondwana underwent widespread rifting from Upper Carboniferous to Lower Jurassic times, due to the tectonic separation of Africa from South America, Madagascar, India, Australia, and Antarctica. The resulting fault-bounded rift basins were filled with 4,000 to 10,000 meters of mainly continental evaporite and clastic sediments, likely after the Triassic.[1] In the Middle Jurassic these basins, which overlay the Pre-Cambrian gneiss basement, were transformed into a passive coastal margin. As the ocean spread inland from the northeast, marine sediments were deposited in shallow water on a continental shelf that became the coastline of present-day East Africa. In the Upper Jurassic, the newly developing Indian Ocean transgressed the Mandawa Basin three times, creating an alternating sequence of marine and terrestrial sediments that thinned out westward to form the 140-meter-thick Tendaguru Beds. The Tendaguru area may have been 3,400 kilometers south of the equator, between 30° and 40° south latitude, according to paleomagnetic studies.[2] Intensified seafloor spreading and rifting may have caused the highest sea levels of Jurassic times during the Kimmeridgian and Tithonian. This resulted in extensive flooding of continental margins and the creation of epicontinental seas.

Stratigraphy, Sedimentology, and
Depositional Environments

Stratigraphic nomenclature relevant to the Tendaguru Beds is under discussion, with suggestions that the conventions of the International Union of Geological Sciences be followed. In this case, the Tendaguru Beds would be considered the Tendaguru Formation, and its horizons would be members.[3] Many authors, however, still refer to the Tendaguru Beds, which consist of the following sequence, from oldest to youngest: Lower Saurian Bed, *Nerinea* Bed, Middle Saurian Bed, *Trigonia smeei* Bed, Upper Saurian Bed, and *Trigonia schwarzi* Bed. Transitional sands have been identified at the base of the *Nerinea*, Middle Saurian, *Trigonia smeei*, and Upper Saurian Beds.

The highly controversial question of dating the sediments is also receiving attention. Despite the discussion about stratigraphic nomenclature, the sequence established by Janensch and Hennig is still accepted without modification. However, no further work has been undertaken to determine whether an unconformity exists between the Upper Saurian Bed and the overlying *Trigonia schwarzi* Bed. In addition to the marine invertebrates used to date the three marine horizons, there are now pollen and calcareous microfossil samples from both marine and terrestrial beds.

At the time of the German Tendaguru Expedition, the Upper Jurassic was thought to date to 5 million years ago (Ma). By Migeod's era, around 1930, a method called helium dating revised this figure to 60 million years. Current methods based on radiometric decay date the Tendaguru Beds at approximately 157 to 127 Ma.

Ammonites such as *Torquatisphinctes* and *Taramelliceras* bespeak a Kimmeridgian age for the *Nerinea* Bed, though an example of *Perisphinctes* indicates a Middle Oxford date.[4] According to recent calculations, the Kimmeridgian stage lasted from 154.7 to 150.5 Ma. The Oxfordian lasted from 156.5 to 154.7 Ma.[5] Palynomorphs like the *Anapiculatisporites-Densoisporites-Trisaccites* assemblage,[6] ostracods, and charophytes like *Mesochara* and *Aclistochara* yield a Kimmeridgian or Tithonian time for the Middle Saurian Bed. Pollen from a marine dinoflagellate assemblage, including *Barbatacysta-Pareodinia*, phytoplankton such as *?Mendicodinium*, and the charophyte *Clavator*, indicate Kimmeridgian to Tithonian ages for the *Trigonia smeei* Bed.[7] The Tithonian stage is thought to have lasted from 150 to 141.8 Ma. However, this bed recently yielded the ammonite *Aspidoceras*, suggesting a Middle Kimmeridgian age.[8] The Upper Saurian Bed is also generally accepted as Kimmeridgian to Tithonian. A fragment of *Brachiosaurus* ilium has been analyzed for $^{235}U/U^{207}P$ ratios, producing a range of 140 to 150 million years before the present.[9] Finally, the *Trigonia schwarzi* Bed is considered Valanginian to Hauterivian, part of the Lower Cretaceous. Current estimates date this period from 137 to 127 million years ago. In addition to marine invertebrates, a marine dinoflagellate pollen assemblage of *Cicatricosisporites, Trisaccites,* and *Ephedripites* adds weight to this assignment.[10]

Lithologically, the terrestrial Saurian Beds consist of fine-grained marls, massive calcite-cemented siltstones, fine-grained sandstones, layers of claystone and carbonate interclasts, and reworked bones. Traces of gypsum are present, as are both in situ and reworked calcretes. High amounts of feldspar occur in all the Saurian Beds, and the clays are predominantly smectite.

The marine horizons (*Nerinea, Trigonia smeei, Trigonia schwarzi*) are formed of sandy limestones and calcite-cemented sandstones rich in bioclasts. A moderately high-energy, near-coast depositional environment is inferred, in part from sedimentological structures in addition to the lithological evidence. The calcareous sandstones of the marine horizons are cross- and trough-bedded, as well as flaser- and ripple trough-bedded, indicating the influence of both wave and tidal action. Hummocky cross-bedded sandstones in the *Nerinea* Bed, along with conglomerate and shell beds in the *Trigonia smeei* Bed, have been interpreted as storm deposits.[11] An unusually formed conglomerate bed may be a tsunami deposit. The *Trigonia smeei* horizon features siliciclastic sediments of coastal barrier systems that interfinger with ooid bars. These ooid structures were colonized by autochthonous coral and oyster assemblages. The presence of allochthonous beds of the bivalve *Trigonia,* oriented convex side up, further supports the inference of current effects. This suggests a near-coast marine region with stable substrates and reduced deposition.[12] Apparently the marine biostrome stopped growing with an increase in sedimentation.[13] Paleontological evidence, such as marine dinoflagellates and marine invertebrates such as bivalves, gastropods, corals, ammonites, belemnites, echinoderms, and foraminifers, confirms the fully marine, shallow, and warm-water nature of these beds.

The main facies of the marine beds are tidal channel, backbarrier tidal flat (in both *Trigonia* Beds), and tidal sand bar or flood and ebb delta. Isolated beach sands also occur.[14] During periods of marine transgression, the barrier systems moved inland, cutting deeply into the tidal flats and coastal plain.[15]

Sedimentologically, the Saurian Beds exhibit ripple cross-bedding and small-scale cross-bedding. The Middle Saurian Bed is composed in part of silty and muddy intertidal sediments. The presence of calcareous peloids, or calcrete horizons, in this level as well as the Upper Saurian Bed suggests that deposition was reduced or absent for periods, and that there may have been erosion and soil formation. The dominance of feldspars in the Saurian Beds also points to erosion. The depositional facies of the Upper Saurian Bed is thought to be a dry, sabkha-like coastal plain, where brackish pools and lakes and small stream and river channels formed. The invertebrate fauna of the Middle Saurian Bed consists of one bivalve and two gastropod genera. The low diversity and dominance of one type may indicate a stress environment, with low-energy, brackish water. The periods during which the oscillating shoreline regressed produced a sequence ranging from fine-grained, nearly horizontal marine sandstones to beach and tidal flat deposits, overlain by coastal plain and sabkha sediments.

Deposition rates for the Middle Saurian Bed have been estimated based on the depth of sediment covering the *Brachiosaurus* carcasses SI and SII, prior to exposure and disarticulation. A figure of 20,000 meters per million years is derived, implying that this horizon accumulated in less than 1,000 years.[16] Evidence from the Mandawa 7 wellbore to the north of Tendaguru points to a depositional rate of four to six centimeters per 1,000 years during Upper Jurassic times.[17]

Additional evidence for water salinity levels in the Middle Saurian Bed is available from bivalved crustaceans known as ostracods. Investigation is ongoing, though preliminary results indicate wide variability from freshwater to brackish conditions (from *Trapezoidella* and *Cypridea* to *Cetacella, Rhinocypris* and *Paracypris*). Though this assemblage may be typical of estuarine conditions, it could just as easily indicate a highly variable degree of environmental salinity, with partly freshwater and partly marine influxes.[18] This assumption is broadly confirmed by microscopic algae known as charophytes. An analysis of gyrogonites, or the calcified portions of the female gametangia, also supports a range from mainly freshwater to slightly brackish conditions (*Mesochara, Adistochara*). However, the gyrogonites are easily transported, and thus may not be authochthonous.[19]

Paleoclimatic inferences may be drawn, with caution, from the lithology of the deposits and their flora and fauna. The presence of several genera of corals in the marine beds points to a warm-weather regime.[20] A subtropical or tropical climate during periods of continental deposition is inferred from the calcretes, the dominance of smectite and illite in the clays, and the presence of gypsum. Further, the entire sequence from the Lower Saurian Bed to the *Trigonia schwarzi* Bed is dominated by *Classopollis* pollen, derived from xerophytic conifers of the family Cheirolepidaceae. The epidermal cuticles of some gymnosperm genera could also point to a regime of limited water availability, high evaporation rates, or elevated soil salinity.[21] Though a variety of other gymnosperm pollens have been identified, at present the relatively low diversity and abundance of conifers adapted to dry upland environments supports the hypothesis of generally dry conditions. Fluctuations in levels of *Classopollis* and *Araucariacites* pollen in the Middle Saurian Bed are evidence for varying moisture levels, temporally or geographically.

Paleoclimatic conditions have been simulated using a general circulation model. CO_2 levels several times the current values are thought to have created a greenhouse effect, warming the planet. Trade winds delivered heavy precipitation to eastern Gondwana from December to February.[22]

Plant Fossils

Despite obvious gaps, the Upper Jurassic biota at Tendaguru can now be reconstructed in greater detail than previously. Our knowledge of the flora, especially, has benefited from detailed modern analyses. Paleobotanical macro- and meso-fossils and pollen also hold promise as paleoclimatic, paleoenvironmental, and paleobiogeographic indicators, once more and better-preserved samples have been collected with proper stratigraphical control.

Taphonomic filters and collection biases have artificially reduced the apparent diversity of the assemblage to coniferous gymnosperms. The German Tendaguru Expedition of 1909–1913 referred to macerated plant remains in the marine *Nerinea* and *Trigonia smeei* Beds and silicified wood in the Upper Saurian and *Trigonia schwarzi* Beds. Only an immature cone, probably from the family Araucariaceae, *Conites araucaroides,* was described. Much of the silicified or dolomitized wood was surface collected from the *Trigonia schwarzi* Bed, and is thus likely of Lower Cretaceous or even Tertiary age. At least four groups of gymnosperms have been recognized, including the coniferous Taxaceae, Taxodiaceae, Cupressaceae, and Ginkgoaceae.[23] Preservation indicates considerable mechanical and biological alteration, through transport, fire, insect attack (burrows and coprolites), and *Teredo*-like infestation. In addition to the remains of stems, an unidentified cone scale and needle have been recognized.

Fusain, or carbonized secondary wood from roots, stems, or branches, has been recovered, along with minute fragments of leaf cuticle, from matrix of the Middle Saurian Bed and the transitional sands between the *Trigonia smeei* and Middle Saurian Beds. Four major gymnosperm families common during Jurassic-Cretaceous boundary times are represented as fusain or cuticle: Taxaceae, Cycadaceae, Cupressaceae, and Ginkgoaceae.[24] Here too the preservation suggests lengthy transport, fungal decay, and insect attack prior to fossilization.

A new species of the coniferous Taxodiaceae, *Glyptostroboxylon tendagurense,* has been identified from a fusain fragment found in the Middle Saurian Bed. It exhibits wood structure typical of small shrubs. Another species of *Glyptostroboxylon* originates in the Upper Saurian Bed. Three new forms, likely assignable to the Taxodiaceae, Podocarpaceae, and an unknown family, are likewise derived from the Upper Saurian Bed. Although systematic assignment is problematical, eight different cuticle types have been recognized to date.[25] They also appear to represent conifers, dominated by the extinct family Cheirolepidaceae, but also possibly include the Ginkgoaceae (*Baiera/Sphenobaiera*) and the Cycadaceae. Unidentified trigonocarpic and platycarpic seeds have been recovered from high in the stratigraphic section. Perithecia of ascomycetous, edaphic fungi have also been recovered.

Palynological data confirms an apparent dominance of coniferous gymnosperms at Tendaguru, but also provides evidence of additional floral diversity. Initial attempts to recover palynomorphs yielded low returns and raised concerns about contamination with recent pollen. *Classopollis torosus,* the pollen of the conifer family Cheirolepidaceae, was found to dominate the samples from the Upper Saurian Bed. It is believed that these dry-adapted trees grew to six meters in height, and populated well-drained soil both in slopes at higher elevations and in lowlands.[26] *Araucariacites australis* pollen was also present, derived from the family Araucariaceae, another extinct conifer, that grew to 60 meters in height. Pteridophyte or fern spores were recovered, which most closely resemble examples from the family Gleicheniaceae.

Since 1990, far more samples have been processed from sediments in field jackets stored in Berlin, as well as from freshly collected Tendaguru matrix. Current indications are that the total section, from the Lower Saurian Bed to the *Trigonia schwarzi* Bed, is dominated by *Classopollis* pollen. This is especially true for the Middle Saurian and *Trigonia smeei* Beds, and indicates a strong terrestrial pollen influx.

Less well known, the Lower Saurian Bed contained less conifer pollen and featured more chorate cysts of marine dinoflagellates than the other Saurian Beds. Marine phytoplankton or dinoflagellates formed a *Barbatacysta-Pareodinia* assemblage in the *Trigonia smeei* Bed, confirming marine deposition. Terrestrially derived pollen is dominated by conifers, but lacks the pteridophytic and bryophytic elements, possibly as a result of habitat flooding.

The Middle Saurian Bed fluctuates between *Classopollis* and *Araucariacites* pollen, but also yields evidence of podocarp pollen (*Podocarpidites*), again supporting the scenario of stands of Cheirolepidaceae, Araucariaceae, and Podocarpaceae at a distance from the shoreline.[27] Spores from a diverse array of pteridophytes (ferns) and bryophytes (mosses, hornworts, and liverworts) demonstrate that these plants colonized moister habitats at Tendaguru, possibly the coastal plains where streams and pools formed. Their spores, characterized as an *Anapiculatisporites-Densoisporites-Trisaccites* assemblage, point to an aquatic environment of deposition. The Middle Saurian Bed also yielded some evidence of phytoplankton or dinoflagellates (*?Mendicodinium*) and freshwater algae of the Zygnemataceae family (*Ovoidites parvus*).

The *Trigonia schwarzi* Bed is characterized by a marine dinoflagellate assemblage of *Cicatricosisporites-Trisaccites-Ephedripites*. Recently, some 57 species of pollen were recovered in total, including 17 marine dinoflagellate palynomorphs, 2 freshwater algal forms, 12 pteridophyte and bryophyte representatives, and 15 gymnosperm taxa.[28]

Animal Fossils

The aquatic environments at Tendaguru hosted a diverse array of life forms. The marine and freshwater invertebrates are dominated by bivalves, gastropods, and corals, but also include cephalopods such as nautiloids and ammonites, belemnites, echinoderms, and brachiopods. Calcareous benthic foraminifera and charophytes are much less common. Crustaceans such as ostracods, crabs, and conchostracans were present.

Insect remains are extremely rare, with a hymenopteric form, possibly of the family Mymaridae, preserved with fusain. It was likely a parasite on insects that attacked wood. Trace fossils have been found in the form of annelid and possibly arthropod burrows, insect coprolites, insect-damaged leaves, and gastroliths. No vertebrate coprolites, skin impressions, or footprints have been identified from the site. An egg, possibly that of a turtle, was collected by Parkinson in 1929 from Quarry M2. Museum records indicate that it was found in the *Trigonia schwarzi* Bed.

The vertebrate fauna at Tendaguru is still being investigated. The original German expedition recovered evidence of the hybodont shark *Orthacodus*. Hennig described a holostean semionotiform fish, *Lepidotus,* from the Upper Saurian Bed. Additional material was examined recently, and a new species,

Lepidotes tendaguruensis, was erected.[29] It is considered a marginal marine form, attaining a length of about 25 centimeters.

Crocodiles are known from teeth found in the Upper Saurian Bed. The closest resemblance is to *?Bernissartensis.* Paramacellodid lizard remains were recently recovered from the Middle Saurian Bed. A sister group to the Scincidae, they resembled such extant African forms as *Gerrhosaurus* and *Zonosaurus.* The genus *Becklesius* may be represented. Sphenodontid lacertilians may also be represented at Tendaguru. Amphibians, possibly frogs, were part of the environment, as their remains have been recovered from freshly collected matrix. A small plantigrade reptile of unknown affinities was collected from the Middle Saurian Bed during the massive excavation at Skeleton S.

The pterosaur material that underwent such a harrowing series of journeys between Africa and Europe has been reassigned taxonomically. Reck originally established one new species of *Rhamphorhynchus* and three of *Pterodactylus.* Found in the Middle and Upper Saurian Beds, they represent smaller forms, with a 2- to 2.5-meter wingspan. The validity of some of the assignments was questioned by Wellnhofer in 1978.[30] The specimen of *Pterodactylus brancai* was assigned to *Dsungaripterus* by Galton in 1980.[31] In 1999, all of Reck's genera and species were declared nomina dubia, as they were based on undiagnostic skeletal elements.[32] A new genus and species from the transitional sands at the base of the Upper Saurian Bed, *Tendaguripterus recki* (Pterodactyloidea, Dsungaripteroidea, Germanodactylidae), was erected. A small individual with a one-meter wingspan, it is considered a dsungaripteroid pterosaur. It was recognized that one of the Tendaguru pterosaurs could be rhamphorhynchoid and another possibly dsungaripteroid, but the remainder are considered unassignable beyond the level of Pterosauria. Another examination of the material in 2001–2002 has led to the interpretation that pterodactyloids, but not rhamphorhynchoids or dsungaripteroids, are present. Furthermore, a tapejaroid pterosaur, possibly a member of the Azhdarchidae, was recognized.[33]

Our understanding of the Mammaliaformes has been greatly expanded by the work of Wolf-Dieter Heinrich at the Museum of Natural History in Berlin. The first Upper Jurassic mammal discovered in Africa, *Brancatherulum tendagurense,* of the Upper Saurian Bed, has been classified in the Pantotheria, Eupantotheria, Amphitheriidae, and Paurodontidae, and, currently, within the ?Peramuridae. Subsequently, matrix from field jackets has yielded three more genera from the Middle Saurian Bed: *Tendagurutherium dietrichi* (Eupantotheria, ?Peramuridae), *Tendagurodon janenschi* (Triconodonta, family incertae sedis), and a member of the little-known allotherid Haramiyidae, *Staffia aenigmatica.*

Tendaguru dinosaurs have been reconsidered from many perspectives, and continue to provide new discoveries and theories. The position of several taxa is still fluid. Phylogenetic taxonomy is an important tool in understanding relationships among groups, and improved character analysis and new fossil discoveries around the world guarantee that current interpretations will change. Members of both the Ornithischia and Saurischia are known from the site.

Dryosaurus lettowvorbecki, the small, swift, bipedal ornithischian herbivore, has been placed in a grouping that includes Ornithischia (Seeley, 1888), Neornithischia (Cooper, 1985), Genasauria (Sereno, 1986), Ornithopoda (Marsh, 1881), Euornithopoda (Sereno, 1986), Iguanodontia (Dollo, 1888), Euiguanodontia (Coria, Salgado, 1996), Dryomorpha (Sereno, 1986), Dryosauroidea (Sereno, 1986), Dryosauridae (Milner and Norman, 1984). (Following current practice, each clade is followed by the name of its establisher and the year of the relevant publication.) The Dryosauridae is considered a monophyletic basal taxon, a sister group to others in the paraphyletic Iguanodontia. Historically, *Dryosaurus* has been variously placed in the Laosauridae, Hypsilophodontidae, Camptosauridae, and Iguanodontidae. In 1981, Galton emended *Dysalotosaurus lettow-vorbecki* to *Dryosaurus lettowvorbecki,* and considered the African and North American forms congeneric.

Kentrosaurus aethiopicus is another ornithischian, an armored quadrupedal herbivore. It is often placed in a hierarchy similar to the following: Ornithischia (Seeley, 1888), Genasauria (Sereno, 1986), Thyreophora (Nopcsa, 1915), Eurypoda (Sereno, 1986), Stegosauria (Marsh, 1877), and Stegosauridae (Marsh, 1880). It is considered a primitive member of the monophyletic Stegosauridae. The name *Kentrosaurus,* chosen by Hennig in 1915, was considered too similar to the North American ceratopian *Centrosaurus,* and was changed to *Kentrurosaurus* in 1916. Nopcsa suggested *Doryphorosaurus* in 1916, but the original name *Kentrosaurus* is today considered valid.

The Saurischia is represented by theropods and sauropods. Theropod remains at Tendaguru, other than teeth, are rare and incomplete, though animals of various size ranges were present. Taxonomic assignment of the theropods represented at Tendaguru is in flux.

Elaphrosaurus bambergi, a swift, bipedal carnivore, is assigned to a number of different groups by various specialists. Some workers would agree with the current placement: Saurischia (Seeley, 1888), Theropoda (Marsh, 1881), Neotheropoda (Bakker, 1986), Ceratosauria (Marsh, 1884), Neoceratosauria (Novas, 1991), Abelisauroidea (Bonaparte, 1991), or Coelophysoidea (Nopcsa, 1928; Holtz, 1994). *Elaphrosaurus* appears to exhibit both abelisauroid and coelophysoid characters. In 1994, a cladistic analysis considered this genus a sister taxon to the Abelisauridae.[34] A year later, another character compilation indicated shared characters with the Coelophysoidea, one of the two lineages within Ceratosauria.[35] In 1999, Sereno retained *Elaphrosaurus* within the Ceratosauroidea.[36] Rauhut's 1998 and 2001 analyses also noted a mix of abelisauroid and ceolophysoid characters.[37] In 2000, Holtz analyzed this genus as a sister group to *Ceratosaurus* plus abelisauroids.[38] Historically, *Elaphrosaurus* has been considered a possible member of the Coeluridae, Ornithomimidae, and Ornithomimosauria. Specimens from North America have been referred to this genus.

Small and large theropods are known from numerous isolated teeth and other skeletal elements. Few of Janensch's initial referrals of 1920 and 1925 have withstood recent scrutiny. *?Megalosaurus ingens,* which Paul listed as a synonym of *?Ceratosaurus ingens* in 1988, is now considered a nomen dubium.

An enormous tooth, upon which the species was named, is not sufficiently diagnostic, but may belong to an imposing ceratosaurian theropod. *?Ceratosaurus roechlingi* and *?Labrosaurus stechowi,* both nomina dubia, were synonymized in 2000 by Madsen and Welles as *?Ceratosaurus stechowi* nomen dubium. *?Ceratosaurus roechlingi* becomes a junior synonym.[39]

The other large theropod, *?Allosaurus tendagurensis,* is considered a nomen dubium by many researchers. If allosaurid, it could be placed in Tetanurae (Gauthier, 1986), Neotenanurae (Sereno et al., 1994), Allosauroidea (Marsh, 1878), Allosauridae (Marsh, 1878).

Among the Saurischia, the Sauropoda, consisting of giant quadrupedal herbivores, is arguably the best known group of Tendaguru dinosaurs. Several lineages of this stem-based, monophyletic clade are present, including Dicraeosauridae, Brachiosauridae, Diplodocidae, Tendaguriidae, and a member whose family is incertae sedis. Various workers throughout the 1990s have revised the relationships among the groups, and the schema presented here is certain to change.

Janensch's two major groups of sauropods, the Bothrosauropodidae and Homalosauropodidae, erected in 1929 on the basis of narial and dental morphology, have been discarded. Within the former family, he placed the subfamily Brachiosaurinae and the genus *Brachiosaurus.* Within the latter family, he placed the subfamily Dicraeosaurinae and the genus *Dicraeosaurus.* Both subfamilies were elevated to family status by other researchers.

Dicraeosaurus has recently been placed in the following relationships: Saurischia (Seeley, 1888), Sauropodomorpha (von Huene, 1932), Sauropoda (Marsh, 1878), Eusauropoda (Upchurch, 1995), Neosauropoda (Bonaparte, 1986), Diplodocoidea (Marsh, 1884, Upchurch, 1995), Dicraeosauridae (von Huene, 1927). In 1995, Calvo and Salgado expressed the opinion that the Dicraeosauridae was a subfamily within Diplodocidae, which, along with *Rebbachisaurus,* comprised the Diplodocomorpha. Other workers consider the Dicraeosauridae a sister group to the Diplodocidae. Salgado posited that the two African species of *Dicraeosaurus* and the South American *Amargasaurus* formed a "lineage sequence." The presence of apomorphies or derived, more specialized characters in *Dicraeosaurus sattleri* and *Amargasaurus cazaui* prompted Salgado's suggestion that the African sauropod might be considered a second species of *Amargasaurus.* Heterochrony, a change in the timing of a trait's appearance or development from ancestors to descendants, is considered to have driven the evolution of the dicraeosaurines toward smaller body size and increased size of bifurcated neural spines.[40]

Brachiosaurus may be classified as Saurischia (Seeley, 1888), Sauropodomorpha (von Huene, 1932), Sauropoda (Marsh, 1878), Eusauropoda (Upchurch, 1995), Neosauropoda (Bonaparte, 1986), Macronaria (Wilson and Sereno, 1998), Camarasauromorpha (Salgado, Coria, and Calvo, 1997), Titanosauriformes (Salgado, Coria, and Calvo, 1997), Brachiosauridae (Riggs, 1904). In 1993 Salgado and in 1997 Salgado, Coria, and Calvo argued that the Brachiosauridae contained various titanosauriforms and other poorly known types, and was therefore not a natural group. In their analysis, they maintained that *Bra-*

chiosaurus belonged in a new monophyletic taxon, Titanosauriformes, a clade that included, among others, the most recent common ancestor of *Brachiosaurus brancai,* the Titanosauria, and all its descendants. Further, they believed that the North American and East African forms showed no unique derived features. In a 1998 analysis, Wilson and Sereno continued to recognize the family Brachiosauridae, as did Upchurch. Wedel, Cifelli, and Sanders also contended that features in the North American genus, *Sauroposeidon,* confirmed the Brachiosauridae as monophyletic.[41] Paul had erected the subgenus *Giraffatitan* to distinguish *Brachiosaurus brancai* in 1988, though subsequent researchers felt that his assignment was incorrectly based on a comparison with possible diplodocid, rather than brachiosaurid, skeletal elements.[42] It is hoped that additional North American brachiosaurid vertebrae from Bone Cabin Quarry, Wyoming, vertebrae from Dry Mesa Quarry, Colorado, and a skull from Garden Park, Colorado, will clarify the status of the family and the relationships between Laurasian and Gondwanan forms.

Fraas's *Gigantosaurus africanus* and Sternfeld's *Tornieria africana* were reassigned to *Barosaurus africanus* by Janensch in 1923. McIntosh did not consider the evidence as completely convincing in 1990. If the placement is accepted, the hierarchy would be Saurischia (Seeley, 1888), Sauropodomorpha (von Huene, 1932), Sauropoda (Marsh, 1878), Eusauropoda (Upchurch, 1995), Neosauropoda (Bonaparte, 1986), Diplodocoidea (Marsh, 1884), Diplodocidae (Marsh, 1884).

Tendaguria was established in its own family, the Tendaguriidae, in 2000.[43] Janensch had briefly described but never classified the vertebrae upon which the new genus and family are based. No additional systematic position has been developed for this highly derived material.

Fraas's *Gigantosaurus robustus,* later Sternfeld's *Tornieria robusta,* was revised to *Janenschia robusta* by Wild in 1991.[44] Long considered a titanosaur, in what today is known as the Titanosauriformes, *Janenschia* was reexamined in 2000 by Bonaparte, Heinrich, and Wild, who refuted the assignment. Exact relationships are unknown, but its closest resemblance is to the Camarasauridae. Furthermore, additional isolated dorsal and caudal vertebrae, formerly referred to *Janenschia,* are now of uncertain taxonomic affinity.

Biogeography

The flora of Tendaguru is of considerable paleogeographic significance. Some researchers recognize three vegetational zones during Upper Jurassic times, of which the North Gondwanan and South Gondwanan Zones are relevant to Tendaguru. The Cheirolepidaceae that produced the abundant *Classopollis* pollen dominated the equatorial North Gondwanan Zone, while the South Gondwanan Zone is characterized by the Araucariaceae and Podocarpaceae, producers of bisaccate and trisaccate pollen. Tendaguru was situated at the border of these two zones, and the pollen assemblage implies a mixture of elements from both.[45] *Podocarpidites* pollen and *Ephedripites* dinoflagellate cysts are typical of the South Gondwanan Zone, but trisaccate pollen was absent from the Kizimbazi well core, a hundred kilometers to the north of Tendaguru, near Kilwa.[46]

Among the macroflora, *Glyptostroboxylon* is now reported for the first time from the Upper Jurassic of Africa. It has previously been recognized from the Tertiary of Europe, Asia, and Antarctica.[47]

The ostracods from Tendaguru are both similar to and different from those of North America. The charophytes appear to be more closely related to those found in parts of Europe than to those in North America.[48] The appearance of *Cypridea* at Tendaguru poses questions, as the taxon is currently known from very late Jurassic strata in North America. Its migration to Gondwana may be related to the development of zonal wind circulation systems in the Upper Jurassic as continental breakup progressed.[49]

Many recent vertebrate discoveries at Tendaguru are important paleobiogeographically, since the flora and fauna of Mesozoic times in Africa are not as well known as those in other parts of the globe. The geographic and temporal range of several groups has been extended. The paramacellodid lizard is the first of its kind to be discovered from this latitude, and is also the oldest known form in Africa. Similarly, if pterosaurs belonging to the Tapejaroidea, Azhdarchidae are confirmed, this group's presence would be extended to the Upper Jurassic of Africa.

Tendaguru remains the only site in Africa where Upper Jurassic pterosaurs have been found. The clades represented at the site are still debated, but if dsungaripteroid forms were present, they are representatives of a group that was well established by the Upper Jurassic and persisted until the Lower Cretaceous. They have been found in Europe, Asia, and possibly North America, as well as South America.

The Sauropoda was a diverse, cosmopolitan, and evolutionarily successful group of dinosaurs whose members appear in the fossil record in the Lower Jurassic and reached a major peak in size and number of genera in the Upper Jurassic. The clade flourished into the Upper Cretaceous. Its relationship to the sister group Prosauropoda implies an Upper Triassic origin for the Sauropoda. Found on nearly every continent, sauropods are thought to have diverged into a number of lineages and dispersed quickly and globally during the Jurassic. Upchurch suggested that while brachiosaurids and diplodocids were able to move between the supercontinents to become the dominant terrestrial sauropod fauna, dicraeosaurids and titanosaurids were specific to Gondwana.[50] Habitat preferences and paleoenvironmental causes are given as explanations. In 1996, Bonaparte theorized that the presence of dicraeosaurids in the Jurassic of Africa (*Dicraeosaurus*) and the Cretaceous of South America (*Amargasaurus*) indicated that this lineage became differentiated in Gondwana, but did not radiate to Laurasia for some paleogeographic reason.[51]

In the 1970s and 1980s the resemblance between dinosaurs from Tendaguru and the North American Morrison Formation was considered paleobiogeographically significant. Comparisons were made between common forms, such as *Brachiosaurus, Barosaurus, Dryosaurus, Elaphrosaurus, ?Allosaurus, ?Ceratosaurus,* and *?Megalosaurus.* The degree of faunal overlap is perhaps not as compelling, as new finds are made or original specimens are redescribed and reassigned. A land connection between North America and Africa, possibly via Europe, was implied to facilitate this exchange.[52] Laurasia and Gondwana were most likely separated before Kimmeridgian times, when *Brachiosaurus* and *Barosaurus* are known from both landmasses. In 1998, Wilson and Sereno contended that the Neosauropoda (diplodocoids and their sister taxa, the camarasaurids, brachiosaurids, and titanosaurids) must have originated earlier than the Upper Jurassic or the split of Pangaea. Consequently, *Brachiosaurus* and *Barosaurus* could have been present in North America and Africa as remnants of an earlier, globally dispersed sauropod fauna, some members of which persisted and diversified, while others went extinct, as the continents became isolated.[53] The identification, by McIntosh and Raath, of coeval *Brachiosaurus* and *Tornieria* (= *Janenschia*), as well as *Camarasaurus* at Kadzi in Zimbabwe, raises interesting questions about the degree of dispersal of the land giants of Tendaguru. In 2002, sauropod and theropod footprints were discovered in the Zambezi Valley of Zimbabwe. Unfortunately, due to political events, this and other faunas, like those at Gokwe, are not likely to be investigated for some time.

The Ornithischia may have split into the Thyreophora and Neornithischia prior to the end of the Triassic. The Iguanodontia, such as *Dryosaurus*, were also present in Gondwana and Laurasia by the Upper Jurassic. Coria and Salgado proposed that a group of basal iguanodontids persisted until the Upper Cretaceous in South America.[54] Another ornithischian group, the Thyreophora, represented at Tendaguru by the stegosaur *Kentrosaurus*, persisted from the Middle Jurassic to the Lower Cretaceous. This cosmopolitan clade has been found in Africa, Europe, Asia, and North and South America.

The Neotheropoda is thought to have split into Ceratosauria and Tetanurae prior to the end of the Triassic. By the Middle Jurassic, the Neotetanurae had diverged into the Allosauroidea and Coelurosauria. Allosauroids survived into the Cretaceous in North America and Africa. Similarly, the Ceratosauria had diverged into Ceratosauroidea and Coelophysoidea before the Triassic ended. Members of the Ceratosauria appear to have been more common until the Middle Jurassic, by which time diverse Tetanurae genera were present on every continent. Around the Upper Jurassic, members of the Ceratosauroidea became the Neoceratosauria. Members of the Neoceratosauria, such as *Elaphrosaurus*, appeared in the Upper Jurassic of Africa and North America. The Neoceratosauria survived into the Cretaceous on Gondwanan continents like South America, Madagascar, and India, as well as Europe.

The taxonomic position of *Elaphrosaurus* is still under revision, but possible assignment within the Ceratosauroidea or Abelisauridae/Coelophysoidea is of interest, as the genus would be a representative of the early radiation of these theropods.

The mammaliaform fauna continues to reinforce the importance of Tendaguru. *Staffia* is the first member of the Allotheria, Haramiyidae, to be found in Gondwana, and extends the presence of this family back to the Upper Jurassic. If *Brancatherulum* and *Tendagurutherium* of the Eupantotheria are correctly assigned to the Peramuridae, they are the oldest representatives of the family, and mark the first occurrence in the Upper Jurassic of Africa. Possibly as a result of collecting or taphonomic bias, or geographic or ecological barriers, no multituberculates have been recovered at Tendaguru. The group is common in the Upper Jurassic of North America and the Lower Cretaceous of Africa.

Certain ancestral anatomical features persisted longer in the Tendaguru mammals than in other Gondwanan or Laurasian mammals. A mandibular eardrum is present in *Tendagurutherium,* with the angular-articular assembly still connected to the dentary. For Upper Jurassic times, this illustrates an ancestral stage in the evolution of the mammalian middle ear, while signifying a derived stage for the Eupantotheria. The Upper Jurassic Eutherian triconodont *Tendagurodon* has an ancestral molar cusp pattern, resembling that found in Upper Triassic to Lower Jurassic triconodonts. These lines of evidence may point to the existence of a unique East African faunal province.[55]

Taphonomy

The question of how the bone accumulations formed at Tendaguru has received renewed attention. A correlation has been hypothesized between high global sea levels and an abundant fossil record. Coastal areas, often populated by numerous dinosaur taxa, received abundant sedimentation from terrestrial and marine sources during these phases. This created optimal conditions for fossilization, and faunas of lowland habitats are frequently preserved in brackish, deltaic, or shallow marine horizons. The late Kimmeridgian to Tithonian was such a period.[56] General habitat preferences have been suggested: diplodocids may have preferred the continental interior; dicraeosaurids are often found on coastal plains; and brachiosaurids, dryosaurids, and stegosaurids ranged over both environments.[57]

There is currently no consensus on a taphonomic regime, as the divergent models of both German and British authors are still cited in the most recent literature. Detailed analyses by Russell, Béland, and McIntosh in 1980, and Heinrich in 1999, have furthered our understanding.

It is generally recognized that there was considerable postmortem disarticulation and transport, leading to a broad range of bone accumulations, from partial or associated skeletons to bonebeds and isolated elements. Of all the specimens collected by the German expeditions, with the exception of those from the more distant sites of Mchuya and Makangaga, 64 percent of the skeletons were one-quarter or less complete, 24 percent were one-half complete, and 10 percent were three-quarters complete.[58] Current and wave sorting aligned limb elements and ribs at the sole *Dryosaurus* site and in at least five monogeneric sauropod quarries in the Middle and Upper Saurian Beds.[59] Abrasion was evident in the delicate processes of well-articulated presacral vertebrae of two genera of sauropods. Water transport preferentially accumulated small and light stegosaur podial elements. The fragmentary preservation of plant remains also indicates transport, water and biological action, and other environmental effects. Analysis of German quarry maps yielded examples of monogeneric, multigeneric, single-individual, and multi-individual sites. The Middle and Upper Saurian Beds produced the richest finds and were dominated by adult sauropods in predominantly single-individual, monospecific quarries.[60] One reason for the rarity of complete sauropod skeletons is the unlikelihood of such large bodies being fully buried quickly enough to prevent disarticulation by water or scavenging by carnivores. Smaller skeletal elements such as the skull and hand and foot bones are easily washed away.

The Middle Saurian Bed was the source of approximately one-quarter of all German and British quarries.[61] The best-preserved semiarticulated sauropods on display in Berlin came from this bed. *Brachiosaurus* was the most common taxon, followed by *Dicraeosaurus* and *Barosaurus*. The most abundant skeletal elements appear to be limb bones of adult individuals. Appendicular sauropod elements, specifically hind limbs of diplodocids and forelimbs of brachiosaurids, predominate at about 30 German-excavated sites.[62] The discovery of upright *Brachiosaurus* hindlimb elements (Skeleton S) and the complete manus for *Janenschia* (Nr. 5, P), *Brachiosaurus* (R), and *Barosaurus* (XIII, 28) suggests that the animals may have been mired. Juvenile sauropods, perhaps less susceptible to entrapment and death in unconsolidated substrates due to their lower body mass, are virtually unknown.

Bonebed accumulations imply a higher-energy depositional environment. Monogeneric bonebeds of *Dryosaurus* and *Kentrosaurus* are a prominent feature of the Middle Saurian Bed. The bipedal iguanodontids also occur in bonebeds in North America, but were only preserved at a single locality in Africa. Two separate but overlapping accumulations are found at the same Kindope site, where the familiar range of disarticulation occurred, from well-articulated partial skeletons to individual vertebrae and limb bones. Reck recognized three size classes, and Heinrich's analysis points to two peaks in the size distribution, based on femur length. Sexual dimorphism and the presence of both juveniles and adults are possible explanations.

Similarly, the *Kentrosaurus* bonebeds featured individuals of various sizes, ages, and likely genders, represented by individual elements and articulated portions of skeletons. Isolated elements were also recovered from several sites. No consensus of opinion exists to explain the accumulation of over 30 individuals at St and HE. The virtual pavement of manual and pedal elements at Quarry X, representing at least six individuals, is thought to be the result of water's transporting these smaller, lighter bones away from decaying carcasses.[63] Scavenging is assumed from the presence of theropod teeth and vertebrae. Stegosaurs appear to have been relatively more abundant in the Middle than in the Upper Saurian Bed, though overall the thyreophorans formed a comparatively minor element of the Tendaguru fauna.[64] Size classes and anatomical differences among similar skeletal elements indicate possible sexual dimorphism.

In contrast, the Upper Saurian Bed was the source of over half the German quarries and one-third of the British excavations. It produced nearly twice as many articulated skeletons as the Middle Saurian Bed.[65] This horizon appears to have supported a more diverse sauropod fauna. *Brachiosaurus* was the most common, and *Janenschia, Tendaguria, Dicraeosaurus,* and *Barosaurus* were also present. The greatest number of barosaurs were found in this bed, with one side of the animals preserved most often.[66] Little can be said about the taphonomy of the Tendaguru theropods or pterosaurs in the absence of quarry maps, notes, or stratigraphic profiles. Extremely delicate pterosaurs and mammal remains indicate low-energy environments or minimal transport prior to burial.

The British expeditions collected primarily from the transitional sands deposited above the Middle Saurian Bed and below the Upper Saurian Bed. Roughly one-half of all British quarries originate in these two horizons.[67] As they are dominated by accumulations of sauropod limbs from multiple individuals, a great deal of postmortem transport and sorting is implied.

Ultimately, a broad range of factors must be considered when interpreting the origin of the Tendaguru sites. Janensch and Hennig proposed that a number of large sauropods became mired in shallow-water lagoon sediments, to be drowned by incoming tides. R. McNeill Alexander's research supports the miring of large sauropods crossing wet clay.[68] Carcasses were then disarticulated, undoubtedly scavenged on occasion, transported, and subjected to further processes prior to burial. In part, the death of the dryosaurs and stegosaurs was explained as a catastrophe of a similar nature.

Reck refuted a mass mortality scenario, pointing to the irregular occurrence of countless bones of many taxa throughout a lengthy accumulation of fine-grained sediment, which was only interrupted during marine transgressions.

Schuchert contended that the extensive ornithischian bonebeds indicated mass drowning episodes in rivers. Carcasses were washed downstream, disarticulated, scavenged, redeposited, and buried. In this view, sauropods inhabited brackish marshes, similar to Reck's saline coastal marshes. This scenario was accepted by British researchers such as Kitchin and Parkinson, who also argued for an estuarine depositional environment.

Both Heinrich and Russell, Béland, and McIntosh drew on additional sedimentological and paleontological data to offer another reconstruction. Pronounced seasonality, inferred from multiple lines of evidence, may have increased in severity and given rise to droughts. Dinosaurs may have concentrated around shrinking resources of food and water. Famine may have induced increased dinosaur mortality, and seasonal flash floods may have transported and concentrated carcasses. Seasonal conditions appear to have persisted throughout the Upper Jurassic, and over the span of many millennia, bones may have accumulated to such an extent that they mirrored sudden mass death events. A theory of long-term attritional deaths emerges, possibly punctuated by occasional natural catastrophes. This theory is similar to Migeod's. Future investigations may further clarify the complex interplay of climate, fauna, flora, and habitat.

Dinosaur Paleobiology

Current research involving Tendaguru dinosaurs has included numerous considerations of paleobiology. New analytical techniques have led to better understandings of the trophic relationships, physiology, biomechanics, and even paleopathology of these intriguing animals. Bone pathologies have been recognized in Tendaguru dinosaurs. These include fused vertebrae in *Dryosaurus*[69] and rehealed ribs in both *Brachiosaurus*[70] and *Dicraeosaurus*.[71] Other aspects of the biology of Tendaguru dinosaurs will be summarized for each group.

Ornithopods. The abundance of well-preserved remains of *Dryosaurus* may indicate some form of social behavior, and has allowed detailed anatomical studies that offer insights into the metabolic regime. These ornithischians possessed a pleurokinetic skull, in which the upper jaws rotated outward as they occluded with the lower jaws. This action produced a transverse power stroke that efficiently processed vegetation in a range of up to two meters from the ground. Histological investigations characterize *Dryosaurus* bone as fibrolamellar and lacking in rest lines and growth rings. These features, among others, can be interpreted as the product of rapid, sustained, indeterminate growth.[72]

Encephalization quotients, the ratios of measured brain volume to brain volume predicted from body mass, have been computed for ornithopods. The values of this lineage are thought to have been much higher than those of sauropods or stegosaurs.

Body mass has been estimated at 104 kilograms, by using polynomial equations to describe body shape.[73] Using the mid-shaft circumference of the femur, a value of 100 to 400 kilograms is derived.[74] Limb proportions indicate an agile and fast-running animal, perhaps capable of achieving 33 to 43 kilometers per hour.[75] Beam theory was employed to analyze ontogenetic changes in *Dryosaurus* bone. It was concluded that long bone architecture underwent the greatest changes at a very early stage in the animal's life. The alteration in mass and distribution of bone indicates that *Dryosaurus* may have been obligatorily quadrupedal as a hatchling, and became bipedal in the first five months as its tail began to function as a counterbalance.[76] This interpretation has been challenged, with the suggestion that the inherent curvature of the femur, maintained throughout all ontogenetic stages, caused the biomechanical alteration.[77] Pathologies have been described in the dorsal vertebrae of this animal.[78]

Stegosaurs. The stegosaurs form a smaller component of the Tendaguru fauna. These animals, also possibly gregarious, were likely low-level browsers that cropped vegetation up to one meter from ground level, or as high as two meters if they reared up on their hind limbs. Fruits, leaves, and seeds would have been stripped or pulled off plants with the hard toothless covering at the front of the jaws known as a rhamphotheca, and roughly processed by nongrinding teeth set further back along the jaws.

Kentrosaurus was graviportal and more slow-moving than most of the Tendaguru dinosaurs. Its body mass has been estimated at 1.4 tonnes, based on the cross-sectional area of the humerus and femur.[79] Using mid-shaft femur circumference, a range of one to four tonnes was calculated.[80] Distinct size ranges within the bone accumulations suggest the possibility of sexual dimorphism.

According to endocast data, *Kentrosaurus* was also among the smallest-brained creatures of the area, with an encephalization quotient of approximately .54.[81] Stegosaurs are noted for a major increase in the size of the endosacrum. The presence of an avian-like glycogen body within the neural canal of the sacrum has been hypothesized, though its physiological function is not clearly understood.[82]

Recent discoveries in China have led researchers to reposition the spines, which Janensch placed on the hip, forward over the shoulders. The combination of plate and spine ornamentation, considered less advanced than other Jurassic stegosaurs, may have served a variety of functions, including agonistic display, thermoregulation, and defense.

Theropods. The Upper Jurassic theropods from this site are not well known, though there is evidence of a range of sizes. The diversity of this group may have been relatively high thanks to a broad range of prey size. It has also been speculated that the unrelieved pressure exerted by carnivores on herbivores, especially sauropods at different growth stages, may have led to maximum size in the sauropods.[83] The mass of the Tendaguru *?Ceratosaurus* is estimated as 1.2 tonnes, based on the cross-sectional area of the femur of *Ceratosaurus* from North America.[84] Based on mid-shaft femur circumference, a figure of 400 to 700 kilograms has been derived.[85] The mass range of *Elaphrosaurus* has been estimated at 100 to 400 kilograms, based on mid-shaft femur circumference.[86]

Sauropods. Bones from Tendaguru sauropods have undergone geochemical analyses. *Barosaurus* bone was thin sectioned as part of a diagenesis study. It is hoped that an understanding of the processes and stages of fossilization will allow inferences about the physical and chemical conditions to which bone has been subjected. It should be possible to interpret certain histological features, the nature of the depositional environment, and the relationship between the internal and external sources of replacement elements such as iron.[87]

Uranium can be precipitated and incorporated into fossil bone, yielding lead isotopes, which may allow dating. A fragment of the left ilium of *Brachiosaurus* from the Upper Saurian Bed was subjected to electron microprobe and isotope analysis. A complex uranium-lead system existed in the sample, and despite correction for radiogenic disturbances such as diagenetic introduction of lead and continuous radon loss, it was not possible to calculate a precise age. The most consistent dates from the sample range from 150 to 140 million years ago.[88]

A cast of the left labyrinth or inner ear of *Brachiosaurus* was preserved, and the dimensions of the three semicircular canals and their ampullae were measured. It was estimated that the labyrinth functioned optimally at a range of .02 to .1 Hertz, lower than that of humans, due to the far greater mass of the animal and correspondingly larger diameter of the semicircular canals.[89]

The volume of a *Brachiosaurus* cranial endocast was adjusted to reflect the possibility that the actual brain was about 50 percent of the endocast volume. When this volume was compared to Colbert's estimate of body volume, an encephalization quotient of .1 was derived.[90] Others used the total endocast volume to arrive at an EQ ranging from .2 to .25 for sauropods as a group.[91] This was the lowest encephalization quotient derived for any group of dinosaurs. Some researchers have concluded that sauropod behavior was not complex.

Massive enlargement of the sacral neural canal, recognized in *Barosaurus* and *Dicraeosaurus,* may indicate the presence of an avian-like glycogen body.

Since their discovery, sauropods have posed intriguing paleobiological questions due to their great size. How their large bodies functioned has been the subject of considerable research and debate. Divergent opinions have been expressed about such matters as the sauropods' neck posture (with consequent implications for hemodynamics and feeding) and metabolic physiology. Biomechanical studies have analyzed limb posture, gait, and speed of locomotion. Tendaguru sauropod genera have featured prominently in discussions of dinosaur physiology, growth rates, and biomechanics.

Fundamentally important for understanding how sauropods worked is knowing how much they weighed. Mass estimates for Tendaguru sauropods vary greatly as new methods are developed. In 1935, Janensch estimated a body volume of 25 and 32 cubic meters for two *Brachiosaurus* skeletons, and a mass of about as many tonnes, respectively.[92] In a later paper, he gave a figure of 40 tonnes for *Brachiosaurus*, based on the mounted skeleton in Berlin, but specified no method of deriving this figure.[93]

Colbert used a scale model and displacement theory to estimate a figure of 78.2 tonnes for *Brachiosaurus*.[94] Béland and Russell measured the cross-sectional area of humeri and femora and derived mass estimates of 14.9 tonnes for *Brachiosaurus*, 8.3 tonnes for *Barosaurus*, and 3.3 tonnes for *Dicraeosaurus*.[95] Anderson, Hall-Martin, and Russell measured the mid-shaft circumference of humeri and femora to arrive at a weight of 29 tonnes for *Brachiosaurus*.[96] Peczkis used mid-shaft femur circumference for a result of 10 to 40 tonnes for *Brachiosaurus, Barosaurus, Dicraeosaurus,* and *Janenschia*.[97] By using a scale model and the water-displacement method, Alexander derived a figure of 46.6 tonnes for *Brachiosaurus*.[98] Paul employed a scale model and water immersion to calculate a mass of 29 tonnes for *Brachiosaurus*.[99] Gunga et al. applied laser stereophotogrammetry to the skeletons mounted in Berlin, to obtain body masses of 74.4 and 12.8 tonnes, respectively, for *Brachiosaurus* and *Dicraeosaurus*.[100] By using a scale model and water immersion, Christiansen developed estimates of 37.4 and 5.4 tonnes for *Brachiosaurus* and *Dicraeosaurus*, respectively.[101] Christian, Heinrich, and Golder adjusted their figure for *Brachiosaurus* to 63 tonnes, to reflect a lower density due to pneumaticity and reduced neck and tail volumes.[102] They considered Henderson's computational method, which estimated volume by calculations on body slices, to be more accurate than their own. Most recently, Seebacher applied polynomial equations to describe body shape, then added volume figures and corrections, to arrive at 28.2 tonnes for the North American *Brachiosaurus,* 20 tonnes for the North American *Barosaurus,* and 4.4 tonnes for *Dicraeosaurus*.[103] Thus, the estimates for *Brachiosaurus* range from 14.9 to 78.2 tonnes, for *Dicraeosaurus* from 3.3 to 40 tonnes, and for *Barosaurus* from 8.3 to 40 tonnes.

Locomotion studies of Tendaguru sauropods substantiated that the structure of the forelimbs and hind limbs was pillar-like.[104] Very large brachiosaurids had relatively lightly built humeri, small olecranon processes, reduced carpals, and columnar forefeet held in an unguligrade position with greatly reduced phalanges.

These features are thought to have resulted in gracile, lightly muscled forelimbs with relatively weak elbows and wrists and inflexible feet. As a result, the thrust of forefeet against the ground did not contribute substantially to propelling the animal forward. The humeral retractor muscles supplied the greatest propulsion, while the forelimbs acted more as support than as locomotor mechanisms. The forelimbs of *Dicraeosaurus* and *Barosaurus,* however, were more robust than those of elephants.

In contrast, the hind limbs of almost all sauropods are massive, more so than those of elephants. Femoral retractor muscles, acting perpendicularly to the long axis of the bone, supplied more propulsion than the forelimbs. The nearly digitigrade hind feet are considered more flexible, and possessed a posterior heel pad. Strong claws would have been important in thrusting off from the ground.

As a result, it is postulated that large animals like *Brachiosaurus* did not have the same locomotor abilities as elephants.[105] They probably moved at a fast walk at maximum, but generally progressed quite slowly, especially when feeding.

Another study also argued that fore- and hind limbs in *Brachiosaurus* were erect, held beneath the body, and that locomotion was constrained in comparison with large modern mammals.[106] Higher walking speeds required either the elbows to flex or the pectoral girdle to shift, in order to compensate for maximum stress loads. Assuming a vertical neck posture, the mass distribution was calculated to place 20 tonnes on the pectoral girdle, or 33 percent of the body total versus 48 percent, as calculated by Alexander.[107]

Locomotion in such a large animal likely proceeded very smoothly and fairly slowly, with minimal differences in joint height as mass was shifted from shoulders to hips and back again. If elbow flexion at the middle of the support phase of forelimb movement is considered minimal, the pectoral joints had to move .2 to .3 meters vertically relative to the center of gravity at every step.

In an effort to understand the energetics of locomotion, limb swinging was studied in large herbivores. Calculating the stride frequency from the natural pendulum frequency of limbs was problematical, due to the great variation observed in both these frequencies among elephants and giraffes. Likewise, estimates of the speed of a track maker from either of these limb frequencies may prove inaccurate.

In dinosaurs, the pectoral and pelvic joints of the swinging limbs were probably accelerated by the ground reaction forces conducted by the supporting limbs. Shorter swing times and longer support times resulted, which may have helped to produce an energy-efficient mode of locomotion for these large terrestrial vertebrates. The average walking speed of adult Tendaguru sauropods was probably three to four kilometers per hour.[108] The maximum speed of a 78-tonne *Brachiosaurus* was calculated at 18 kilometers per hour.[109] Assuming some form of social behavior from trackway evidence, groups of sauropods were probably obliged to move over considerable distances when foraging for food, as they would have had a great impact on the local vegetation.

Recently, neck posture has been debated for a number of lineages, based on anatomical and hemodynamic arguments. Computer modeling of the cervical vertebrae of North American diplodocids implied maximum vertical feeding heights considerably reduced from previous estimates. These forms browsed low, though some sauropods occasionally reared up on hind legs.[110] The tall neural spines of *Dicraeosaurus* also presumably restricted the vertical range of motion of the neck, but lateral flexibility seems to have been enhanced. In this sauropod, the neck appears to have been straight at the transition from dorsal to cervical vertebrae, and then curved downward toward the skull.[111]

Computer simulations of the skeletons of macronarian sauropods such as *Camarasaurus* and *Brachiosaurus* reveal that the necks were not held vertically, but closer to horizontal.[112] This conclusion is supported by the lack of keystoning in the centra of posterior cervical vertebrae in *Brachiosaurus*. The angle of transition from dorsal to cervical vertebrae is nearly straight, continuing throughout the neck with minimal curvature in a neutral pose. The head of *Brachiosaurus* is thought to have been held about six meters above the ground.[113]

In 1998, two divergent opinions emerged regarding sauropod neck biomechanics. One argued that vertebral structure and articulation were analogous to those of segmented beams. Depending on the morphology of neural spines, transverse processes, and cervical ribs, the neck could be braced ventrally against compression by cervical ribs that could not compress, or dorsally by ligaments, epaxial muscles, and tendons. A combination of these features was also recognized in some sauropods.[114] The conclusion was a rigid neck, carried nearly horizontally. Primarily ventral bracing through overlapping cervical ribs was characteristic of *Brachiosaurus*. Dorsal bracing was evident in *Dicraeosaurus,* with its tall, bifurcated neural spines and short cervical ribs.

The second viewpoint calculated forces of compression and bending moments, or torque, on discs situated between cervical vertebrae in *Brachiosaurus*. In this model, dorsal muscles, ligaments, and tendons provided support for a neck held nearly vertically. Tension on the cervical ribs restricted the neck from moving backward. The center of mass was close to the base of the neck. Further, the cervical ribs were considered to be positioned too close to the intervertebral joints to habitually support the neck in a horizontal position.[115] Other researchers have also posited intervertebral discs in sauropods, producing a steeply inclined neck in brachiosaurids.

Pennate muscles, capable of exerting greater forces than striated muscles, may have been associated with features resembling tendons in the cervical vertebrae of the brachiosaurid skeleton M23 from Tendaguru.[116] This support mechanism may have provided additional strength or stability to the neck. It is, however, generally considered that powerful elastin tissues, like a ligamentum nuchae, would have been required to lift long sauropod necks.

The discovery of *Sauroposeidon,* an enormous brachiosaurid in the Lower Cretaceous of North America, led others to argue that the extremely long cervical ribs were held in bundles by bands that would allow the ribs to slide past one another like tendons. Accordingly, the immobile ventral bracing system proposed previously was not considered valid. Epaxial muscles and articulated vertebrae provided control over tension and compression respectively. Tension was maintained ventrally by cervical ribs. It was noted that the neck of brachiosaurids might have been held in a shallower S-curve, because there was a point in the cervical sequence where neural spines became lower.[117] Discussions of neck posture in *Brachiosaurus* are hampered by the incomplete state of the neural spines of the first few cervical vertebrae.

Strong objections have been raised to greatly elevated necks in sauropods on the basis of a line of evidence other than the construction of the neck skeleton itself. In 1973, it was estimated that the hydrostatic or gravitational pressure of a column of blood extending 6.5 meters from the heart to the brain of *Brachiosaurus* required a systemic arterial pressure of 568 mm Hg if the animal was in a quadrupedal stance.[118] To cope with such high pressures, adaptations such as a small radius ventricle, a thick myocardium, thick-walled arteries, and some form of pressure reduction before blood reached the capillaries were hypothesized.

A rearing *Barosaurus,* its head held 12 meters above the ground, was estimated to require a systolic or contraction-phase blood pressure of 800 mm Hg.[119] To cope with such pressures, it was calculated that a 50-tonne sauropod's heart would have to weigh over 1.6 tonnes.[120] The oxygen consumption of such thick ventricles would have been very high, and the mechanical efficiency low. An alternative mechanism was suggested, whereby a four-chambered primary heart pumped blood to a secondary heart which in turn pushed blood to a series of three pairs of single-chambered hearts along the subclavian arteries in the neck. A thick, rigid skin and check valves in the cerebral arteries complemented the system.

Other researchers disputed this solution, estimating that aortic pressures of 740 mm Hg were created to push blood eight meters from heart to head.[121] To sustain such pressures, a ventricle 90 centimeters thick, in a heart 2.3 meters in diameter and weighing 7.3 tonnes, would have been necessary.[122] This was considered mechanically unlikely. It was suggested that sauropods were amphibious, so that surrounding water pressure reduced stress on the circulatory system. Alternately, an enlargement at the base of the aorta would reduce the effort of the heart and still maintain high blood pressure.

An alternative was proposed in 1996: a siphon loop in which blood ascended in the arteries, passed through capillaries in the neck and head, and descended through the veins. The siphon loop suggestion would overcome a column of blood with a hydrostatic pressure of 640 mm Hg or 740 mm Hg at the aorta, but equally high negative pressure would exist at the skull. It was argued that blood vessel collapse could be prevented if vessels were surrounded by thick, rigid skin and the brain and cranial vessels were surrounded with cerebrospinal fluid at similar negative pressure.[123] Lower aortic pressures could theoretically be sustained if the blood vessels were supported by muscles and extremely dense connective tissue attached to a nearly rigid skin, creating higher pressure in the vessels.

In 1995 and 1999, a wide variety of physiological functions were calculated for Tendaguru sauropods, including the cardio-circulatory system of *Brachiosaurus.* The distance from heart to brain was measured as eight meters, implying a length of 9.8 meters for the carotid arteries. A four-chambered heart was hypothesized, weighing 386 kilograms and capable of moving 17.4 liters per stroke at a frequency of 14.6 beats per minute. To overcome hydrostatic pressure, the heart had to pump at a pressure of 600 to 750 mm Hg.[124] A rete mirabile along the neck was proposed to attenuate this pressure at the brain. A total blood volume of 3,600 liters created tremendous hydrostatic pressure in the limbs, calling for adaptations like venous pumps and valves, rigid connective tissue, and thick skin to prevent edema. This blood volume was estimated to be six times that of *Dicraeosaurus*, whose heart may have weighed 48 kilograms and beaten 23 times per minute.[125] Another estimate suggested a one-tonne heart for a 30- to 32-tonne tachymetabolic *Brachiosaurus*.[126]

In 2000, it was suggested that a 40-tonne endothermic *Barosaurus* would require a 2-tonne left ventricle that was 52 centimeters thick, in order to pump blood at a pressure of 700 mm Hg.[127] During the diastolic or dilation phase, the ventricle would expand to 1.6 meters in diameter. The siphon loop was criticized as likely to produce fatal gas bubbles in blood at low pressures, and the difficulty of making all blood vessels extremely rigid to prevent collapse was cited. Smaller stroke volumes would allow for a smaller heart beating more quickly or a lower blood flow rate. If output of the heart were reduced to 50 percent of the level of an endotherm, a 530-kilogram ventricle would be required to maintain 700 mm Hg. This was still considered mechanically inefficient, and would result in a heart rate of less than 20 beats per minute. It was suggested that, if sauropod metabolism was ectothermic, the "blood flow requirement" would be reduced, allowing for a more mechanically efficient heart and, possibly, vertical necks.

The ways that sauropods held their necks have a bearing on the feeding envelopes utilized by different sauropod species, especially if resource partitioning was a necessity. If *Brachiosaurus* held its neck nearly vertically, it could have browsed at a height of about 14 meters, whereas *Dicraeosaurus* would have fed at roughly four meters, and *Barosaurus* at about 12 meters.[128] Estimates of the feeding area of *Brachiosaurus* vary from 20 to 30 cubic meters in a two-meter vertical range[129] to 150 cubic meters.[130]

Long necks and forelimbs would have allowed foraging over wide vertical and horizontal ranges to take advantage of different vegetation types. This ability is considered important—with a number of sauropod genera co-existing at Tendaguru sympatrically, some degree of resource partitioning is likely. Competition for food resources would have developed with other herbivorous dinosaurs in the area, like stegosaurs and iguanodontids. Perhaps particular plant types were targeted by certain sauropods at different times in their life cycle.

Teeth, jaw structures, and jaw musculature are major factors in feeding. Earlier assumptions of two tooth forms, spatulate for cropping coarse vegetation and peg-like for raking somewhat softer plant food, have been supplanted by recent models of tooth morphology. Sauropods are not thought to have had muscular cheeks, though sturdy hyoid bones imply a strong tongue, which would have positioned food as it was processed.

The teeth of *Brachiosaurus* grew along the entire margin of the upper and lower jaws and are considered cone-chisel and spatulate in morphology.[131] Some researchers maintain that the teeth of the upper and lower jaws aligned and the position of jaw muscles and shape of the articular appear to have produced an orthal or simple vertical bite.[132] The resulting wear facets on the crowns formed chisel edges suited to an accurate shearing motion that cut rather than tore tough plants. Another study observed wear patterns consistent with a diplodocid-like raking motion that stripped foliage off branches.[133] Fine scratches have been found on the enamel of *Brachiosaurus* teeth.[134] Other researchers maintain that the microwear pattern on *Brachiosaurus* teeth resembles the coarser scratching and pitting observed on *Camarasaurus* teeth.[135] It is implied that brachiosaurs fed on different vegetation than diplodocids.

It is believed that the teeth of Tendaguru sauropods did relatively little food processing, but gastroliths have been found in association with *Barosaurus* and *Dicraeosaurus,* suggesting a ventriculus or grinding organ. It has also been suggested that bacteria in an extended digestive system, combined with prolonged passage, helped extract maximum value from low-energy-value foliage.[136]

Diplodocoids like *Dicraeosaurus* possessed peg-file teeth that were restricted to the anteriormost portion of the upper and lower jaws. Jaw action may have been propalinal, moving forward and backward. Some workers have interpreted the position of wear facets as indicating that a tooth of the upper jaw touched one or more teeth in the lower jaw, allowing a precise shearing action.[137] Others believe the feeding mechanism would have resembled the foliage-stripping action of *Diplodocus.*[138] The comparatively short neck and forelimbs and tall neural spines imply low-level browsing. Digestion was aided by gastroliths.

Diplodocids like the very long necked *Barosaurus* also possessed only anterior teeth. The cranial fragments of *Barosaurus* from Tendaguru do not allow a detailed analysis and there are no *Barosaurus* skulls from North America. Lower jaw motion may have been palinal, or backward. This sauropod's teeth may have worn like those of *Diplodocus,* an unusual pattern in which facets formed on the labial or "outer" faces of the upper and lower teeth. Several feeding strategies have been proposed as the cause of this wear pattern, the most recent suggesting that softer foliage like ferns was stripped off branches independently by the upper or the lower jaw. Tooth-to-tooth contact has been postulated by some workers.[139] Other researchers argue that the wear patterns were caused by contact with food material.[140] Classified as high browsers thanks to a long neck and the possibility that they reared up, they may also have browsed at less than maximum levels thanks to a laterally flexible neck.

The amount of food that sauropods required would depend on their metabolic rates, and as we have already seen, this has been a contentious issue among paleontologists and biologists. Large dinosaurs have frequently been compared to modern large reptiles, whose great volume or mass and comparatively small surface area allow them to maintain a fairly constant internal body temperature with reptilian metabolic levels. Such maintenance is known as inertial homeothermy or gigantothermy. Some workers have suggested that higher metabolic rates could be adjusted to lower levels seasonally or upon attaining maturity.[141]

Some researchers consider endothermy in sauropods unlikely or impossible. Béland and Russell calculated ecological efficiency ratios for the dinosaurs of the Upper Saurian Bed at Tendaguru, using both endothermic and ectothermic models. They concluded that it was unlikely that sauropods at Tendaguru were endothermic.

The energetics of *Brachiosaurus* was discussed with respect to food consumption in 1983. Using assumptions about the caloric content of extant cycads, ferns, and conifers, digestive efficiency, and rates of free-living metabolism compared to standard rates, it was concluded that endothermy was likely impossible in a 36-tonne *Brachiosaurus*.[142]

The question of how these gigantic animals sustained themselves on vegetation still remains unanswered. A number of skeletal adaptations have been examined to understand sauropod feeding mechanisms and behaviors. Sauropod skulls and jaws were considered very small for the task, until recent comparisons by Christiansen indicated that the ratio of sauropod muzzle width to skull length is greater than in modern herbivorous mammals. Sauropods' ability to efficiently crop vegetation with little oral processing may have enabled them to maintain a metabolic rate exceeding that of reptiles.[143]

It has even been suggested that the reason sauropods were so successful on land was that they possessed both large tachyaerobic muscles and the high-pressure circulatory systems and efficient lungs needed to supply their elevated oxygen requirements. This condition has been called terramegathermy. High-capacity internal organs also produced heat, implying an elevated resting metabolic level. Thermoregulation in this model was achieved by storing heat within the tremendous body mass, releasing it at night.[144]

The application of laser scanning and stereophotogrammetry to the Tendaguru sauropod skeletons has allowed detailed extrapolations of volume, mass distribution, and physiological functions.[145] Using allometric formulae derived from endotherms, it was calculated that the oxygen consumption of a 74.6-tonne *Brachiosaurus* was 50 liters per minute or 3,000 liters per hour, four times that of *Dicraeosaurus*. The lungs expanded to a volume of just over 5,800 liters, about six times that of *Dicraeosaurus*. The respiration rate was estimated at three per minute for *Brachiosaurus* and 4.6 for *Dicraeosaurus*. Tidal volume for *Brachiosaurus* was thought to be 516 liters.

The volume of air in the bronchial tubes, trachea, larynx, nose, and mouth during each breath is called "dead air space," since it does not contribute to oxygen exchange or carbon dioxide removal in the lungs. In *Brachiosaurus*, it amounts to 20 liters. Consequently, according to some researchers, a 30-tonne sauropod with an extremely long neck would most likely have had an air sac system and reptilian metabolic levels for energy-efficient respiration.[146]

Brachiosaurus lungs may have consisted of multiple tubular chambers that branched asymmetrically and consisted of large, cavernous, ventrally located sacs and dorsally located parenchyma.[147] Because of the presence of highly pneumatic vertebrae and ribs, an air sac system is postulated. Difficulties were envisioned in the relative position of the heart and the lungs. If the air passage and blood vessels lay ventral to the heart, high pulmonary artery pressure would be required to supply blood to the lungs. If the heart were dorsal to the lungs, the pulmonary vein pressure might have been unacceptably high, unless the air passages were also situated dorsally.

From the oxygen consumption assumed above, a basal metabolism rate of 1.46 million kilojoules per day was extrapolated for *Brachiosaurus*, 3.5 times that of *Dicraeosaurus*.[148] A free-living metabolism could easily have exceeded this by 15 to 20 percent. This rate is thought to have led to a daily caloric requirement of 344,000 kilocalories. The caloric content of the postulated diet of cycads, conifers, and ferns amounted to perhaps eight kilojoules per gram, which as a dry mass has a nutritive value of 30 percent. Of this, it is assumed that 50 percent was effectively absorbed. To meet its caloric requirements, then, *Brachiosaurus* would have had to consume 360 kilograms of vegetation per day. Using similar assumptions about food value and digestive efficiency, other researchers estimated that a 30-tonne, tachymetabolic *Brachiosaurus* required a food intake of 500 kilograms per 12- to 14-hour foraging period.[149] At a rate of one to six bites per minute, each mouthful would have amounted to 100 to 600 grams. Another reconstruction posits that a 15-tonne ectothermic *Brachiosaurus* would have consumed 60 kilograms of vegetation per day. Assuming an ectothermic fauna, the Tendaguru biomass has been estimated at 250 tonnes per square kilometer.[150]

Animals the size of *Brachiosaurus* would have produced a great deal of heat internally through the action of large muscle masses on limbs and the action of large hearts. The surface area of *Brachiosaurus* skin without folds may have been 139 square meters, and that of *Dicraeosaurus* just over 46 square meters. It is hypothesized that a combination of behavioral adaptations, such as immersion in mud or water, and the naturally conductive and convective heat exchangers like the neck, tail, and limbs, comprising 50 percent of the surface area, could have adequately dispersed heat produced by basal metabolism.

Sixty-four years after Gross's pioneering bone histology study, the extremity bones of four genera of Tendaguru sauropods at various growth stages were analyzed. *Brachiosaurus, Dicraeosaurus, Janenschia,* and *Barosaurus* had fibrolamellar cortical bone that showed few lines of arrested growth. This implied rapid, uninterrupted, but determinate growth, through three recognizable stages of life, from hatchling to juvenile to adult. In all cases, growth slowed substantially as the animals neared their greatest size. The four genera could be distinguished by varying degrees of bone remodeling and growth markers.

Polish lines, the growth markers in the fibrolamellar bone, allowed ages to be calculated. One individual of *Janenschia* was sexually mature at approximately 11 years, reached its maximum size at 26 years, and died at age 38.[151]

Barosaurus bone exhibited two distinct structural types, indicative of both rapid uninterrupted growth and slower, interrupted, partly cyclical growth. Sexual dimorphism or distinct taxa are possible explanations. Sexual maturity probably occurred at 70 percent of maximum size, and fast growth continued well into adulthood.

Dicraeosaurus bone also showed evidence of cyclical growth, and this genus may have reached sexual maturity at about 80 percent of its greatest size.

Brachiosaurus growth appears to have been swift and continued late into the animal's life, but was also subject to periodic interruptions and considerable slowing after sexual maturity. This genus became sexually mature at about 40 percent of maximum body size. These different growth strategies may be a function of different adult body sizes or may indicate a response to seasonality or reproductive cycles.[152]

Sauropod dinosaurs are thought to have laid eggs. Paul has described large dinosaurs as r-strategists, whose reproductive level was high. Possibly hundreds or thousands of eggs were produced by a sauropod in a 40-year interval, and breeding may have occurred every one or two years.[153] Sauropod growth rates may have been several hundred grams daily, so that a hatchling weighing a few kilograms would exceed 100 kilograms in the first year. In this model, sexual maturity was achieved at one-third of full adult size. Prolific birth rates and rapid growth rates were essential to ensure a viable, rapidly increasing population capable of surviving accidents, disease, and high levels of predation on juveniles.[154] A relatively high number of juveniles may have allowed for a higher predator-prey ratio than in recent ecosystems.

Plainly, there is great potential for further significant discoveries both at Tendaguru and within the material already collected from the site. It is a tremendous good fortune that this gift of geological processes was investigated, interpreted, and preserved. The impressive body of work generated by Tendaguru offers a rare insight into the remote and exotic world of Jurassic Africa. It is an enduring tribute to all those who labored with such admirable dedication and persistence, in the face of great adversity, for almost a century.

NOTES

1. 1907: Something Curious in the African Bush

1. Sattler's discovery of Tendaguru in 1906 is supported by references in Tendaguru-Akten I Rep. 76 Va Sekt. 2, Tit. 10, Nr. 21, Adhib. A, Geheimes Staatsarchiv Preussischer Kulturbesitz, Berlin.

2. R. W. Howard, *The Dawn Seekers: The First History of American Paleontology* (New York: Harcourt Brace and Jovanovich, 1975), p. 263.

3. C. Gillman, "Dar es Salaam, 1860 to 1940: A story of growth and change," *Tanganyika Notes and Records,* no. 20 (1945), p. 22.

4. J. Iliffe, *Tanganyika under German Rule, 1905–1912* (Cambridge: Cambridge University Press, 1969), p. 82.

5. Fraas to administration, October 17 and 19, 1907, Royal Natural History Collection, Stuttgart. Courtesy of Dr. R. Wild, Staatliches Museum für Naturkunde, Schloss Rosenstein, Stuttgart.

6. V. Harlow and E. M. Chilver, eds., *History of East Africa,* vol. 2 (London: Oxford University Press, 1965), p. 207.

7. J. Iliffe, *A Modern History of Tanzania* (Cambridge: Cambridge University Press, 1979), pp. 165, 200.

8. E. Fraas, "Dinosaurier in Deutsch-Ostafrika," *Umschau* 12, no. 48 (1908), p. 947.

9. E. Fraas, "Entdeckung von Überresten grosser Dinosaurier in Ostafrika," *Gaea* 44 (1907), p. 55.

10. Ibid.

11. Fraas, "Dinosaurier in Deutsch-Ostafrika," p. 946.

12. E. Fraas, "Die ostafrikanischen Dinosaurier," *Verhandlungen der Gesellschaft deutscher Naturforscher und Ärzte* 83, no. 1 (1911), p. 39.

2. 1908: A Matter of National Honor

1. E. Fraas, "Ostafrikanische Dinosaurier," *Palaeontographica* 55 (1908), p. 119.

2. W. O. Dietrich, "Geschichte der Sammlungen des Geologisch-Paläontologischen Institut und Museums der Humboldt-Universität zu Berlin," *Berichte der Geologischen Gesellschaft der DDR* 5 (1960), p. 279.

3. *Das Museum für Naturkunde der Königlichen Friedrich-Wilhelm-Universität in Berlin zur Eröffnungsfeier* (Berlin: Ernst & Korn, 1889), p. 6.

4. W. von Branca, "Allgemeines über die Tendaguru-Expedition," *Archiv für Biontologie* 3, no. 1 (1914), p. 3.

5. Ibid., p. 4.

6. Ibid.

7. G. Tornier, "Bericht des Vorsitzenden: Festsitzung zur Berichterstattung über Werden, Verlauf und bisherige Ergebnisse der Tendaguru-Expedition," *Sitzungsberichte der Gesellschaft Naturforschender Freunde zu Berlin* (1912), p. 121.

8. Sattler to Janensch, December 23, 1908, Paläontologisches Institut, Museum für Naturkunde, Berlin (hereafter MfN). Courtesy of Dr. W.-D. Heinrich.

9. E. Hennig, "Dr. Phil. Werner Janensch," *Berichte der Geologischen Gesellschaft der DDR* 9 (1964), p. 711.

10. Hennig correspondence, March 4, 1909, Universitätsarchiv, Eberhard Karls Universität, Tübingen (hereafter UAT) 407/2.

11. E. Hennig, *Gewesene Welten: Auf Saurierjagd im ostafrikanischen Busch* (Zürich: Albert Müller, 1955), p. 65.

3. 1909: A Cemetery of Giants

1. Janensch field notes, April 2, 1909, MfN.
2. Ibid., April 6, 1909, MfN.
3. Ibid.
4. W. Janensch and E. Hennig, "Erster Bericht über die Tendaguru-Expedition," *Sitzungsberichte der Gesellschaft Naturforschender Freunde zu Berlin* (1909), p. 358.
5. W. Janensch, "Bericht über den Verlauf der Tendaguru-Expedition," *Archiv für Biontologie* 3, no. 1 (1914), p. 20.
6. Janensch field notes, April 10, 1909, MfN.
7. E. Hennig, *Am Tendaguru: Leben und Wirken einer deutschen Forschungsexpedition zur Ausgrabung vorweltlicher Riesensaurier in Deutsch-Ostafrika* (Stuttgart: E. Schweizerbart'sche Verlagsbuchhandlung, 1912), p. 12.
8. Ibid., p. 19.
9. Hennig field notes, April 17, 1909, UAT 407/80.
10. Janensch field notes, April 16, 1909, MfN.
11. Janensch and Hennig, "Erster Bericht," p. 359.
12. Ibid.
13. Janensch, "Bericht über den Verlauf," p. 21.
14. Janensch and Hennig, "Erster Bericht," p. 359.
15. Hennig field notes, April 23, 1909, UAT 407/80.
16. Janensch and Hennig, "Erster Bericht," p. 360.
17. Hennig field notes, April 25, 1909, UAT 407/80.
18. Hennig field notes, April 27, 1909, UAT 407/80.
19. W. von Branca, "Allgemeines über die Tendaguru-Expedition," *Archiv für Biontologie* 3, no. 1 (1914), p. 9.
20. Janensch, "Bericht über den Verlauf," p. 27.
21. Hennig, *Am Tendaguru*, p. 41.
22. Deutsche Kolonialgesellschaft, *Deutscher Kolonial-Atlas* (Berlin: Dietrich Reimer, 1910), p. 24.
23. M. von Eckenbrecher, *Im Dichten Pori: Reise- und Jagdbilder aus Deutsch-Ostafrika* (Berlin: Mittler und Sohn, 1912), pp. 42–43.
24. Hennig field notes, July 19, 1909, UAT 407/80.
25. Hennig field notes, July 20, 1909, UAT 407/80.
26. Hennig field notes, July 23, 1909, UAT 407/80.
27. Hennig, *Am Tendaguru*, p. 66.
28. Ibid., p. 122.
29. Ibid., p. 67.
30. Hennig field notes, August 3, 1909, UAT 407/80.
31. Ibid.
32. Hennig field notes, August 26, 1909, UAT 407/80.
33. W. Janensch, "Zweiter Bericht über die Tendaguru-Expedition," *Sitzungsberichte der Gesellschaft Naturforschender Freunde zu Berlin* (1909), pp. 500–501.
34. Ibid., p. 502.
35. Hennig field notes, September 15, 1909, UAT 407/80.
36. Hennig field notes, October 30, 1909, UAT 407/80.
37. Von Branca, "Allgemeines über die Tendaguru-Expedition," p. 9.
38. Hennig field notes, November 23, 1909, UAT 407/80.
39. Hennig field notes, November 28, 1909, UAT 407/80.
40. Hennig field notes, December 1, 1909, UAT 407/80.

41. Hennig field notes, December 2, 1909, UAT 407/80.
42. Hennig field notes, December 3, 1909, UAT 407/80.
43. Ibid.
44. Hennig field notes, December 23, 1909, UAT 407/80.
45. Hennig field notes, December 24, 1909, UAT 407/80.
46. Hennig field notes, December 31, 1909, UAT 407/80.
47. Hennig field notes, January 9, 1910, UAT 407/80.
48. W. Janensch, "Nachtrag zum zweiten Bericht über die Deutsche Tendaguru-Expedition," *Sitzungsberichte der Gesellschaft Naturforschender Freunde zu Berlin* (1909), p. 631.
49. Janensch, "Bericht über den Verlauf," p. 52.
50. H. Reck, "3. Bericht über den weiteren Verlauf der Tendaguru-Expedition," *Sitzungsberichte der Gesellschaft Naturforschender Freunde zu Berlin* (1910), p. 374.
51. Janensch, "Nachtrag zum zweiten Bericht," p. 631.
52. Von Branca, "Allgemeines über die Tendaguru-Expedition," p. 11.
53. Ibid., p. 6.

4. 1909–1910: Geology in the Rain

1. E. Hennig, "Beiträge zur Geologie und Stratigraphie Deutsch-Ostafrikas," *Archiv für Biontologie* 3, no. 3 (1913), p. 55.
2. H. Schnee, *Deutsches Kolonial-Lexikon*, vol. 3 (Leipzig: Quelle & Meyer, 1920), p. 455.
3. Hennig correspondence, April 1, 1910, UAT 407/2.
4. The information on African tribes that follows is taken from *Militärisches Orientierungsheft für Deutsch-Ostafrika*, 1911, section 8, pp. 6–8.
5. W. Janensch, "Nachtrag zum zweiten Bericht über die Deutsche Tendaguru-Expedition," *Sitzungsberichte der Gesellschaft Naturforschender Freunde zu Berlin* (1909), p. 631.
6. Hennig field notes, May 3, 1910, UAT 407/81.
7. Hennig field notes, May 9, 1910, UAT 407/81.
8. E. Hennig, *Am Tendaguru: Leben und Wirken einer deutschen Forschungsexpedition zur Ausgrabung vorweltlicher Riesensaurier in Deutsch-Ostafrika* (Stuttgart: E. Schweizerbart'sche Verlagsbuchhandlung, 1912), pp. 96–97.
9. Hennig field notes, May 10, 1910, UAT 407/81.
10. Hennig, *Am Tendaguru*, p. 96.
11. B. Harpur, *The Official Halley's Comet Book* (London: Hodder and Stoughton, 1985), p. 31.
12. W. Janensch, "Bericht über den Verlauf der Tendaguru-Expedition," *Archiv für Biontologie* 3, no. 1 (1914), p. 51.
13. Hennig field notes, June 20, 1910, UAT 407/81.
14. Hennig field notes, July 6, 1910, UAT 407/81.
15. E. Hennig, "*Kentrosaurus aethiopicus* der Stegosauride des Tendaguru," *Sitzungsberichte der Gesellschaft Naturforschender Freunde zu Berlin* 6 (1915), p. 220.
16. Hennig field notes, July 27, 1910, UAT 407/81.
17. H. Reck, "3. Bericht über den weiteren Verlauf der Tendaguru-Expedition," *Sitzungsberichte der Gesellschaft Naturforschender Freunde zu Berlin* (1910), p. 373.
18. Hennig field notes, August 10, 1910, UAT 407/81.
19. Hennig field notes, August 17, 1910, UAT 407/81.
20. Hennig field notes, September 1, 1910, UAT 407/81.
21. Hennig field notes, September 27, 1910, UAT 407/81.
22. Hennig field notes, October 12, 1910, UAT 407/81.

23. H. Reck, "Die Ausgrabungen fossiler Riesentiere in Deutsch-Ostafrika," *Umschau* 14 (1910), p. 1043.

24. Hennig field notes, October 26, 1910, UAT 407/81.

25. Insurance claim on file at Paläontologisches Institut, MfN. Courtesy of Dr. W.-D. Heinrich.

26. Hennig field notes, December 1, 1910, UAT 407/81.

27. W. Janensch, "Bericht über den Verlauf," p. 51.

28. Hennig field notes, December 10, 1910, UAT 407/81.

29. Hennig field notes, December 24, 1910, UAT 407/81.

30. Hennig field notes, January 1, 1911, UAT 407/81.

31. Reck, "Die Ausgrabung fossiler Riesentiere," p. 1043.

32. Hennig field notes, January 2, 1911, UAT 407/81.

33. Hennig, *Am Tendaguru,* p. 10.

34. Hennig correspondence, January 6, 1911, UAT 407/2.

35. Ibid.

36. Great Britain, Naval Intelligence Division, *A Handbook of German East Africa* (London: H.M.S.O., 1920), p. 186.

37. Hennig, *Am Tendaguru,* p. 71.

38. Janensch, "Bericht über den Verlauf," p. 52.

39. H. Reck, "4. Bericht über die Ausgrabungen und Ergebnisse der Tendaguru-Expedition (Grabungsperiode 1911)," *Sitzungsberichte der Gesellschaft Naturforschender Freunde zu Berlin* (1911), p. 385.

40. Janensch, "Bericht über den Verlauf," p. 51.

41. Ibid., p. 49.

42. W. von Branca, "Allgemeines über die Tendaguru-Expedition," *Archiv für Biontologie* 3, no. 1 (1914), p. 11.

5. 1911: Along the Railway

1. H. Reck, "4. Bericht über die Ausgrabungen und Ergebnisse der Tendaguru-Expedition (Grabungsperiode 1911)," *Sitzungsberichte der Gesellschaft Naturforschender Freunde zu Berlin* (1911), p. 385.

2. Hennig field notes, May 29, 1911, UAT 407/81.

3. W. Janensch, "Bericht über den Verlauf der Tendaguru-Expedition," *Archiv für Biontologie* 3, no. 1 (1914), p. 55.

4. Reck, "4. Bericht über die Ausgrabungen und Ergebnisse," p. 388.

5. Hennig field notes, July 21, 1911, UAT 407/81.

6. Hennig field notes, July 24, 1911, UAT 407/82.

7. Janensch, "Bericht über den Verlauf," p. 56.

8. Hennig field notes, August 28, 1911, UAT 407/82.

9. Hennig field notes, September 5, 1911, UAT 407/82.

10. Hennig field notes, September 17, 1911, UAT 407/82.

11. H. Schnee, *Deutsches Kolonial-Lexikon* (Leipzig: Quelle & Meyer, 1920), vol. 2, p. 300.

12. Janensch, "Bericht über den Verlauf," p. 57.

13. Ibid., p. 56.

14. Ibid., p. 57.

15. Reck, "4. Bericht über die Ausgrabungen und Ergebnisse," p. 391.

16. Ibid., p. 389.

17. Ibid., p. 388.

18. Janensch, "Bericht über den Verlauf," p. 57.

19. Ibid., p. 58.

20. Ibid.

21. W. von Branca, "Allgemeines über die Tendaguru-Expedition," *Archiv für Biontologie* 3, no. 1 (1914), p. 13.

22. Hennig, *Am Tendaguru,* 1912, p. 33.

6. 1911–1912: A Museum Overflows

1. Hennig field notes, November 1, 1911, UAT 407/82.

2. G. Masur, *Imperial Berlin* (New York: Basic Books, 1971), p. 132.

3. W. von Branca, "Allgemeines über die Tendaguru-Expedition," *Archiv für Biontologie* 3, no. 1 (1914), p. 7.

4. H. Reck, "4. Bericht über die Ausgrabungen und Ergebnisse der Tendaguru-Expedition (Grabungsperiode 1911)," *Sitzungsberichte der Gesellschaft Naturforschender Freunde zu Berlin* (1911), p. 396.

5. Ibid.

6. Von Branca, "Allgemeines über die Tendaguru-Expedition," p. 11.

7. Ibid., p. 9.

8. Reichs-Kolonialamt, *Die Deutschen Schutzgebiete in Afrika und der Südsee: Amtliche Jahresberichte* (Berlin: Mittler und Sohn, 1912–1913), p. 44.

9. Von Branca, "Allgemeines über die Tendaguru-Expedition," pp. 10–11.

10. W. von Branca, "Kurzer Bericht über die von Dr. Reck erzielten Ergebnisse im vierten Grabungsjahre 1912," *Archiv für Biontologie* 3, no. 1 (1914), p. 62.

11. Hennig field notes, December 10, 1911, UAT 407/82.

12. Hennig field notes, December 31, 1911, UAT 407/82.

13. K. Sapper, "Hans Reck," *Zeitschrift für Vulkanologie* 17 (1936–1938), p. 227.

14. Von Branca, "Kurzer Bericht," p. 61.

15. Ibid.

16. Reck correspondence, February 19, 1913, MfN.

17. I. Reck, *Mit der Tendaguru-Expedition im Süden von Deutsch-Ostafrika* (Berlin: Dietrich Reimer, 1924), pp. 40–41.

18. Ibid., p. 48.

19. E. Hennig, "*Kentrurosaurus aethiopicus:* Die Stegosaurierfunde von Tendaguru, Deutsch-Ostafrika," *Palaeontographica,* Suppl. 7, Reihe 1, Teil 1 (1925), p. 110.

20. Ibid., p. 108.

21. Ibid., p. 112.

22. Von Branca, "Kurzer Bericht," p. 62.

23. Reck correspondence, "2 Bericht," June 12, 1912, MfN.

24. Reck correspondence, "3 Bericht," June 22, 1912, MfN.

25. Ibid.

26. Reck correspondence, "4 Bericht," July 5, 1912, MfN.

27. Reck correspondence, "3 Bericht," June 22, 1912, MfN.

28. Von Branca, "Kurzer Bericht," p. 62.

29. H. Reck, "Die deutschostafrikanischen Flugsaurier," *Centralblatt für Mineralogie, Geologie und Paläontologie,* Abt. B (1931), p. 325.

30. Reck correspondence, "4 Bericht," July 5, 1912, MfN.

31. I. Reck, *Mit der Tendaguru-Expedition,* p. 47.

32. Reck correspondence, "5 Bericht," July 18, 1912, MfN.

33. Reck correspondence, "8 Bericht," September 15, 1912, MfN.

34. Reck correspondence, "7 Bericht," September 10, 1912, MfN.

35. Reck correspondence, "9 Bericht," October 7, 1912, MfN.

36. Ibid.

37. Ibid.

38. Ibid.

39. Reck correspondence, "10 Bericht," November 2, 1912, MfN.

40. Von Branca, "Kurzer Bericht," p. 63.

41. Reck correspondence, "10 Bericht," November 2, 1912, MfN.
42. I. Reck, *Mit der Tendaguru-Expedition,* p. 73.
43. Reck correspondence, "10 Bericht," November 2, 1912, MfN.
44. Reck correspondence, "11 Bericht," November 18, 1912, MfN.
45. Ibid.
46. H. Schnee, *Deutsches Kolonial-Lexikon* (Leipzig: Quelle & Meyer, 1920), vol. 2, p. 459.
47. Reck, "Safari nach Liwale," Dec. 8–9, 1912, Archiv, MfN; correspondence "12 Bericht," December 28, 1912, MfN.
48. Reck correspondence, "12 Bericht," December 28, 1912, MfN.
49. Ibid.
50. I. Reck, *Mit der Tendaguru-Expedition,* p. 60.
51. Reck correspondence, "12 Bericht," December 28, 1912, MfN.
52. Reck correspondence, "13 Bericht," January 26, 1913, MfN.
53. Reck correspondence, February 19, 1913, MfN.
54. Reck correspondence, "13 Bericht," January 26, 1913, MfN.
55. Ibid.
56. Ibid.
57. I. Reck, *Mit der Tendaguru-Expedition,* p. 49.
58. Ibid., p. 7.

7. 1913–1918: Fresh Discoveries and a Bitter War

1. Reck correspondence, February 19, 1913, MfN.
2. I. Reck, *Auf einsamen Märschen im Norden von Deutsch-Ostafrika* (Berlin: Dietrich Reimer, 1925), p. 43.
3. Reck correspondence, May 30, 1913, MfN.
4. Reck field notes, June 23–27, 1913, MfN.
5. Reck field notes, June 24, 1913, MfN.
6. Reck correspondence, July 14, 1914, MfN.
7. Reck field notes, July 12, 1913, MfN.
8. H. Reck, *Oldoway, die Schlucht des Urmenschen: Die Entdeckung des altsteinzeitlichen Menschen in Deutsch-Ostafrika* (Leipzig: F. A. Brockhaus, 1933), p. 16.
9. Ibid., p. 59.
10. Ibid., p. 74.
11. Reck correspondence, Christmas 1913, MfN.
12. H. Reck, *Oldoway,* p. 179.
13. Von Branca, "Allgemeines über die Tendaguru-Expedition," *Archiv für Biontologie* 3, no. 1 (1914), pp. 7–8.
14. Ibid.
15. Ibid.
16. Ibid., p. 13.
17. Reck correspondence, February 10, 1914, MfN.
18. Reck correspondence, December 11, 1919, MfN.
19. Tendaguru-Expedition Akten, March 23, 1914, I/76, Geheimes Staatsarchiv Preussischer Kulturbesitz, Berlin.
20. Reichs-Kolonialamt, *Die Deutschen Schutzgebiete in Afrika und der Südsee: Amtliche Jahresberichte* (Berlin: Mittler und Sohn, 1912–1913), pp. 17–18.
21. H. F. Osborn, "Eberhard Fraas," *Science* 41, no. 1059 (1915), p. 572.
22. C. Miller, *Battle for the Bundu: The First World War in East Africa* (New York: Macmillan, 1974), p. 143.
23. Reck, *Auf einsamen Märschen,* p. 89.
24. J. R. Sibley, *Tanganyikan Guerrilla: East African Campaign, 1914–1918* (London: Pan Books, 1973), p. 115.

25. C. Deppe and L. Deppe, *Um Ostafrika* (Dresden: Beutelspacher, 1925), p. 93.
26. Miller, *Battle for the Bundu*, p. 256.
27. Sibley, *Tanganyikan Guerrilla*, p. 135.
28. Reck, *Auf einsamen Märschen*, p. 113.
29. J. Goebel, *Afrika zu unsern Füssen* (Berlin: K. F. Koehler, 1925), p. 106.
30. L. Boell, *Die Operationen in Ostafrika* (Hamburg: Walter Dachert, 1951), p. 424.
31. Ibid., p. 429.
32. Ibid., p. 428–29.
33. J. Iliffe, *A Modern History of Tanzania* (Cambridge: Cambridge University Press, 1979), p. 270.

8. 1919–1924: The British Museum in Tanganyika Territory

1. Orde-Browne to BM(NH), July 28, 1918, 1004/391/1, British Museum East Africa Expedition Archive, Archives Section, Natural History Museum, London (hereafter DF).
2. F. H. W. Sheppard, ed., *Survey of London: The Museums Area of South Kensington and Westminster,* vol. 38 (London: The Athlone Press, 1975), pp. 213–14.
3. Woodward to trustees, BM(NH), August 12, 1918, DF 1004/391/1.
4. W. T. Stearn, *The Natural History Museum at South Kensington: A History of the British Museum (Natural History), 1753–1980* (London: Heinemann, 1981), p. 110.
5. Cutler to Calgary Natural History Society, June 16, 1913, Glenbow Archives, Calgary.
6. W. E. Cutler, "The Badlands of Alberta," *Canadian Illustrated Monthly,* January 1922, p. 22.
7. Attestation paper for W. E. Cutler, Canadian Overseas Expeditionary Force, June 28, 1915, Public Archives of Canada, Ottawa.
8. "Late Prof. Cutler navigated rivers to city Winnipeg. Scientist who died of malaria in Africa was lover of novelty and adventure," *Calgary Daily Herald,* September 3, 1925, p. 11.
9. Cutler, "The Badlands of Alberta," p. 51.
10. Smith, BM(NH) Museum Report, July 21, 1929, DF 1004/391/8.
11. Secretary to the administration, Dar es Salaam, to secretary of state for the colonies, February 11, 1920, DF 1004/391/1.
12. Woodward to trustees, October 20, 1920, DF 1004/391/1.
13. R. Blythe, *The Age of Illusion: England in the Twenties and Thirties, 1919–1940* (London: H. Hamilton, 1963), p. 7.
14. D. H. Aldcroft, *The Inter-War Economy: Britain, 1919–1939* (London: Batsford, 1970), p. 37.
15. S. Glynn and J. Oxborrow, *Interwar Britain: A Social and Economic History* (London: Allen and Unwin, 1976), p. 126.
16. Byatt to Harmer, March 25, 1923, DF 1004/391/1.
17. Peringuey to Harmer, April 17, 1923, DF 1004/391/1.
18. Harmer to Cutler, July 19, 1923, DF 1004/391/1.
19. Ibid.
20. Harmer to Cutler, November 8, 1923, DF 1004/391/1.
21. Gray, Dawes & Co. to Smith, December 24, 1923, January 1, 1924, DF 1004/391/1.
22. Woodward to Harmer, January 8, 1924, DF 1004/391/2.
23. Hichens to Harmer, January 7, 1924, DF 1004/391/2.
24. Smith to Leakey, February 6, 1924, DF 1004/391/2.
25. Receipt of monies for Cutler and Leakey, February 12/13, 1924, and Smith to Cooke, February 25, 1924, DF 1004/391/2.

9. 1924–1925: Cutler, Leakey, and a Difficult Start

1. Army & Navy Co-operative Society to Smith, January 31, 1924, DF 1004/391/2.
2. Wray to Stahlschmidt & Co., March 3, 1924, DF 1004/391/2.
3. Richman, Symes & Co. to BM(NH), February 26, 1924, DF 1004/391/2.
4. Jonathan Fallowfield to Smith, February 19, 1924, DF 1004/391/2.
5. Wray to Cutler, March 24, 1924, DF 1004/391/5.
6. L. S. B. Leakey, *White African* (London: Hodder and Stoughton, 1937), p. 106.
7. Ibid., p. 107.
8. Ibid.
9. Ibid., pp. 114–15.
10. Ibid.
11. Ibid., p. 119.
12. Cutler field notes, June 5, 1924, DF 5000/7.
13. Cutler to Harmer, June 18, 1924, DF 1004/391/4.
14. Cutler field notes, June 11, 1924, DF 5000/7.
15. Leakey, *White African,* pp. 120–21.
16. Cutler to Harmer, June 18, 1924, DF 1004/391/4.
17. Leakey, *White African,* pp. 128–29.
18. Cutler geology notes, Trench I, p. 1, DF 5000/8.
19. Cutler geology notes, Ditch II, p. 2, DF 5000/8.
20. Ibid., p. 35, "Bones from Ditches."
21. Cutler to Harmer, October 5, 1924, DF 1004/391/3.
22. Cutler geology notes, Trench 3, p. 3, DF 5000/8.
23. Cutler geology notes, Ditch VI, p. 49, DF 5000/8.
24. Cutler geology notes, Ditch VII, p. 7, DF 5000/8.
25. Cutler geology notes, Ditch VIII, p. 55, DF 5000/8.
26. Cutler geology notes, Ditch IX, p. 16, DF 5000/8.
27. Cutler geology notes, Ditch X, p. 57, DF 5000/8.
28. Cutler geology notes, Ditch XI, p. 60, DF 5000/8.
29. Cutler field notes, June 24, 1924, DF 5000/7.
30. Cutler field notes, July 8, 1924, DF 5000/7.
31. Leakey, *White African,* pp. 124–25.
32. Cutler field notes, July 18, 1924, DF 5000/7.
33. Cutler field notes, July 23, 1924, DF 5000/7.
34. Cutler to Harmer, 15 August, 1924, DF 1004/391/4.
35. Ibid.
36. Cutler to Harmer, September 8, 1924. DF 1004/391/3.
37. Cutler field notes, December 15, 1924, DF 5000/7.
38. Cutler field notes, December 25, 1924, DF 5000/7.
39. Cutler field notes, December 26, 1924, DF 5000/7.
40. Cutler field notes, January 1, 1925, DF 5000/7.
41. Cutler field notes, January 2, 1925, DF 5000/7.
42. Smith to Price, January 7, 1925, DF 1004/391/3.
43. Harmer to trustees, January 19, 1925, DF 1004/391/3.
44. Ibid.
45. Harmer to Colonial Office, December 19, 1924, DF 1004/391/3.
46. Cutler to Smith, November 24, 1924, DF 1004/391/3.
47. Ibid.
48. Ibid.
49. Ibid.
50. Harmer to Cutler, January 2, 1925, DF 1004/391/3.
51. Ibid.
52. Ibid.

53. Cutler field notes, January 5, 1925, DF 5000/7.
54. Cutler field notes, January 7, 1925, DF 5000/7.
55. Cutler field notes, January 8, 1925, DF 5000/7.
56. Cutler field notes, January 24, 1925, DF 5000/7.
57. Cutler field notes, February 15, 1925, DF 5000/7.
58. Cutler field notes, March 7, 1925, DF 5000/7.

10. 1925: Berlin Builds Dinosaurs

1. T. Friedrich, *Berlin between the Wars* (New York: Vendome Press, 1991), p. 46.
2. W. D. Matthew, "Notes on the scientific museums of Europe," *Natural History Magazine* 21 (1921), p. 186.
3. W. Freydank, *Museum für Naturkunde der Humboldt-Universität zu Berlin: 100 Jahre Museumsgebäude in der Invalidenstraße 43* (Berlin: Museum für Naturkunde, 1989), p. 14.
4. W. D. Matthew, "Jurassic dinosaurs of Utah and East Africa," *Bulletin of the Geological Society of America* 34 (1923), p. 406.
5. Matthew, "Notes on the scientific museums of Europe," pp. 186–87.
6. P. Fritzsche, *Reading Berlin 1900* (Cambridge, Mass.: Harvard University Press, 1996), pp. 53, 17.
7. Royal Museum of Natural History, Geological-Paleontological Collection, *Acta betreffend den Kustos Prof. Dr. Janensch,* June 15, 1920, Archiv, MfN.
8. Hermann Jaeger, personal communication, 1985.
9. W. Mann, *Berlin zur Zeit der Weimarer Republik* (Berlin: Das Neue Berlin, 1957), appendix for 1923, timetable, p. 1.
10. W. Von Eckardt and S. L. Gilman, *Bertolt Brecht's Berlin: A Scrapbook of the Twenties* (Garden City, N.Y.: Anchor Press, 1975), p. 13.
11. W. Janensch, "Ein aufgestelltes Skelett des Stegosauriers *Kentrurosaurus aethiopicus* E. Hennig aus den Tendaguru-schichten Deutsch-Ostafrikas," *Palaeontographica,* Suppl. 7, Reihe 1, Teil 1 (1925), pp. 261–62.
12. Ibid., p. 263.
13. Reck, "Das erste rekonstruierte Skelett der Tendagurusaurier-Lagerstätte in Deutsch-Ostafrika," *Afrika-Nachrichten* 17–18 (1924), p. 260.
14. Janensch, "Das erste aufgestellte Skelett eines Dinosauriers von Tendaguru in Deutsch-Ostafrika," *Naturforscher* 1 (1924), p. 252.
15. Reck, "Das erste rekonstruierte Skelett," pp. 259–60.
16. Hennig, "Ein Drache aus Deutsch-Ostafrika," *Umschau* 29 (1925), p. 110.

11. 1925: A Death in Africa

1. Harmer to Arnold, March 30, 1925, DF 1004/391/5.
2. Cutler to Harmer, March 20, 1925, DF 1004/391/5.
3. Cutler to Harmer, May 16, 1925, DF 1004/391/5.
4. Cutler field notes, May 23 and 25, 1925, DF 5000/7.
5. Cutler notes: Tendaguru 1925: catalogue of work done, May 29, 1925, DF 5000/12.
6. Cutler field notes, June 30, 1925, DF 5000/7.
7. Cutler field notes, July 1, 1925, DF 5000/7.
8. Cutler to Anderson, June 8, 1925, DF 5000/10.
9. Cutler field notes, June 9, 1925, DF 5000/7.
10. Cutler notes: Tendaguru 1925: catalogue of work done, July 3, 1925, DF 5000/12.
11. Cutler field notes, July 14, 1925, DF 5000/7.
12. Cutler notes: Tendaguru 1925: catalogue of work done, July 9, 1925, DF 5000/12.
13. Cutler to Anderson, July 13, 1925, DF 5000/10.

14. Cutler to Anderson, July 24, 1925, DF 5000/10.
15. Cutler to Harmer, July 25, 1925, DF 1004/391/4.
16. Cutler notes: Tendaguru 1925: catalogue of work done, July 23, 1925, DF 5000/12.
17. Ibid., July 22, 1925, DF 5000/12.
18. Cutler to Anderson, July 25, 1925, DF 5000/10.
19. Cutler to Manning, July 25, 1925, DF 5000/10.
20. Cutler field notes, July 27, 1925, DF 5000/7.
21. Cutler field notes, August 1, 1925, DF 5000/7.
22. Cutler to Harmer, August 10, 1925, DF 5000/10.
23. Cutler field notes, August 20, 1921, DF 5000/7.
24. Cameron to Harmer, September 1, 1925, DF 1004/391/4.
25. Wyatt to Harmer, BM(NH), August 31, 1925, DF 1004/391/5.
26. Lachlan to Harmer, BM(NH), September 6, 1925, DF 1004/391/5.
27. Wyatt to Harmer, BM(NH), August 31, 1925, DF 1004/391/5.

12. 1925: A New Recruit

1. Tanganyika Territory government to secretary of state for the colonies, telegram, September 1, 1925, DF 1004/391/4.
2. Smith to Hobley, September 1, 1925, DF 1004/391/4.
3. Keith to Bather, September 7, 1925, DF 1004/391/4.
4. Leakey to Smith, September 7, 1925, DF 1004/391/4.
5. F. W. H. Migeod, *A View of Sierra Leone* (New York: Brentano's, 1927), p. 126.
6. W. E. Swinton, "F. W. H. Migeod," *Nature* 170, no. 4318 (1952), p. 134.
7. W. E. Swinton, "Frederick William Hugh Migeod," *Proceedings of the Geological Association* 64 (1953), p. 65.
8. G. G. Simpson, *Simple Curiosity: Letters from George Gaylord Simpson to His Family, 1921–1970,* ed. L. F. Laporte (Berkeley and Los Angeles: University of California Press, 1987), p. 75.
9. Migeod to Smith, September 14, 1925, DF 1004/391/5.
10. Smith to Rothschild, September 15, 1925, DF 1004/391/5.
11. Leakey to Smith, September 17, 1925, DF 1004/391/5.
12. Smith to Harmer, September 29, 1925, DF 1004/391/5.
13. Lachlan to Harmer, September 9, 1925, DF 1004/391/5.
14. Lachlan to Harmer, September 23, 1925, DF 1004/391/5.
15. Lachlan to Harmer, September 9, 1925, DF 1004/391/5.
16. Ibid.
17. Lachlan to Harmer, September 6, 1925, DF 1004/391/5.
18. Lachlan to Harmer, October 1925, DF 1004/391/5.

13. 1925–1926: An Expedition Saved

1. Migeod expedition diary, November 4, 1925, DF 5000/16.
2. Migeod to Harmer, November 6, 1925, DF 1004/391/5.
3. Migeod expedition diary, November 7, 1925, DF 5000/16.
4. Migeod to Harmer, November 22, 1925, DF 1004/391/5.
5. Ibid.
6. Migeod expedition diary, November 14, 1925, DF 5000/16.
7. Migeod to Harmer, November 29, 1925, DF 1004/391/5.
8. Deacon to Smith and Migeod, February 13, 1926, and Migeod to Harmer, February 14, 1926, DF 1004/391/5.
9. Migeod expedition diary, January 25, 1926, DF 5000/16.
10. Migeod expedition diary, March 1, 1926, DF 5000/16.

11. Migeod expedition diary, July 16, 1926, DF 5000/16.
12. Lachlan to Harmer, November 30, 1925, DF 1004/391/5.
13. Migeod to Harmer, November 22, 1925, DF 1004/391/5.
14. Migeod expedition diary, November 30, 1925, DF 5000/16.
15. Migeod expedition diary, December 7, 1925, DF 5000/16.
16. Migeod to Harmer, December 26, 1925, DF 1004/391/5.
17. Migeod expedition diary, February 4, 1926, DF 5000/16.
18. H. Reck, "Die britische Tendaguru-Expedition 1924," *Afrika-Nachrichten* 1 (1925), p. 3.
19. Migeod expedition diary, February 5, 1926, DF 5000/16.
20. Migeod expedition diary, December 12, 1925, DF 5000/16.
21. Migeod to Harmer, January 10, 1926, DF 1004/391/5.
22. Migeod to Harmer, May 26, 1926, DF 1004/391/6.
23. Migeod to Harmer, June 30, 1926, DF 1004/391/6.
24. Migeod, "The dinosaurs of Tendaguru," *Journal of the African Society* 26 (1927), p. 333.
25. Migeod to Harmer, December 26, 1925, DF 1004/391/5.
26. Migeod expedition diary, December 31, 1925, DF 5000/16.
27. Migeod expedition diary, December 21, 1925, DF 5000/16.
28. Migeod expedition diary, December 31, 1925, DF 5000/16.
29. Migeod expedition diary, January 9, 1926, DF 5000/16.
30. Migeod expedition diary, February 2, 1926, DF 5000/16.
31. L. S. B. Leakey, *White African* (London: Hodder and Stoughton, 1937), p. 159.
32. Ibid., p. 177.
33. Reck, "Die britische Tendaguru-Expedition 1924," pp. 3–4.
34. Burton to Smith, January 21, 1926, DF 1004/391/5.
35. Higgs and Warris to Colonial Office of East Africa, June 14, 1926, DF 1004/391/6.
36. Migeod expedition diary, February 17, 1926, DF 5000/16.
37. Migeod expedition diary, February 20, 1926, DF 5000/16.
38. Migeod expedition diary, February 18, 1926, DF 5000/16.
39. Migeod expedition diary, August 24, 1926, DF 5000/16.
40. Migeod to Harmer, March 14, 1926, DF 1004/391/7.
41. Migeod expedition diary, March 13, 1926, DF 5000/16.
42. Migeod expedition diary, May 26, 1926, DF 5000/16.
43. Migeod expedition diary, May 31, 1926, DF 5000/16.
44. Deacon to Migeod, June 11, 1926, DF 1004/391/6.
45. Migeod to Harmer, June 6, 1926, DF 1004/391/5.
46. Deacon to Migeod, July 8 and 9, 1926, DF 1004/391/6.
47. Migeod expedition diary, July 28, 1926, DF 5000/16.
48. Migeod expedition diary, July 31, 1926, DF 5000/16.
49. Migeod expedition diary, August 5, 1926, DF 5000/16.
50. Migeod expedition diary, August 22, 1926, DF 5000/16.
51. Migeod to Harmer, October 8, 1926, DF 1004/391/6.
52. Migeod to Parlett and Kershaw, October 14, 1926, DF 1004/391/6.
53. Bather to Migeod, January 20, 1926, DF 1004/391/5.
54. Bather to Migeod, June 21, 1926.
55. Smith to Migeod, July 15, 1926, DF 1004/391/6.
56. Bather to Migeod, July 13, 1926, DF 1004/391/6.
57. Migeod to Bather, August 29, 1926, DF 1004/391/6.
58. Bather to Harmer, July 20, 1926, DF 1004/391/7.
59. Leakey to Smith, August 28, 1926, DF 1004/391/6.
60. Migeod to Smith, October 17, 1926, DF 1004/391/6.
61. Migeod expedition diary, October 21, 1926, DF 5000/16.
62. Deacon to Migeod, October 3, 1926, DF 1004/391/6.

63. Migeod expedition diary, November 10, 1926, DF 5000/16.

64. Bather to trustees, October 21, 1926, DF 1004/391/6.

65. Bather to Cooke, November 11, 1926, DF 1004/391/6.

66. Migeod, "The dinosaurs of Tendaguru," p. 339.

67. F. W. H. Migeod, "British Museum East Africa Expedition: Progress in the year 1926," *Natural History Magazine* 1, no. 2 (1927), p. 35.

68. Migeod expedition diary, 1926, DF 5000/16.

14. 1926–1927: Berlin in Chaos

1. T. Friedrich, *Berlin between the Wars* (New York: Vendome Press, 1991), p. 63.

2. A. Gill, *A Dance between Flames: Berlin between the Wars* (London: J. Murray, 1993), p. 157.

3. W. Janensch, "Ueber *Elaphrosaurus bambergi* und die Megalosaurier aus den Tendaguru-Schichten DeutschOstafrikas," *Sitzungsberichte der Gesellschaft Naturforschender Freunde zu Berlin* (1920), p. 225.

4. Janensch, "Die Coelurosaurier und Theropoden der Tendaguru-Schichten Deutsch-Ostafrikas," *Palaeontographica*, Suppl. 7, Reihe 1, Teil 1 (1925), p. 7.

5. Ibid., p. 31.

6. Ibid., p. 80.

7. Ibid., p. 97.

8. Janensch, "Ein aufgestelltes und rekonstruiertes Skelett von *Elaphrosaurus bambergi*: Mit einem Nachtrag zur Osteologie dieses Coelurosauriers," *Palaeontographica*, Suppl. 7, Reihe 1, Teil 1 (1929), p. 284.

9. Ibid.

10. Berckhemer to Stille, January 26, 1933, MfN.

11. Von Huene to Janensch, March 31, 1927, MfN.

12. Smith to Leakey, December 9, 1926, DF 1004/391/6.

13. Harmer to BM(NH) trustees, February 25, 1927, DF 1004/391/7.

14. Parkinson to colonial secretary, Nairobi, February 2, 1928, DF 1004/391/7.

15. Kershaw to Migeod, November 28, 1926, DF 5000/19.

16. Parlett and Kershaw to Migeod, January 2, 1927, DF 5000/19.

17. Parlett to Migeod, February 3, 1927, DF 5000/19.

18. Parlett to Migeod, March 31, 1927, DF 5000/20.

19. Harmer to Migeod, March 4, 1927, DF 5000/20.

20. Parlett to Migeod, April 26, 1927, DF 1004/391/7.

21. Deacon to Smith, May 14, 1927, DF 1004/391/7.

22. Parlett to Migeod, June 29, 1927, DF 5000/20.

23. Parkinson to BM(NH), July 28, 1927, DF 1004/391/7.

24. Parkinson to Bather, July 20, 1927, DF 1004/391/7.

25. Bather, probably to Harmer, July 20, 1927, DF 1004/391/7.

26. Bather to Parkinson, July 25, 1927, DF 1004/391/7.

27. Parkinson to Bather, October 8, 1927, DF 1004/391/7.

28. Bather to BM(NH) trustees, October 13, 1927, DF 1004/391/7.

29. Deacon to Smith, December 23, 1927, DF 1004/391/7.

30. J. Parkinson, *The Dinosaur in East Africa: An Account of the Giant Reptile Beds of Tendaguru, Tanganyika Territory* (London: H. F. & G. Witherby, 1930), p. 23.

15. 1927–1929: Geology at Tendaguru

1. Smith to Green, Colonial Office, April 24, 1928, DF 1004/391/8.

2. Lang memorandum, May 2, 1928, DF 1004/391/8.

9. Grunberger, *A Social History of the Third Reich*, p. 24.

10. Ibid., p. 29.

11. H. Reck, "Auf der Suche nach Saurier-Skeletten in der Karru Natals," *Umschau* 37, no. 17 (1933), p. 328.

12. Grunberger, *A Social History of the Third Reich*, p. 27.

13. Whiting, *The Home Front: Germany*, p. 27.

14. Grunberger, *A Social History of the Third Reich*, p. 36.

15. Hennig, July 10, 1934, UAT 407/9.

16. Hennig, July 30, 1934, UAT 407/9.

17. E. Hennig, "Der Sedimentstreifen des Lindi-Kilwa-Hinterlandes (Deutsch-Ostafrika)," *Palaeontographica*, Suppl. 7, Reihe 2, Teil 2 (1937), p. 102.

18. Hennig, September 20, 1934, UAT 407/9.

19. Hennig, October 1, 1934, UAT 407/9.

20. Hennig, October 6, 1934, UAT 407/9.

21. E. Hennig, *Gewesene Welten: Auf Saurierjagd im ostafrikanischen Busch* (Zürich: Albert Müller, 1955), p. 140.

22. Hennig, "Der Sedimentstreifen des Lindi-Kilwa-Hinterlandes," p. 101.

23. Hennig, *Gewesene Welten*, p. 155.

24. Hennig, "Der Sedimentstreifen des Lindi-Kilwa-Hinterlandes," p. 101.

25. Hennig, *Gewesene Welten*, pp. 140–41.

26. Whiting, *The Home Front: Germany*, p. 36.

27. Grunberger, *A Social History of the Third Reich*, p. 181.

28. Whiting, *The Home Front: Germany*, p. 32.

29. Ibid., p. 30.

30. A. T. Hopwood, "Prof. Hans Reck," *Nature* 140 (1937), p. 351.

31. W. Gross, "Die Typen des mikroskopischen Knochenbaues bei fossilen Stegocephalen und Reptilien," *Zeitschrift für Anatomie* 103 (1934), pp. 731–64.

32. W. Janensch, "Die Schädel der Sauropoden *Brachiosaurus, Barosaurus,* und *Dicraeosaurus* aus den Tendaguru-Schichten Deutsch-Ostafrikas," *Palaeontographica,* Suppl. 7, Reihe 1, Teil 2 (1935–1936), p. 255.

33. W. Janensch, "Die Wirbelsäule von *Brachiosaurus brancai,*" *Palaeontographica,* Suppl. 7, Reihe 1, Teil 2 (1950), p. 60.

34. W. Janensch, "Die Skelettrekonstruktion von *Brachiosaurus brancai,*" *Palaeontographica,* Suppl. 7, Reihe 1, Teil 3 (1950), p. 97.

35. Ibid.

36. W. Janensch, "Gestalt und Grösse von *Brachiosaurus* und anderen riesenwüchsigen Sauropoden," *Der Biologe* 7, no. 4 (1938), p. 131.

37. Brown to Janensch, January 21, 1938, MfN.

38. Schuchert to Janensch, July 27, 1938, MfN.

39. Grunberger, *A Social History of the Third Reich*, pp. 245, 257.

19. 1939–1976: Destruction and Renewal

1. M. Middlebrook, *The Berlin Raids: RAF Bomber Command Winter, 1943–44* (New York: Viking, 1988), p. 24.

2. K. Adam, "Die Württembergische Naturaliensammlung zu Stuttgart im Zweiten Weltkrieg," *Stuttgarter Beiträge zur Naturkunde,* series C, 30 (1991), pp. 96–97.

3. I. Jahn, "Der neue Museumsbau und die Entwicklung neuer museologischer Konzeptionen und Aktivitäten seit 1890," *Wissenschaftliche Zeitschrift der Humboldt-Universität* 38 (1989), pp. 303–304.

4. M. Middlebrook and C. Everitt, *The Bomber Command Diaries: An Operational Reference Book, 1939–1945* (New York: Viking, 1985), pp. 452–53.

5. C. Whiting, *The Home Front: Germany* (Chicago: Time-Life Books, 1982), p. 148.

6. Jeletzky to author, December 21, 1984.

7. O. Groehler, *Bombenkrieg gegen Deutschland* (Berlin: Akademie-Verlag, 1990), pp. 398, 400.

8. Ibid., p. 423.

9. A. Tully, *Berlin: Story of a Battle, April–May 1945* (New York: Macfadden-Bartell, 1964), p.117.

10. Middlebrook, *The Berlin Raids,* p. 320.

11. T. LeTissier, *The Battle of Berlin, 1945* (London: Jonathan Cape, 1988), p. 119.

12. M. Wyman, *DPs: Europe's Displaced Persons, 1945–1951* (Ithaca: Cornell University Press, 1998), pp. 17–20.

13. D. Botting, *In the Ruins of the Reich* (London: Grafton Books, 1986), p. 163.

14. Middlebrook, *The Berlin Raids,* p. 328.

15. Botting, *In the Ruins of the Reich,* p. 167.

16. Adam, "Die Württembergische Naturaliensammlung," p. 84.

17. W. T. Stearn, *The Natural History Museum at South Kensington: A History of the British Museum (Natural History), 1753–1980* (London: Heinemann, 1981), p. 149.

18. W. E. Swinton, "John Parkinson," *Nature* 160, no. 4064 (1947), pp. 390–91.

19. A. Tusa and J. Tusa, *The Berlin Blockade* (London: Hodder and Stoughton, 1988), p. 190.

20. W. E. Swinton, "F. W. H. Migeod," *Nature* 170, no. 4318 (1952), p. 184.

21. W. Janensch, "Der Ornithopode *Dysalotosaurus* der Tendaguruschichten," *Palaeontographica,* Suppl. 7, Reihe 1, Teil 3 (1955), p. 137.

22. Ibid., p. 151; Janensch, "Skelettrekonstruktion von *Dysalotosaurus lettow-vorbecki,*" *Palaeontographica,* Suppl. 7, Reihe 1, Teil 3 (1961), p. 239.

23. Janensch, "Skelettrekonstruktion von *Dysalotosaurus lettow-vorbecki,*" p. 239.

24. H. Jahnke to author, September 1, 1998.

25. H. Kinder and W. Hilgemann, *Penguin Atlas of World History,* vol. 2 (Harmondsworth: Penguin, 1978), p. 248.

26. J. Man, *Berlin Blockade* (New York: Ballantine Books, 1973), p. 130.

27. H. Jaeger, "Werner Janensch, 1878–1969," *Berichte der Deutschen Gesellschaft für Geologische Wissenschaften* 16, no. 2 (1971), p. 154.

28. L. S. Kurtz, *Historical Dictionary of Tanzania* (Metuchen, N.J.: The Scarecrow Press, 1978), p. 107.

29. U. Oberli, "Der 'Saurierfriedhof' am Tendaguru: Auf den Spuren einer fast vergessenen Ausgrabungsexpedition in Tansania," *Schweizer Strahler* 6, no. 8 (1983), p. 380.

20. 1971–2001: Russell to Africa, *Brachiosaurus* to Tokyo, Berlin to Tendaguru

1. Mturi to National Museum of Natural Sciences, document NS/5850-1, October 20, 1976, Archives, Canadian Museum of Nature, Ottawa (hereafter CMN).

2. D. A. Russell expedition diary, July 20, 1977, CMN.

3. Ibid., July 21, 1977.

4. Russell to Lemieux, November 2, 1977, CMN.

5. Lemieux to Mturi, August 25, 1978, CMN.

6. Mturi to Lemieux, document NS/1465-2, September 21, 1978, CMN.

7. D. A. Russell expedition diary, October 31, 1978, CMN.

8. G. I. Smith, *Ghosts of Kampala* (London: Weidenfeld and Nicolson, 1980), p. 182.

9. D. A. Russell expedition diary, November 3, 1978, CMN.

10. D. M. Jarzen, "A preliminary report of the palynomorphs recovered from Tendaguru hill (Tanzania)," *Pollen et Spores* 23, no. 1 (1981), p. 159.

11. M. Barthel, "Mit dem Riesensaurier in Japan," *Humboldt-Universität,* no. 2 (1984–1985), p. 5.

12. Both quotations are from "DDR schickt unersetzliche Saurier-Skelette nach Japan: Ost-Berlin nimmt irreparable Transportschaden in Kauf," *Berliner Morgenpost,* May 27, 1984, p. 4.

13. "Urvogel und Saurier," *Frankfurter Allgemeine Zeitung,* July 19, 1984, p. 21.

14. Barthel, "Mit dem Riesensaurier in Japan," p. 5, and "Urvogel und Saurier," p. 21.

15. "40,000 Tokioter sahen bisher Saurier aus Berlin: Lehr-und Forschungsschau findet grossen Anklang," *Berliner Zeitung,* July 22–23, 1984, p. 21.

16. Hasegawa, personal communication, December 1, 1998.

17. L. L. Jacobs, D. A. Winkler, W. R. Downs, and E. M. Gomani, "New material of an Early Cretaceous titanosaurid sauropod dinosaur from Malawi," *Paleontology* 36 (1993), p. 523; L. L. Jacobs, D. A. Winkler, Z. M. Kaufulu, and W. R. Downs, "The Dinosaur Beds of northern Malawi, Africa," *National Geographic Research* 6, no. 2 (1990), p. 196.

18. S. M. Ball et al., *Final Report of Tanzanian Mpingo 96, Cambridge Mpingo Project,* 1998, p. 32.

19. C. Werner and W. Zils, "Tendaguru-Expedition Dezember 1996," *Terra Nostra,* 7, no. 97, p. 30.

21. A Significant Contribution

1. P. E. Kent, "Continental margin of East Africa—a region of vertical movements," in *The Geology of Continental Margins,* ed. C. A. Burk and C. L. Drake (New York: Springer-Verlag, 1974), pp. 313–20.

2. D. A. Russell, P. Béland, and J. S. McIntosh, "Paleoecology of the dinosaurs of Tendaguru (Tanzania)," *Memoires de Société géologique de France* 59, no. 139 (1980), pp. 169–75.

3. M. E. Schudack, "Some charophytes from the Middle Dinosaur Member of the Tendaguru Formation (Upper Jurassic of Tanzania)," *Mitteilungen aus dem Museum für Naturkunde in Berlin, Geowissenschaftliche Reihe* 2 (1999), pp. 201–205.

4. W. Zils et al., "Orientierende Tendaguru-Expedition 1994," *Berliner Geowissenschaftliche Abhandlungen* 16.2 (1995), pp. 483–531.

5. J. Pálfry, P. L. Smith, and J. K. Mortensen, "A U-Pb and ^{40}Ar/^{39}Ar time scale for the Jurassic," *Canadian Journal of Earth Sciences* 37 (2000), p. 931.

6. E. Schrank, "Age and biogeographic links of palynomorph assemblages from the dinosaur beds of Tendaguru (Late Jurassic, Tanzania)," *Geo-Eco-Trop* 22 (2000): 141–46, Actes du 4éme Symposium de Palynologie africaine, Sousse, Tunisia, April 25–30, 1999.

7. E. Schrank, "Eine Expedition zur Dinosaurierlagerstätte Tendaguru (Tansania) und erste palynologische Ergebnisse," *29. Jahrestagung des Arbeitskreises für Paläobotanik und Palynologie,* program and abstracts, Utrecht, The Netherlands, 2001, p. 27.

8. Zils et al., "Orientierende Tendaguru-Expedition 1994," p. 495.

9. R. L. Romer, "Isotopically heterogeneous initial Pb and continuous ^{222}Rn loss in fossils: The U-Pb systematics of *Brachiosaurus brancai,*" *Geochimica et Cosmochimica Acta* 65, no. 22 (2001), pp. 4201–13.

10. Schrank, "Eine Expedition zur Dinosaurierlagerstätte Tendaguru," p. 27.

11. R. Bussert and M. Aberhan, "Sedimentation and palaeoecology of a tide, storm, and tsunami(?)-influenced coast at a passive margin: An example from the Upper Jurassic-Lower Cretaceous of SE Tanzania," 2001 Margins Meeting, program with abstracts, *Schriftenreihe der deutschen geologischen Gesellschaft* 14 (2001), pp. 36–37.

12. R. Bussert, "Palaeoenvironmental interpretation of the Upper Jurassic-Lower Cretaceous Tendaguru Beds of southeast Tanzania, based on petrographic investigations," 18th Colloquium of African Geology, Graz, Austria, abstracts, *Journal of African Earth Sciences* 30, no. 4a (2000), p. 18.

13. W. O. Dietrich, "Zur Stratigraphie und Paläontologie der Tendaguruschichten," *Palaeontographica,* Suppl. 7, Reihe 2, Teil 2 (1933), pp. 1–86.

14. Bussert and Aberhan, "Sedimentation and palaeoecology of a tide, storm, and tsunami(?)-influenced coast at a passive margin," pp. 36–37.

15. Bussert, "Die flachmarine und kontinentale Sedimentation an einer durch Gezeiten, Stürme und Tsunamis(?)-geprägten Küste eines passiven Kontinentalrandes (Oberjura-Unterkreide, Südost-Tansania), Sediment 2001, Jena, abstracts, *Schriftenreihe der deutschen geologischen Gesellschaft* 13 (2001), pp. 28–29.

16. Russell, Béland, and McIntosh, "Paleoecology of the dinosaurs of Tendaguru (Tanzania)."

17. P. E. Kent, J. A. Hunt, and D. W. Johnstone, *The Geology and Geophysics of Coastal Tanzania,* Geophysical Paper no. 6, Natural Environment Research Council, Institute of Geological Sciences (London: H.M.S.O., 1971), p. 88.

18. M. Schudack and U. Schudack, "Ostracods from the Middle Dinosaur Member of the Tendaguru Formation (Upper Jurassic of Tanzania)," *Neues Jahrbuch für Geologie und Paläontologie,* Monatshefte 6 (2002), pp. 321–33.

19. Schudack, "Some charophytes from the Middle Dinosaur Member of the Tendaguru Formation," pp. 201–205.

20. Dietrich, "Zur Stratigraphie und Paläontologie der Tendaguruschichten," pp. 1–86.

21. W.-D. Heinrich et al., "The German-Tanzanian Tendaguru Expedition 2000," *Mitteilungen aus dem Museum für Naturkunde in Berlin, Geowissenschaftliche Reihe* 4 (2001), pp. 223–37.

22. G. T. Moore et al., "Paleoclimate of the Kimmeridgian/Tithonian (Late Jurassic) world: I. Results using a general circulation model," *Palaeogeography, Palaeoclimatology, Palaeoecology* 93 (1992), pp. 113–50.

23. E. Kahlert, S. Schultka, and H. Süss, "Die mesophytische Flora der Saurierlagerstätte am Tendaguru (Tansania): Erste Ergebnisse," *Mitteilungen aus dem Museum für Naturkunde in Berlin, Geowissenschaftliche Reihe* 2 (1999), pp. 185–99.

24. R. Grube, S. Schultka, and H. Süss, "Kutikulen und Fusite—Hinweise auf eine oberjurassische Flora von Tendaguru (Tansania)," *29 Jahrestagung des Arbeitskreises für Paläobotanik und Palynologie,* program and abstracts, Utrecht, The Netherlands, 2001, p. 8.

25. Ibid.

26. D. M. Jarzen, "A preliminary report of the palynomorphs recovered from Tendaguru hill (Tanzania)," *Pollen et Spores* 23, no. 1 (1981), pp. 149–63.

27. Schrank, "Age and biogeographic links of palynomorph assemblages from the dinosaur beds of Tendaguru," pp. 141–46.

28. E. Schrank, "Palynology of the dinosaur beds of Tendaguru (Tanzania)—preliminary results," *Mitteilungen aus dem Museum für Naturkunde in Berlin, Geowissenschaftliche Reihe* 2 (1999), pp. 171–83.

29. G. Arratia and H.-P. Schultze, "Semionotiform fish from the Upper Jurassic of Tendaguru (Tanzania)," *Mitteilungen aus dem Museum für Naturkunde in Berlin, Geowissenschaftliche Reihe* 2 (1999), pp. 135–53.

30. P. Wellnhofer, ed., *Handbuch der Paläoherpetologie,* Teil 19, Pterosauria (Stuttgart: Gustav Fischer Verlag, 1978).

31. P. M. Galton, "Avian-like tibiotarsi of pterodactyloids (Reptilia: Pterosauria) from the Upper Jurassic of East Africa," *Paläontologische Zeitschrift* 54, nos. 3–4 (1980), pp. 331–42.

32. D. M. Unwin and W.-D. Heinrich, "On a pterosaur jaw from the Upper Jurassic of Tendaguru (Tanzania)," *Mitteilungen aus dem Museum für Naturkunde in Berlin, Geowissenschaftliche Reihe* 2 (1999), pp. 121–34.

33. J. M. Sayao and A. W. A. Kellner, "New data on the pterosaur fauna from Tendaguru (Tanzania), Upper Jurassic, Africa," *Journal of Vertebrate Paleontology* 21, supplement to no. 3 (2001), Abstracts of Papers, Sixty-first Annual Meeting, p. 97A.

34. T. R. Holtz, "The phylogenetic position of the Tyrannosauridae: Implications for theropod systematics," *Journal of Palaeontology* 68, no. 5 (1994), pp. 1100–17.

35. T. R. Holtz, "A new phylogeny of the Theropoda," *Journal of Vertebrate Paleontology* 15, supplement to no. 3 (1995), Abstracts of Papers, Fifty-fifth Annual Meeting, p. 35A.

36. P. C. Sereno, "The evolution of dinosaurs," *Science* 284 (1999), pp. 2137–47.

37. O. Rauhut, "*Elaphrosaurus bambergi* and the early evolution of theropod dinosaurs," *Journal of Vertebrate Paleontology* 18, supplement to no. 3 (1998), Abstracts of Papers, Fifty-eighth Annual Meeting, p. 71A; O. Rauhut, "The interrelationships and evolution of basal theropods (*Dinosauria, Saurischia*)" (Ph.D. diss., University of Bristol, 2000).

38. T. R. Holtz, Jr., "A new phylogeny of the carnivorous dinosaurs," *Gaia* 15 (2000), p. 15.

39. J. H. Madsen and S. P. Welles, Ceratosaurus *(Dinosauria, Theropoda): A Revised Osteology* (Salt Lake City: Utah Geological Survey, 2000).

40. L. Salgado, "The macroevolution of the Diplodocomorpha (Dinosauria; Sauropoda): A developmental model," *Ameghiniana* 36, no. 2 (1999), pp. 203–16.

41. M. J. Wedel, R. L. Cifelli, and R. K. Sanders, "Osteology, paleobiology, and relationships of the sauropod dinosaur *Sauroposeidon*," *Acta Palaeontologica Polonica* 45, no. 4 (2000), pp. 343–88.

42. J. A. Wilson and P. C. Sereno, "Early evolution and higher-level phylogeny of sauropod dinosaurs," Society of Vertebrate Paleontology Memoir 5, *Journal of Vertebrate Paleontology* 18, suppl. to no. 2 (1998).

43. J. F. Bonaparte, W.-D. Heinrich, and R. Wild, "Review of *Janenschia* Wild, with the description of a new sauropod from the Tendaguru Beds of Tanzania and a discussion on the systematic value of procoelous caudal vertebrae in the Sauropoda," *Palaeontographica* 256, Abt. A (2000), pp. 25–76.

44. R. Wild, "*Janenschia* n. g. robusta (E. Fraas 1908) pro *Tornieria robusta* (E. Fraas 1908) (Reptilia, Saurischia, Sauropodomorpha)," *Stuttgarter Beiträge zur Naturkunde,* series B, no. 173 (1991), pp. 1–4.

45. Schrank, "Age and biogeographic links of palynomorph assemblages from the dinosaur beds of Tendaguru," pp. 141–46.

46. E. Schrank, "Palynomorphs from dinosaur-bearing Upper Jurassic strata of Tanzania in a palaeoenvironmental and phytogeographic context," Tenth International Palynological Conference, Nanjing, China, 2000, Abstracts, p. 147.

47. H. Süss and S. Schultka, "First record of *Glyptostroboxylon* from the Upper Jurassic of Tendaguru, Tanzania," *Botanical Journal of the Linnean Society* 135 (2001), pp. 421–29.

48. Schudack, "Some charophytes from the Middle Dinosaur Member of the Tendaguru Formation," pp. 201–205.

49. Schudack and Schudack, "Ostracods from the Middle Dinosaur Member of the Tendaguru Formation, (Upper Jurassic of Tanzania)," *Neues Jahrbuch für Geologie und Paläontologie,* Monatshefte 6 (2002), pp. 321–33.

50. P. Upchurch, "The evolutionary history of sauropod dinosaurs," *Philosophical Transactions of the Royal Society of London B,* 349 (1995), pp. 365–90.

51. J. F. Bonaparte, "Cretaceous tetrapods of Argentina," *Münchner Geowissenschaftliche Abhandlungen* A 30 (1996), pp. 73–130.

52. P. M. Galton, "The ornithopod dinosaur *Dryosaurus* and a Laurasia-Gondwanaland connection," *Nature* 268 (1977), pp. 230–32.

53. Wilson and Sereno, "Early evolution and higher-level phylogeny of sauropod dinosaurs."

54. R. A. Coria and L. Salgado, "A basal iguanodontian (Ornithischia: Ornithopoda) from the Late Cretaceous of South America," *Journal of Vertebrate Paleontology* 16, no. 3 (1996), pp. 445–57.

55. W.-D. Heinrich, "First haramiyid (Mammalia, Allotheria) from the Mesozoic of Gondwana," *Mitteilungen aus dem Museum für Naturkunde in Berlin, Geowissenschaftliche Reihe* 2 (1999), pp. 169–70.

56. H. Haubold, "Dinosaurs and fluctuating sea levels during the Mesozoic," *Historical Biology* 4 (1990), pp. 75–106.

57. Hunt et al., "The global sauropod fossil record," *Gaia* 10 (1994), pp. 261–79.

58. Russell, "Progress Report #2," unpublished paper, National Museum of Natural Sciences, 1977, p. 2, CMN.

59. W.-D. Heinrich, "The taphonomy of dinosaurs from the Upper Jurassic of Tendaguru (Tanzania), based on field sketches of the German Tendaguru Expedition (1909–1913)," *Mitteilungen aus dem Museum für Naturkunde in Berlin, Geowissenschaftliche Reihe* 2 (1999), pp. 25–61.

60. Ibid.

61. Russell, "Progress Report #2," p. 5.

62. Heinrich, "The taphonomy of dinosaurs from the Upper Jurassic of Tendaguru," pp. 25–61.

63. Ibid.

64. Russell, Béland, and McIntosh, "Paleoecology of the dinosaurs of Tendaguru (Tanzania)," pp. 169–75.

65. Russell, "Progress Report #2," p. 5.

66. Russell, Béland, and McIntosh, "Paleoecology of the dinosaurs of Tendaguru (Tanzania)," pp. 169–75.

67. Russell, "Progress Report #2," p. 5.

68. Alexander, *Dynamics of Dinosaurs and Other Extinct Giants* (New York: Columbia University Press, 1989), pp. 30–33.

69. W. Janensch, "Eine halbseitige überzählige Wirbelbildung bei einem Dinosaurier," *Sitzungsberichte der Gesellschaft Naturforschender Freunde zu Berlin* (1934), pp. 458–62.

70. H. E. Kaiser, "Eine Rippe mit pathologischen Erscheinungen des Sauropods *Brachiosaurus brancai* Janensch, des grössten Saurischier (Dinosaurier) des Tendagurumaterials aus der jüngeren Jura," *Monatshefte für Veterinär Medizin* 9, no. 4 (1954), pp. 373–74.

71. W. Janensch, "Die Wirbelsäule der Gattung *Dicraeosaurus,*" *Palaeontographica,* Suppl. 7, Reihe 1, Teil 2, (1929), pp. 35–133.

72. A. Chinsamy, "Ontogenetic changes in the bone histology of the Late Jurassic ornithopod *Dryosaurus lettowvorbecki,*" *Journal of Vertebrate Paleontology* 15, no. 1 (1995), pp. 96–104.

73. F. Seebacher, "A new method to calculate allometric length-mass relationships of dinosaurs," *Journal of Vertebrate Paleontology* 21, no. 1 (2001), pp. 51–60.

74. J. Peczkis, "Implications of body-mass estimates for dinosaurs," *Journal of Vertebrate Paleontology* 14, no. 4 (1994), pp. 520–33.

75. R. A. Thulborn, "Speeds and gaits of dinosaurs," *Palaeogeography, Palaeoclimatology, Palaeoecology* 38 (1982), pp. 227–56.

76. R. E. Heinrich, C. B. Ruff, and D. B. Weishampel, "Femoral ontogeny and locomotor biomechanics of *Dryosaurus lettowvorbecki* (Dinosauria, Iguanodontidae)," *Zoological Journal of the Linnean Society* 108 (1993), pp. 179–96.

77. D. W. Dilkes, "An ontogenetic perspective on locomotion in the Late Cretaceous dinosaur *Maiasaura peeblesorum* (Ornithischia: Hadrosauridae)," *Canadian Journal of Earth Sciences* 38, no. 8 (2001), pp. 1205–27.

78. Janensch, "Eine halbseitige überzählige Wirbelbildung bei einem Dinosaurier," pp. 458–62.

79. P. Béland and D. A. Russell, "Dinosaur metabolism and predator-prey ratios in the fossil record," in *A Cold Look at the Warm-blooded Dinosaurs,* ed. R. D. K. Thomas and E. C. Olson (Boulder, Colo.: Westview Press for the American Association for the Advancement of Science, 1980), 85–102.

80. Peczkis, "Implications of body-mass estimates for dinosaurs," pp. 520–33.

81. J. A. Hopson, "Relative brain size and behavior in archosaurian reptiles," *Annual Review of Ecology and Systematics* 8 (1977), pp. 429–48.

82. E. B. Giffin, "Endosacral enlargements in dinosaurs," *Modern Geology* 16 (1991), pp. 101–12.

83. Russell, Béland, and McIntosh, "Paleoecology of the dinosaurs of Tendaguru (Tanzania)," pp. 169–75.

84. Béland and Russell, "Dinosaur metabolism and predator-prey ratios in the fossil record," pp. 85–102.

85. J. F. Anderson, A. Hall-Martin, and D. A. Russell, "Long-bone circumference and weight in mammals, birds and dinosaurs," *Journal of Zoology* 207 (1985), pp. 53–61.

86. Peczkis, "Implications of body-mass estimates for dinosaurs," pp. 520–33.

87. O. Wings, "Early diagenetic processes and cementation in the fossilization of tetrapod bones," project in progress at University of Bristol, 2001, <http://www.Paleontology.uni-bonn.de/mitarbeiter/wings/bristol.htm>, accessed September 7, 2002.

88. Romer, "Isotopically heterogeneous initial Pb and continuous ^{222}Rn loss in fossils," pp. 4201–13.

89. F. Thoss and P. Schwartze, "Über mögliche funktionelle Eigenschaften des Labyrinths von *Brachiosaurus brancai*," *Acta Biologica et Medica Germanica* 34, no. 5 (1975), pp. 899–906.

90. C. McGowan, *The Successful Dragons: A Natural History of Extinct Reptiles* (Toronto: Samuel Stevens, 1983).

91. Hopson, "Relative brain size and behavior in archosaurian reptiles," pp. 429–48.

92. W. Janensch, "Die Schädel der Sauropoden *Brachiosaurus, Barosaurus* und *Dicraeosaurus* aus den Tendaguru-Schichten Deutsch-Ostafrikas," *Palaeontographica,* Suppl. 7, Reihe 1, Teil 2 (1935–1936), pp. 249–98.

93. W. Janensch, "Gestalt und Grösse von *Brachiosaurus* und anderen riesenwüchsigen Sauropoden," *Der Biologe* 7, no. 4 (1938), pp. 130–34.

94. E. H. Colbert, "The weights of dinosaurs," *American Museum Novitates,* no. 2076 (1962), pp. 1–16.

95. Béland and Russell, "Dinosaur metabolism and predator-prey ratios in the fossil record," pp. 85–102.

96. Anderson, Hall-Martin, and Russell, "Long-bone circumference and weight in mammals, birds and dinosaurs," pp. 53–61.

97. Peczkis, "Implications of body-mass estimates for dinosaurs," pp. 520–33.

98. R. McN. Alexander, "Mechanics of posture and gait of some large dinosaurs," *Zoological Journal of the Linnean Society* 83, no. 1 (1985), pp. 1–25.

99. G. S. Paul, "The brachiosaur giants of the Morrison and Tendaguru with a description of a new subgenus, *Giraffatitan,* and a comparison of the world's largest dinosaurs," *Hunteria* 2, no. 3 (1988), pp. 1–14.

100. H.-C. Gunga, K. A. Kirsch, F. Baartz, et al., "New data on the dimensions of *Brachiosaurus brancai* and their physiological implications," *Naturwissenschaften* 82 (1995), pp. 190–92; H.-C. Gunga, K. Kirsch, J. Rittweger, et al., "Body size and body volume distribution in two sauropods from the Upper Jurassic of Tendaguru (Tanzania)," *Mitteilungen aus dem Museum für Naturkunde in Berlin, Geowissenschaftliche Reihe* 2 (1999), pp. 91–102.

101. P. Christiansen, "Locomotion in sauropod dinosaurs," *Gaia* 14 (1997), pp. 45–75.

102. A. Christian, W.-D. Heinrich, and W. Golder, "Posture and mechanics of the forelimbs of *Brachiosaurus brancai* (Dinosauria: Sauropoda)," *Mitteilungen aus dem Museum für Naturkunde in Berlin, Geowissenschaftliche Reihe* 2 (1999), pp. 63–73.

103. Seebacher, "A new method to calculate allometric length-mass relationships of dinosaurs," pp. 51–60.

104. Christiansen, "Locomotion in sauropod dinosaurs," pp. 45–75.

105. W. P. Coombs, "Sauropod habits and habitats," *Palaeogeography, Palaeoclimatology, Palaeoecology* 17 (1975), pp. 1–33.

106. Christian, Heinrich, and Golder, "Posture and mechanics of the forelimbs of *Brachiosaurus brancai* (Dinosauria: Sauropoda)," pp. 63–73.

107. Alexander, *The Dynamics of Dinosaurs and Other Extinct Giants,* p. 55.

108. A. Christian et al., "Limb swinging in elephants and giraffes and implications for the reconstruction of limb movements and speed estimates in large dinosaurs," *Mitteilungen aus dem Museum für Naturkunde in Berlin, Geowissenschaftliche Reihe* 2 (1999), pp. 81–90.

109. Thulborn, "Speeds and gaits of dinosaurs," pp. 227–56.

110. K. A. Stevens and J. M. Parrish, "Neck posture and feeding habits of two Jurassic sauropod dinosaurs," *Science* 284 (1999), pp. 798–800.

111. K. A. Stevens and J. M. Parrish, "The intrinsic curvature of sauropod necks (Saurischia: Dinosauria)," in preparation.

112. K. A. Stevens, "DinoMorph: parametric modeling of skeletal structures," *Senckenbergiana Lethaea* 82, no. 1 (2002), pp. 23–34.

113. K. A. Stevens and J. Parrish, "Biological implications of digital reconstructions of the whole body of sauropod dinosaurs," in press.

114. J. Martin, V. Mártin-Rolland, and E. Frey, "Not cranes or masts, but beams: The biomechanics of sauropod necks," *Oryctos* 1 (1998), pp. 113–20.

115. A. Christian and W.-D. Heinrich, "The neck posture of *Brachiosaurus brancai*," *Mitteilungen aus dem Museum für Naturkunde in Berlin, Geowissenschaftliche Reihe* 1 (1998), pp. 73–80.

116. Alexander, "Mechanics of posture and gait of some large dinosaurs," pp. 1–25.

117. Wedel, Cifelli, and Sanders, "Osteology, paleobiology, and relationships of the sauropod dinosaur *Sauroposeidon*," pp. 343–88.

118. L. A. Hohnke, "Haemodynamics in the Sauropoda," *Nature* 244 (1973), pp. 309–10.

119. D. S. J. Choy and P. Altman, "The cardiovascular system of *Barosaurus*: An educated guess," *Lancet* 340 (1992), pp. 534–36.

120. R. S. Seymour, "Dinosaurs, endothermy and blood pressure," *Nature* 262 (1976), pp. 207–208.

121. R. W. Millard, H. B. Lillywhite, and A. R. Hargens, "Cardiovascular system design and *Barosaurus*," *Lancet* 340 (1992), p. 914.

122. J. M. Dennis, "*Barosaurus* and its circulation," *Lancet* 340 (1992), p. 1228.

123. H. S. Badeer and J. W. Hicks, "Circulation to the head of *Barosaurus* revisited: Theoretical considerations," *Comparative Biochemistry and Physiology* 114A, no. 3 (1996), pp. 197–203.

124. Gunga, Kirsch, Rittweger, et al., "Body size and body volume distribution in two sauropods," pp. 91–102.

125. Gunga, Kirsch, Baartz, et al., "New data on the dimensions of *Brachiosaurus brancai* and their physiological implications," pp. 190–92.

126. G. S. Paul, "Terramegathermy and Cope's Rule in the land of the titans," *Modern Geology* 23 (1998), pp. 179–217.

127. R. S. Seymour and H. B. Lillywhite, "Hearts, neck posture and metabolic intensity of sauropod dinosaurs," *Proceedings of the Royal Society of London B* 267 (2000), pp. 1883–87.

128. P. Upchurch and P. M. Barrett, "The evolution of sauropod feeding mechanisms," in *Evolution of Herbivory in Terrestrial Vertebrates: Perspectives from the Fossil Record*, ed. H.-D. Sues (Cambridge: Cambridge University Press, 2000), pp. 79–122.

129. Christian and Heinrich, "The neck posture of *Brachiosaurus brancai*," pp. 73–80.

130. H.-C. Gunga and K. Kirsch, "Von Hochleistungsherzen und wackeligen Hälsen," *Forschung: Das Magazin der deutschen Forschungsgemeinschaft* 2–3 (2001), p. 8.

131. J. O. Calvo, "Jaw mechanics in sauropod dinosaurs," *Gaia* 10 (1994), pp. 183–93.

132. Upchurch and Barrett, "The evolution of sauropod feeding mechanisms," pp. 79–122.

133. P. Christiansen, "Feeding mechanisms of the sauropod dinosaurs *Brachiosaurus, Camarasaurus, Diplodocus,* and *Dicraeosaurus*," *Historical Biology* 14, no. 3 (2000), pp. 137–52.

134. A. R. Fiorillo, "Enamel microstructure in *Diplodocus, Camarasaurus,* and *Brachiosaurus* (Dinosauria: Sauropoda) and its lack of influence on resource partitioning by sauropods in the Late Jurassic," 1995, in *Sixth Symposium on Mesozoic Terrestrial Ecosystems and Biota: Short Papers,* ed. Ailing Sun and Yuanquing Wang (Beijing: China Ocean Press, 1995), pp. 147–49.

135. A. R. Fiorillo, "Dental microwear patterns of the sauropod dinosaurs *Camarasaurus* and *Diplodocus:* Evidence for resource partitioning in the Late Jurassic of North America," *Historical Biology* 13 (1998), pp. 1–16.

136. J. O. Farlow, "Speculations about the diet and digestive physiology of herbivorous dinosaurs," *Paleobiology* 13, no. 1 (1987), pp. 60–72.

137. Upchurch and Barrett, "The evolution of sauropod feeding mechanisms," pp. 79–122.

138. Christiansen, "Feeding mechanisms of the sauropod dinosaurs *Brachiosaurus, Camarasaurus, Diplodocus,* and *Dicraeosaurus,*" pp. 137–52.

139. Calvo, "Jaw mechanics in sauropod dinosaurs," pp. 183–93.

140. Upchurch and Barrett, "The evolution of sauropod feeding mechanisms," pp. 79–122, and Christiansen, "Feeding mechanisms of the sauropod dinosaurs *Brachiosaurus, Camarasaurus, Diplodocus,* and *Dicraeosaurus,*" pp. 137–52.

141. J. O. Farlow, "Dinosaur energetics and thermal biology," in D. B. Weishampel, P. Dodson, and H. Osmolska, *The Dinosauria* (Berkeley and Los Angeles: University of California Press, 1990), pp. 43–55.

142. J. C. Weaver, "The improbable endotherm: The energetics of the sauropod dinosaur *Brachiosaurus,*" *Paleobiology* 9, no. 2 (1983), pp. 173–82.

143. P. Christiansen, "On the head size of sauropodomorph dinosaurs: Implications for ecology and physiology," *Historical Biology* 13 (1999), pp. 269–97.

144. Paul, "Terramegathermy and Cope's Rule in the land of the titans," pp. 179–217.

145. Gunga, Kirsch, Baartz, et al., "New data on the dimensions of *Brachiosaurus brancai* and their physiological implications," pp. 191–92.

146. C. B. Daniels and J. Pratt, "Breathing in long necked dinosaurs; did the sauropods have bird lungs?" *Comparative Biochemistry and Physiology* 101A, no. 1 (1992), pp. 43–46.

147. S. F. Perry and C. Reuter, "Hypothetical lung structure of *Brachiosaurus* (Dinosauria: Sauropoda) based on functional constraints," *Mitteilungen aus dem Museum für Naturkunde in Berlin, Geowissenschaftliche Reihe* 2 (1999), pp. 75–79.

148. Gunga, Kirsch, Baartz, et al., "New data on the dimensions of *Brachiosaurus brancai* and their physiological implications," pp. 191–92.

149. Paul, "Terramegathermy and Cope's Rule in the land of the titans," pp. 179–217.

150. Russell, Béland, and McIntosh, "Paleoecology of the dinosaurs of Tendaguru (Tanzania)," pp. 169–75.

151. P. M. Sander, "Longbone histology of the Tendaguru sauropods: Implications for growth and biology," *Paleobiology* 26, no. 3 (2000), pp. 466–88.

152. P. M. Sander, "Life history of Tendaguru sauropods as inferred from long bone histology," *Mitteilungen aus dem Museum für Naturkunde in Berlin, Geowissenschaftliche Reihe* 2 (1999), pp. 103–12.

153. G. S. Paul, "Dinosaur reproduction in the fast lane: Implications for size, success, and extinction," in *Dinosaur Eggs and Babies,* ed. K. Carpenter, K. Hirsch, and J. R. Horner (Cambridge: Cambridge University Press, 1994), pp. 244–55.

154. Paul, "Terramegathermy and Cope's Rule in the land of the titans," pp. 179–217.

REFERENCES

Archives

British Museum East Africa Expedition Archive. Archives Section. Natural History Museum, London. By permission of the Trustees of the Natural History Museum.
Canadian Museum of Nature. Archives. Ottawa.
Geheimes Staatsarchiv Preussischer Kulturbesitz, Berlin.
Hennig, Edwin. Nachlass. 407. Eberhard Karls Universität, Tübingen.
Museum für Naturkunde an der Humboldt-Universität zu Berlin. Documents are held both in the Archiv and at the Paläontologisches Institut.
Staatliches Museum für Naturkunde. Schloss Rosenstein, Stuttgart.

Publications

Adam, K. 1991. "Die Württembergische Naturaliensammlung zu Stuttgart im Zweiten Weltkrieg." *Stuttgarter Beiträge zur Naturkunde*, series C, 30: 81–97.
Aldcroft, D. H. 1970. *The Inter-War Economy: Britain, 1919–1939*. London: Batsford.
Alexander, R. McN. 1985. "Mechanics of posture and gait of some large dinosaurs." *Zoological Journal of the Linnean Society* 83, no. 1: 1–25.
———. 1989. *Dynamics of Dinosaurs and Other Extinct Giants*. New York: Columbia University Press.
Anderson, J. F., A. Hall-Martin, and D. A. Russell. 1985. "Long-bone circumference and weight in mammals, birds, and dinosaurs." *Journal of Zoology* 207: 53–61.
Arratia, G., and H.-P. Schultze. 1999. "Semionotiform fish from the Upper Jurassic of Tendaguru (Tanzania)." *Mitteilungen aus dem Museum für Naturkunde in Berlin, Geowissenschaftliche Reihe* 2: 135–53.
Badeer, H. S., and J. W. Hicks. 1996. "Circulation to the head of *Barosaurus* revisited: Theoretical considerations." *Comparative Biochemistry and Physiology* 114A, no. 3: 197–203.
Bakker, R. T. 1986. *The Dinosaur Heresies: New Theories Unlocking the Mystery of the Dinosaurs and Their Extinction*. New York: William Morrow.
Ball, S. M. J., A. S. Smith, N. S. Keylock, et al. 1998. *Final Report of Tanzanian Mpingo 96*. Cambridge: Cambridge Mpingo Project.
Barthel, M. 1984–85. "Mit dem Riesensaurier in Japan." *Humboldt-Universität*, no. 2: 5.
Béland, P., and D. A. Russell. 1980. "Dinosaur metabolism and predator-prey ratios in the fossil record." In *A Cold Look at the Warm-blooded Dinosaurs*, ed. R. D. K. Thomas and E. C. Olson, 85–102. Boulder, Colo.: Westview Press for the American Association for the Advancement of Science.
Blythe, R. 1963. *The Age of Illusion: England in the Twenties and Thirties, 1919–1940*. London: H. Hamilton.
Boell, L. 1951. *Die Operationen in Ostafrika*. Hamburg: Walter Dachert.
Bonaparte, J. F. 1986. "Les dinosaures (carnosaures, allosauridés, sauropodes, cétiosauridés) du Jurassique moyen de Cerro Cóndor (Chubut, Argentina)." *Annales de Paléontologie* 72: 325–86.

————. 1991. "The Gondwanan theropod families Abelisauridae and Noasauridae." *Historical Biology* 5: 1–25.

————. 1996. "Cretaceous tetrapods of Argentina." *Münchner Geowissenschaftliche Abhandlungen* A 30: 73–130.

Bonaparte, J. F., W.-D. Heinrich, and R. Wild. 2000. "Review of *Janenschia* Wild, with the description of a new sauropod from the Tendaguru Beds of Tanzania and a discussion on the systematic value of procoelous caudal vertebrae in the Sauropoda." *Palaeontographica* 256, Abt. A: 25–76.

Botting, D. 1986. *In the Ruins of the Reich.* London: Grafton Books.

Bussert, R. 2000. "Palaeoenvironmental interpretation of the Upper Jurassic-Lower Cretaceous Tendaguru Beds of southeast Tanzania, based on petrographic investigations." 18th Colloquium of African Geology, Graz, Austria, abstracts, *Journal of African Earth Sciences* 30, no. 4a: 18.

————. 2001. "Die flachmarine und kontinentale Sedimentation an einer durch Gezeiten, Stürme und Tsunamis(?)-geprägten Küste eines passiven Kontinentalrandes (Oberjura-Unterkreide, Südost-Tansania)." Sediment 2001, Jena, abstracts, *Schriftenreihe der deutschen geologischen Gesellschaft* 13: 28–29.

Bussert, R., and M. Aberhan. 2001. "Sedimentation and palaeoecology of a tide-, storm, and tsunami(?)-influenced coast at a passive margin: An example from the Upper Jurassic-Lower Cretaceous of SE Tanzania." 2001 Margins Meeting, program with abstracts, *Schriftenreihe der deutschen geologischen Gesellschaft* 14: 36–37.

Calvo, J. O. 1994. "Jaw mechanics in sauropod dinosaurs." *Gaia* 10: 183–93.

Calvo, J. O., and L. Salgado. 1995. "*Rebbachisaurus tessonei* sp. nov., a new Sauropoda from the Albian-Cenomanian of Argentina; new evidence on the origin of the Diplodocidae." *Gaia* 11: 13–33.

Charig, A. 1990. "Francis Rex Parrington." *Biographical Memoirs of the Royal Society* 36: 361–78.

Chinsamy, A. 1995. "Ontogenetic changes in the bone histology of the Late Jurassic ornithopod *Dryosaurus lettowvorbecki.*" *Journal of Vertebrate Paleontology* 15, no. 1: 96–104.

Choy, D. S. J., and P. Altman. 1992. "The cardiovascular system of *Barosaurus*: An educated guess." *Lancet* 340: 534–36.

Christian, A., and W.-D. Heinrich. 1998. "The neck posture of *Brachiosaurus brancai.*" *Mitteilungen aus dem Museum für Naturkunde in Berlin, Geowissenschaftliche Reihe* 1: 73–80.

Christian, A., W.-D. Heinrich, and W. Golder. 1999. "Posture and mechanics of the forelimbs of *Brachiosaurus brancai* (Dinosauria: Sauropoda)." *Mitteilungen aus dem Museum für Naturkunde in Berlin, Geowissenschaftliche Reihe* 2: 63–73.

Christian, A., R. H. G. Müller, G. Christian, and H. Preuschoft. 1999. "Limb swinging in elephants and giraffes and implications for the reconstruction of limb movements and speed estimates in large dinosaurs." *Mitteilungen aus dem Museum für Naturkunde in Berlin, Geowissenschaftliche Reihe* 2: 81–90.

Christiansen, P. 1997. "Locomotion in sauropod dinosaurs." *Gaia* 14: 45–75.

————. 1999. "On the head size of sauropodomorph dinosaurs: Implications for ecology and physiology." *Historical Biology* 13: 269–97.

————. 2000. "Feeding mechanisms of the sauropod dinosaurs *Brachiosaurus, Camarasaurus, Diplodocus,* and *Dicraeosaurus.*" *Historical Biology* 14, no. 3: 137–52.

Colbert, E. H. 1962. "The weights of dinosaurs." *American Museum Novitates,* no. 2076: 1–16.

Coombs, W. P. 1975. "Sauropod habits and habitats." *Palaeogeography, Palaeoclimatology, Palaeoecology* 17: 1–33.

Cooper, M. R. 1985. "A revision of the ornithischian dinosaur *Kangnasaurus coetzeei* Haughton, with a classification of the Ornithischia." *Annals of the South African Museum* 95: 281–317.

Coria, R. A., and L. Salgado. 1996. "A basal iguanodontian (Ornithischia: Ornithopoda) from the Late Cretaceous of South America." *Journal of Vertebrate Paleontology* 16, no. 3: 445–57.

Cutler, W. E. 1922. "The Badlands of Alberta." *Canadian Illustrated Monthly*, January, 19–25, 50–51.

Daniels, C. B., and J. Pratt. 1992. "Breathing in long necked dinosaurs; did the sauropods have bird lungs?" *Comparative Biochemistry and Physiology* 101A, no. 1: 43–46.

"DDR schickt unersetzliche Saurier-Skelette nach Japan: Ost-Berlin nimmt irreparable Transportschaden in Kauf." 1984. *Berliner Morgenpost*, May 27, p. 4.

Dennis, J. M. 1992. "*Barosaurus* and its circulation." *Lancet* 340: 1228.

Deppe, C., and L. Deppe. 1925. *Um Ostafrika*. Dresden: Beutelspacher.

Deutsche Kolonialgesellschaft. 1910. *Deutscher Kolonial-Atlas*. Berlin: Dietrich Reimer.

Dietrich, W. O. 1933. "Zur Stratigraphie und Paläontologie der Tendaguruschichten." *Palaeontographica*, Suppl. 7, Reihe 2, Teil 2: 1–86.

———. 1960. "Geschichte der Sammlungen des Geologisch-Paläontologischen Institut und Museums der Humboldt-Universität zu Berlin." *Berichte der Geologischen Gesellschaft der DDR* 5: 247–89.

Dilkes, D. W. 2001. "An ontogenetic perspective on locomotion in the Late Cretaceous dinosaur *Maiasaura peeblesorum* (Ornithischia: Hadrosauridae)." *Canadian Journal of Earth Sciences* 38, no. 8: 1205–27.

Dollo, L. 1888. "Iguanodontidae et Camptonotidae." *Comptes rendus hebdomadaires des séances de l'Academie des Sciences, Paris* 106: 775–77.

Farlow, J. O. 1987. "Speculations about the diet and digestive physiology of herbivorous dinosaurs." *Paleobiology* 13, no. 1: 60–72.

———. 1990. "Dinosaur energetics and thermal biology." In *The Dinosauria*, ed. D. B. Weishampel, P. Dodson, and H. Osmolska, 43–55. Berkeley and Los Angeles: University of California Press.

Fiorillo, A. R. 1995. "Enamel microstructure in *Diplodocus, Camarasaurus,* and *Brachiosaurus* (Dinosauria: Sauropoda) and its lack of influence on resource partitioning by sauropods in the Late Jurassic." In *Sixth Symposium on Mesozoic Terrestrial Ecosystems and Biota: Short Papers*, ed. Ailing Sun and Yuanqing Wang, 147–49. Beijing: China Ocean Press.

———. 1998. "Dental microwear patterns of the sauropod dinosaurs *Camarasaurus* and *Diplodocus*: Evidence for resource partitioning in the Late Jurassic of North America." *Historical Biology* 13: 1–16.

Fraas, E. 1907. "Entdeckung von Überresten grosser Dinosaurier in Ostafrika." *Gaea* 44: 55–56.

———. 1908a. "Dinosaurier in Deutsch-Ostafrika." *Umschau* 12, no. 48: 943–48.

———. 1908b. "Ostafrikanische Dinosaurier." *Palaeontographica* 55: 105–44.

———. 1911. "Die ostafrikanischen Dinosaurier." *Verhandlungen der Gesellschaft deutscher Naturforscher und Ärzte* 83, no. 1: 27–41.

Freydank, W. 1989. *Museum für Naturkunde der Humboldt-Universität zu Berlin: 100 Jahre Museumsgebäude in der Invalidenstraße 43*. Berlin: Museum für Naturkunde.

Friedrich, T. 1991. *Berlin between the Wars*. New York: Vendome Press.

Fritzsche, P. 1996. *Reading Berlin, 1900*. Cambridge, Mass.: Harvard University Press.

Galton, P. M. 1977. "The ornithopod dinosaur *Dryosaurus* and a Laurasia-Gondwana-land connection." *Nature* 268: 230–32.

———. 1980. "Avian-like tibiotarsi of pterodactyloids (Reptilia: Pterosauria) from the Upper Jurassic of East Africa." *Paläontologische Zeitschrift 54*, nos. 3–4: 331–42.

Gauthier, J. A. 1986. "Saurischian monophyly and the origin of birds." *Memoirs of the California Academy of Sciences* 8: 1–55.

Giffin, E. B. 1991. "Endosacral enlargements in dinosaurs." *Modern Geology* 16: 101–12.

Gill, A. 1993. *A Dance between Flames: Berlin between the Wars.* London: J. Murray.

Gillman, C. 1945. "Dar es Salaam, 1860 to 1940: A story of growth and change." *Tanganyika Notes and Records* 20: 1–23.

Glynn, S., and J. Oxborrow. 1976. *Interwar Britain: A Social and Economic History.* London: Allen and Unwin.

Goebel, J. 1925. *Afrika zu unsern Füssen.* Berlin: K. F. Koehler.

Great Britain. Naval Intelligence Division. 1920. *A Handbook of German East Africa.* Compiled by the Geographical Section of the Naval Intelligence Division, Naval Staff, Admiralty. London: H.M.S.O.

Groehler, O. 1990. *Bombenkrieg gegen Deutschland.* Berlin: Akademie-Verlag.

Gross, W. 1934. "Die Typen des mikroskopischen Knochenbaues bei fossilen Stego-cephalen und Reptilien." *Zeitschrift für Anatomie* 103: 731–64.

Grube, R., S. Schultka, and H. Süss. 2001. "Kutikulen und Fusite—Hinweise auf eine oberjurassische Flora von Tendaguru (Tansania)." *29 Jahrestagung des Arbeitskreises für Paläobotanik und Palynologie,* program and abstracts, Utrecht, The Netherlands, p. 8.

Grunberger, R. 1974. *A Social History of the Third Reich.* Harmondsworth: Penguin.

Gunga, H.-C., and K. Kirsch. 2001. "Von Hochleistungsherzen und wackeligen Hälsen." *Forschung: Das Magazin der deutschen Forschungsgemeinschaft* 2–3: 4–9.

Gunga, H.-C., K. A. Kirsch, F. Baartz, L. Röcker, W.-D. Heinrich, W. Lisowski, A. Wiede-mann, and J. Albertz. 1995. "New data on the dimensions of *Brachiosaurus brancai* and their physiological implications." *Naturwissenschaften* 82: 190–92.

Gunga, H.-C., K. Kirsch, J. Rittweger, L. Röcker, A. Clarke, J. Albertz, A. Wiedemann, S. Mokry, T. Suthau, A. Wehr, W.-D. Heinrich, and H.-P. Schultze. 1999. "Body size and body volume distribution in two sauropods from the Upper Jurassic of Tendaguru (Tanzania)." *Mitteilungen aus dem Museum für Naturkunde in Berlin, Geowissen-schaftliche Reihe* 2: 91–102.

Harlow, V., and E. M. Chilver, eds. 1965. *History of East Africa.* Volume 2. London: Oxford University Press.

Harpur, B. 1985. *The Official Halley's Comet Book.* London: Hodder and Stoughton.

Haubold, H. 1990. "Dinosaurs and fluctuating sea levels during the Mesozoic." *Historical Biology* 4: 75–106.

Heinrich, R. E., C. B. Ruff, and D. B. Weishampel. 1993. "Femoral ontogeny and loco-motor biomechanics of *Dryosaurus lettowvorbecki* (Dinosauria, Iguanodontidae)." *Zoological Journal of the Linnean Society* 108: 179–96.

Heinrich, W.-D. 1999a. "First haramiyid (Mammalia, Allotheria) from the Mesozoic of Gondwana." *Mitteilungen aus dem Museum für Naturkunde in Berlin, Geowis-senschaftliche Reihe* 2: 159–70.

———. 1999b. "The taphonomy of dinosaurs from the Upper Jurassic of Tendaguru (Tanzania), based on field sketches of the German Tendaguru Expedition (1909–1913)." *Mitteilungen aus dem Museum für Naturkunde in Berlin, Geowissenschaft-liche Reihe* 2: 25–61.

Heinrich, W.-D., R. Bussert, M. Aberhan, O. Hampe, S. Kapilima, E. Schrank, S. Schultka, G. Maier, E. Msaky, B. Sames, and R. Chami. 2001. "The German-Tanzanian Tendaguru Expedition 2000." *Mitteilungen aus dem Museum für Naturkunde in Berlin, Geowissenschaftliche Reihe* 4: 223–37.

Hennig, E. 1912. *Am Tendaguru: Leben und Wirken einer deutschen Forschungsexpedition zur Ausgrabung vorweltlicher Riesensaurier in Deutsch-Ostafrika.* Stuttgart: E. Schweitzerbart'sche Verlagsbuchhandlung.

———. 1913. "Beiträge zur Geologie und Stratigraphie Deutsch-Ostafrikas." *Archiv für Biontologie* 3, no. 3: 1–72.

———. 1915. "*Kentrosaurus aethiopicus* der Stegosauride des Tendaguru." *Sitzungsberichte der Gesellschaft Naturforschender Freunde zu Berlin*: 219–47.

———. 1925a. "Ein Drache aus Deutsch-Ostafrika." *Umschau* 29: 108–10.

———. 1925b. "*Kentrurosaurus aethiopicus*: Die Stegosaurierfunde von Tendaguru, Deutsch-Ostafrika." *Palaeontographica*, Suppl. 7, Reihe 1, Teil 1: 100–253.

———. 1937. "Der Sedimentstreifen des Lindi-Kilwa-Hinterlandes (Deutsch-Ostafrika)." *Palaeontographica*, Suppl. 7, Reihe 2, Teil 2: 99–186.

———. 1955. *Gewesene Welten: Auf Saurierjagd im ostafrikanischen Busch.* Zürich: Albert Müller.

———. 1964. "Dr. Phil. Werner Janensch." *Berichte der Geologischen Gesellschaft der DDR* 9: 711–17.

Hohnke, L. A. 1973. "Haemodynamics in the Sauropoda." *Nature* 244: 309–10.

Holtz, T. R. 1994. "The phylogenetic position of the Tyrannosauridae: Implications for theropod systematics." *Journal of Palaeontology* 68, no. 5: 1100–17.

———. 1995. "A new phylogeny of the Theropoda." *Journal of Vertebrate Paleontology* 15, supplement to no. 3, Abstracts of Papers, Fifty-fifth Annual Meeting, p. 35A.

———. 2000. "A new phylogeny of the carnivorous dinosaurs." *Gaia* 15: 5–61.

Hopson, J. A. 1977. "Relative brain size and behavior in archosaurian reptiles." *Annual Review of Ecology and Systematics* 8: 429–48.

Hopwood, A. T. 1937. "Prof. Hans Reck." *Nature* 140: 351.

Howard, R. W. 1975. *The Dawnseekers: The First History of American Paleontology.* New York: Harcourt Brace Jovanovich.

Hunt, A. P., M. G. Lockley, S. G. Lucas, and C. A. Meyer. 1994. "The global sauropod fossil record." *Gaia* 10: 261–79.

Iliffe, J. 1969. *Tanganyika under German Rule, 1905–1912.* Cambridge: Cambridge University Press.

———. 1979. *A Modern History of Tanzania.* Cambridge: Cambridge University Press.

Jacobs, L. L., D. A. Winkler, W. R. Downs, and E. M. Gomani. 1993. "New material of an Early Cretaceous titanosaurid sauropod dinosaur from Malawi." *Paleontology* 36: 523–34.

Jacobs, L. L., D. A. Winkler, Z. M. Kaufulu, and W. R. Downs. 1990. "The Dinosaur Beds of northern Malawi, Africa." *National Geographic Research* 6, no. 2: 196–204.

Jaeger, H. 1971. "Werner Janensch, 1878–1969." *Berichte der Deutschen Gesellschaft für Geologische Wissenschaften* 16, no. 2: 149–54.

Jahn, I. 1989. "Der neue Museumsbau und die Entwicklung neuer museologischer Konzeptionen und Aktivitäten seit 1890." *Wissenschaftliche Zeitschrift der Humboldt-Universität* 38: 287–307.

Janensch, W. 1909a. "Nachtrag zum zweiten Bericht über die Deutsche Tendaguru-Expedition." *Sitzungsberichte der Gesellschaft Naturforschender Freunde zu Berlin*: 631.

———. 1909b. "Zweiter Bericht über die Tendaguru-Expedition." *Sitzungsberichte der Gesellschaft Naturforschender Freunde zu Berlin*: 500–503.

———. 1914. "Bericht über den Verlauf der Tendaguru-Expedition." *Archiv für Biontologie* 3, no. 1: 17–58.

———. 1920. "Über *Elaphrosaurus bambergi* und die Megalosaurier aus den Tendaguru-Schichten Deutsch-Ostafrikas." *Sitzungsberichte der Gesellschaft Naturforschender Freunde zu Berlin*: 225–35.

———. 1924. "Das erste aufgestellte Skelett eines Dinosauriers von Tendaguru in Deutsch-Ostafrika." *Naturforscher* 1: 251–52, 281–82.

———. 1925a. "Ein aufgestelltes Skelett des Stegosauriers *Kentrurosaurus aethiopicus* E. Hennig aus den Tendaguru-schichten Deutsch-Ostafrikas." *Palaeontographica*, Suppl. 7, Reihe 1, Teil 1: 255–76.

———. 1925b. "Die Coelurosaurier und Theropoden der Tendaguru-Schichten Deutsch-Ostafrikas." *Palaeontographica*, Suppl. 7, Reihe 1, Teil 1: 1–97.

———. 1929a. "Ein aufgestelltes und rekonstruiertes Skelett von *Elaphrosaurus bambergi*: Mit einem Nachtrag zur Osteologie dieses Coelurosauriers." *Palaeontographica*, Suppl. 7, Reihe 1, Teil 1: 277–86.

———. 1929b. "Die Wirbelsäule der Gattung *Dicraeosaurus*." *Palaeontographica*, Suppl. 7, Reihe 1, Teil 2: 35–133.

———. 1934. "Eine halbseitige überzählige Wirbelbildung bei einem Dinosaurier." *Sitzungsberichte der Gesellschaft Naturforschender Freunde zu Berlin*: 458–62.

———. 1935–36. "Die Schädel der Sauropoden *Brachiosaurus, Barosaurus*, und *Dicraeosaurus* aus den Tendaguru-Schichten Deutsch-Ostafrikas." *Palaeontographica*, Suppl. 7, Reihe 1, Teil 2: 145–298.

———. 1936. "Ein aufgestelltes Skelett von *Dicraeosaurus hansemanni*." *Palaeontographica*, Suppl. 7, Reihe 1, Teil 2: 299–308.

———. 1938. "Gestalt und Grösse von *Brachiosaurus* und anderen riesenwüchsigen Sauropoden." *Biologe* 7, no. 4: 130–34.

———. 1950a. "Die Skelettrekonstruktion von *Brachiosaurus brancai*." *Palaeontographica*, Suppl. 7, Reihe 1, Teil 3: 95–103.

———. 1950b. "Die Wirbelsäule von *Brachiosaurus brancai*." *Palaeontographica*, Suppl. 7, Reihe 1, Teil 3: 27–93.

———. 1955. "Der Ornithopode *Dysalotosaurus* der Tendaguruschichten." *Palaeontographica*, Suppl. 7, Reihe 1, Teil 3: 105–76.

———. 1961. "Skelettrekonstruktion von *Dysalotosaurus lettow-vorbecki*." *Palaeontographica*, Suppl. 7, Reihe 1, Teil 3: 237–40.

Janensch, W., and E. Hennig. 1909. "Erster Bericht über die Tendaguru-Expedition." *Sitzungsberichte der Gesellschaft Naturforschender Freunde zu Berlin*: 358–60.

Jarzen, D. M. 1981. "A preliminary report of the palynomorphs recovered from Tendaguru hill (Tanzania)." *Pollen et Spores* 23, no. 1: 149–63.

Kahlert, E., S. Schultka, and H. Süss. 1999. "Die mesophytische Flora der Saurierlagerstätte am Tendaguru (Tansania): Erste Ergebnisse." *Mitteilungen aus dem Museum für Naturkunde in Berlin, Geowissenschaftliche Reihe* 2: 185–99.

Kaiser, H. E. 1954. "Eine Rippe mit pathologischen Erscheinungen des Sauropods *Brachiosaurus brancai* Janensch, des grössten Saurischier (Dinosaurier) des Tendagurumaterials aus der jüngeren Jura." *Monatshefte für Veterinär Medizin* 9, no. 4: 373–74.

Kent, P. E. 1974. "Continental margin of East Africa—a region of vertical movements." In *The Geology of Continental Margins*, ed. C. A. Burk and C. L. Drake, 313–20. New York: Springer-Verlag.

Kent, P. E., J. A. Hunt, and D. W. Johnstone. 1971. *The Geology and Geophysics of Coastal Tanzania*. Geophysical Paper no. 6. Natural Environment Research Council, Institute of Geological Sciences. London: H.M.S.O.

Kinder, H., and W. Hilgemann. 1978. *Penguin Atlas of World History.* Volume 2. Harmondsworth: Penguin.

Kurtz, L. S. 1978. *Historical Dictionary of Tanzania.* Metuchen, N.J.: The Scarecrow Press.

"Late Prof. Cutler navigated rivers to city Winnipeg. Scientist who died of malaria in Africa was lover of novelty and adventure." 1925. *Calgary Daily Herald,* September 3, p. 11.

Leakey, L. S. B. 1937. *White African.* London: Hodder and Stoughton.

LeTissier, T. 1988. *The Battle of Berlin, 1945.* London: Jonathan Cape.

MacDonough, G. 1997. *Berlin: A Portrait of its History, Politics, Architecture, and Society.* New York: St. Martin's Press.

Madsen, J. H., and S. P. Welles. 2000. Ceratosaurus *(Dinosauria, Theropoda): A Revised Osteology.* Salt Lake City: Utah Geological Survey.

Man, J. 1973. *Berlin Blockade.* New York: Ballantine Books.

Mann, W. 1957. *Berlin zur Zeit der Weimarer Republik.* Berlin: Das Neue Berlin.

Marsh, O. C. 1877. "New order of extinct Reptilia (Stegosauria) from the Jurassic of the Rocky Mountains." *American Journal of Science,* series 3, 14: 513–14.

———. 1878. "Principal characters of American Jurassic dinosaurs. Pt. I." *American Journal of Science,* series 3, 16: 411–16.

———. 1880. "Principal characters of American Jurassic dinosaurs. Pt. III." *American Journal of Science,* series 3, 19: 253–59.

———. 1881. "Principal characters of American Jurassic dinosaurs. Pt. V." *American Journal of Science,* series 3, 21: 417–23.

———. 1884. "Principal characters of American Jurassic dinosaurs. Pt. VIII. The order Theropoda." *American Journal of Science,* series 3, 27: 329–41.

Martin, J., V. Mártin-Rolland, and E. Frey. 1998. "Not cranes or masts, but beams: The biomechanics of sauropod necks." *Oryctos* 1: 113–20.

Masur, G. 1971. *Imperial Berlin.* New York: Basic Books.

Matthew, W. D. 1921. "Notes on the scientific museums of Europe." *Natural History Magazine* 21: 185–90.

———. 1923. "Jurassic dinosaurs of Utah and East Africa." *Bulletin of the Geological Society of America* 34: 405–406.

McGowan, C. 1983. *The Successful Dragons: A Natural History of Extinct Reptiles.* Toronto: Samuel Stevens.

Middlebrook, M. 1988. *The Berlin Raids: RAF Bomber Command Winter, 1943–44.* New York: Viking.

Middlebrook, M., and C. Everitt. 1985. *The Bomber Command Diaries: An Operational Reference Book, 1939–1945.* New York: Viking.

Migeod, F. W. H. 1927a. "British Museum East Africa Expedition: Progress in the year 1926." *Natural History Magazine* 1, no. 2: 34–43.

———. 1927b. "The dinosaurs of Tendaguru." *Journal of the African Society* 26: 323–40.

———. 1927c. *A View of Sierra Leone.* New York: Brentano's.

———. 1931. "Digging for dinosaurs: An East African giant. Sixty million years." *Times* (London), February 21, p. 13.

Militärisches Orientierungsheft für Deutsch-Ostafrika. 1911. Dar es Salaam: Deutsch-Ostafrikanische Rundschau.

Millard, R. W., H. B. Lillywhite, and A. R. Hargens. 1992. "Cardiovascular system design and *Barosaurus.*" *Lancet* 340: 914.

Miller, C. 1974. *Battle for the Bundu: The First World War in East Africa.* New York: Macmillan.

Milner, A. R., and D. B. Norman. 1984. "The biogeography of advanced ornithopod dinosaurs (Archosauria: Ornithischia)—a cladistic-vicariance model." In *Third Symposium on Mesozoic Terrestrial Ecosystems: Short Papers,* ed. W.-E. Reif and F. Westphal, 145–50. Tübingen: Attempto Verlag.

Moore, G. T., D. N. Hayashida, C. A. Ross, and S. R. Jacobson. 1992. "Paleoclimate of the Kimmeridgian/Tithonian (Late Jurassic) world: I. Results using a general circulation model." *Palaeogeography, Palaeoclimatology, Palaeoecology* 93: 113–50.

Das Museum für Naturkunde der Königlichen Friedrich-Wilhelm-Universität in Berlin zur Eröffnungsfeier. 1889. Berlin: Ernst & Korn.

Nopcsa, F. 1915. "Die Dinosaurier der siebenbürgischen Landesteile Ungarns." *Mitteilungen Jahrbuch des Königlichen Ungarische Geologische Reichsanstalt* 23: 1–26.

———. 1928. "Paleontological notes on reptiles." *Geologica Hungarica* 1: 1–84.

Oberli, U. 1983. "Der 'Saurierfriedhof' am Tendaguru: Auf den Spuren einer fast vergessenen Ausgrabungsexpedition in Tansania." *Schweizer Strahler* 6, no. 8: 374–80.

Osborn, H. F. 1915. "Eberhard Fraas." *Science* 41, no. 1059: 571–72.

Pálfry, J., P. L. Smith, and J. K. Mortensen. 2000. "A U-Pb and ^{40}Ar/^{39}Ar time scale for the Jurassic." *Canadian Journal of Earth Sciences* 37: 923–44.

Parkinson, J. 1930. *The Dinosaur in East Africa: An Account of the Giant Reptile Beds of Tendaguru, Tanganyika Territory.* London: H. F. & G. Witherby.

Paul, G. S. 1988a. "The brachiosaur giants of the Morrison and Tendaguru with a description of a new subgenus, *Giraffatitan,* and a comparison of the world's largest dinosaurs." *Hunteria* 2, no. 3: 1–14.

———. 1988b. *Predatory Dinosaurs of the World: A Complete Illustrated Guide.* New York: Simon and Schuster.

———. 1994. "Dinosaur reproduction in the fast lane: Implications for size, success, and extinction." In *Dinosaur Eggs and Babies,* ed. K. Carpenter, K. Hirsch, and J. R. Horner, 244–55. Cambridge: Cambridge University Press.

———. 1998. "Terramegathermy and Cope's Rule in the land of the titans." *Modern Geology* 23: 179–217.

Peczkis, J. 1994. "Implications of body-mass estimates for dinosaurs." *Journal of Vertebrate Paleontology* 14, no. 4: 520–33.

Perry, S. F., and C. Reuter. 1999. "Hypothetical lung structure of *Brachiosaurus* (Dinosauria: Sauropoda) based on functional constraints." *Mitteilungen aus dem Museum für Naturkunde in Berlin, Geowissenschaftliche Reihe* 2: 75–79.

Rauhut, O. 1998. "*Elaphrosaurus bambergi* and the early evolution of theropod dinosaurs." *Journal of Vertebrate Paleontology* 18, supplement to no. 3, Abstracts of Papers, Fifty-eighth Annual Meeting, p. 71A.

———. 2000. "The interrelationships and evolution of basal theropods (*Dinosauria, Saurischia*)." Ph.D. diss., University of Bristol.

Reck, H. 1910a. "Die Ausgrabungen fossiler Riesentiere in Deutsch-Ostafrika." *Umschau* 14: 1041–48.

———. 1910b. "3. Bericht über den weiteren Verlauf der Tendaguru-Expedition." *Sitzungsberichte der Gesellschaft Naturforschender Freunde zu Berlin*: 372–75.

———. 1911. "4. Bericht über die Ausgrabungen und Ergebnisse der Tendaguru-Expedition (Grabungsperiode 1911)." *Sitzungsberichte der Gesellschaft Naturforschender Freunde zu Berlin*: 385–97.

———. 1924. "Das erste rekonstruierte Skelett der Tendagurusaurier-Lagerstätte in Deutsch-Ostafrika." *Afrika-Nachrichten* 17–18: 259–60.

————. 1925. "Die britische Tendaguru-Expedition 1924." *Afrika-Nachrichten* 1: 3–4.

————. 1931. "Die deutschostafrikanischen Flugsaurier." *Centralblatt für Mineralogie, Geologie und Paläontologie,* Abt. B: 321–36.

————. 1933a. "Auf der Suche nach Saurier-Skeletten in der Karru Natals." *Umschau* 37, no. 17: 324–28.

————. 1933b. *Oldoway, die Schlucht des Urmenschen: Die Entdeckung des altstein-zeitlichen Menschen in Deutsch-Ostafrika.* Leipzig: F. A. Brockhaus.

Reck, I. 1924. *Mit der Tendaguru-Expedition im Süden von Deutsch-Ostafrika.* Berlin: Dietrich Reimer.

————. 1925. *Auf einsamen Märschen im Norden von Deutsch-Ostafrika.* Berlin: Dietrich Reimer.

Reichs-Kolonialamt. 1912–13. *Die Deutschen Schutzgebiete in Afrika und der Südsee: Amtliche Jahresberichte.* Berlin: Mittler und Sohn.

Riggs, E. S. 1904. "Structure and relationships of opisthocoelian dinosaurs. Part II: The Brachiosauridae." *Field Columbian Museum* 2, no. 6: 229–48.

Romer, R. L. 2001. "Isotopically heterogeneous initial Pb and continuous ^{222}Rn loss in fossils: The U-Pb systematics of *Brachiosaurus brancai.*" *Geochimica et Cosmochimica Acta* 65, no. 22: 4201–13.

Russell, D. A., P. Béland, and J. S. McIntosh. 1980. "Paleoecology of the dinosaurs of Tendaguru (Tanzania)." *Memoires de Société géologique de France* 59, no. 139: 169–75.

Salgado, L. 1993. "Comments on *Chubutisaurus insignis* Del Corro (Saurischia, Sauropoda)." *Ameghiniana* 30, no. 3: 265–70.

————. 1999. "The macroevolution of the Diplodocomorpha (Dinosauria; Sauropoda): A developmental model." *Ameghiniana* 36, no. 2: 203–16.

Salgado, L., R. Coria, and J. O. Calvo. 1997. "Evolution of titanosaurid sauropods. I: Phylogenetic analysis based on the postcranial evidence." *Ameghiniana* 34, no. 1: 3–32.

Sander, P. M. 1999. "Life history of Tendaguru sauropods as inferred from long bone histology." *Mitteilungen aus dem Museum für Naturkunde in Berlin, Geowissenschaftliche Reihe* 2: 103–12.

————. 2000. "Longbone histology of the Tendaguru sauropods: Implications for growth and biology." *Paleobiology* 26, no. 3: 466–88.

Sapper, K. 1936–38. "Hans Reck." *Zeitschrift für Vulkanologie* 17: 225–32.

Sayao, J. M., and A. W. A. Kellner. 2001. "New data on the pterosaur fauna from Tendaguru (Tanzania), Upper Jurassic, Africa." *Journal of Vertebrate Paleontology* 21, supplement to no. 3, Abstracts of Papers, Sixty-first Annual Meeting, p. 97A.

Schnee, H. 1920. *Deutsches Kolonial-Lexikon.* 3 vols. Leipzig: Quelle & Meyer.

Schrank, E. 1999. "Palynology of the dinosaur beds of Tendaguru (Tanzania)—preliminary results." *Mitteilungen aus dem Museum für Naturkunde in Berlin, Geowissenschaftliche Reihe* 2: 171–83.

————. 2000a. "Age and biogeographic links of palynomorph assemblages from the dinosaur beds of Tendaguru (Late Jurassic, Tanzania)." Actes du 4éme Symposium de Palynologie africaine, Sousse, Tunisia, April 25–30, 1999. *Geo-Eco-Trop* 22: 141–46.

————. 2000b. "Palynomorphs from dinosaur-bearing Upper Jurassic strata of Tanzania in a palaeoenvironmental and phytogeographic context." In *Abstracts,* Tenth International Palynological Conference, Nanjing, China, June 24–30, ed. Wang Weiming, Ouyang Shu, Sun Xiangjun, and Yu Ge, p. 147.

————. 2001. "Eine Expedition zur Dinosaurierlagerstätte Tendaguru (Tansania) und erste palynologische Ergebnisse." *29. Jahrestagung des Arbeitskreises für Paläobotanik und Palynologie,* program and abstracts, Utrecht, The Netherlands, p. 27.

Schudack, M. E. 1999. "Some charophytes from the Middle Dinosaur Member of the Tendaguru Formation (Upper Jurassic of Tanzania)." *Mitteilungen aus dem Museum für Naturkunde in Berlin, Geowissenschaftliche Reihe* 2: 201–205.

Schudack, M., and U. Schudack. 2002. "Ostracods from the Middle Dinosaur Member of the Tendaguru Formation (Upper Jurassic of Tanzania)." *Neues Jahrbuch für Geologie und Paläontologie,* Monatshefte 6: 321–33.

Seebacher, F. 2001. "A new method to calculate allometric length-mass relationships of dinosaurs." *Journal of Vertebrate Paleontology* 21, no. 1: 51–60.

Seeley, H. G. 1888. "On the classification of the fossil animals commonly named Dinosauria." *Proceedings of the Royal Society of London* 43: 165–71.

Sereno, P. C. 1986. "Phylogeny of the bird-hipped dinosaurs (Order Ornithischia)." *National Geographic Research* 2: 234–56.

———. 1999. "The evolution of dinosaurs." *Science* 284: 2137–47.

Sereno, P. C., J. A. Wilson, H. C. Larsson, D. B. Dutheil, and H.-D. Suess. 1994. "Early Cretaceous dinosaurs from the Sahara." *Science* 266: 267–71.

Seymour, R. S. 1976. "Dinosaurs, endothermy, and blood pressure." *Nature* 262: 207–208.

Seymour, R. S., and H. B. Lillywhite. 2000. "Hearts, neck posture and metabolic intensity of sauropod dinosaurs." *Proceedings of the Royal Society of London B* 267: 1883–87.

Sheppard, F. H. W., ed. 1975. *Survey of London: The Museums Area of South Kensington and Westminster.* Volume 38. London: The Athlone Press.

Sibley, J. R. 1973. *Tanganyikan Guerrilla: East African Campaign, 1914–18.* London: Pan Books.

Simpson, G. G. 1987. *Simple Curiosity: Letters from George Gaylord Simpson to His Family, 1921–1970.* Edited by L. F. Laporte. Berkeley and Los Angeles: University of California Press.

Smith, G. I. 1980. *Ghosts of Kampala.* London: Weidenfeld and Nicolson.

Stearn, W. T. 1981. *The Natural History Museum at South Kensington: A History of the British Museum (Natural History), 1753–1980.* London: Heinemann.

Stevens, K. A. 2002. "DinoMorph: Parametric modeling of skeletal structures." *Senckenbergiana Lethaea* 82, no. 1: 23–34.

Stevens, K. A., and J. M. Parrish. 1999. "Neck posture and feeding habits of two Jurassic sauropod dinosaurs." *Science* 284: 798–800.

———. 2001. "Biological implications of digital reconstructions of the whole body of sauropod dinosaurs." *Journal of Vertebrate Paleontology* 21, supplement to no. 3, Abstracts of Papers, Sixty-first Annual Meeting, p. 104A. To be published in a volume on the Sauropod Symposium in honor of Jack McIntosh, SVP, Bozeman, Montana, October 2001, by University of California Press.

———. In preparation. "The intrinsic curvature of sauropod necks (Saurischia: Dinosauria)."

Süss, H., and S. Schultka. 2001. "First record of *Glyptostroboxylon* from the Upper Jurassic of Tendaguru, Tanzania." *Botanical Journal of the Linnean Society* 135: 421–29.

Swinton, W. E. 1947. "John Parkinson." *Nature* 160, no. 4064: 390–91.

———. 1952. "F. W. H. Migeod." *Nature* 170, no. 4318: 184.

———. 1953. "Frederick William Hugh Migeod." *Proceedings of the Geological Association* 64: 64–65.

Thoss, F., and P. Schwartze. 1975. "Über mögliche funktionelle Eigenschaften des Labyrinths von *Brachiosaurus brancai.*" *Acta Biologica et Medica Germanica* 34, no. 5: 899–906.

Thulborn, R. A. 1982. "Speeds and gaits of dinosaurs." *Palaeogeography, Palaeoclimatology, Palaeoecology* 38: 227–56.

Tornier, G. 1912. "Bericht des Vorsitzenden: Festsitzung zur Berichterstattung über Werden, Verlauf und bisherige Ergebnisse der Tendaguru-Expedition." *Sitzungsberichte der Gesellschaft Naturforschender Freunde zu Berlin*: 120–23, 150–52.

Tully, A. 1964. *Berlin: Story of a Battle, April–May 1945*. New York: Macfadden-Bartell.

Tusa, A., and J. Tusa. 1988. *The Berlin Blockade*. London: Hodder and Stoughton.

Unwin, D. M., and W.-D. Heinrich. 1999. "On a pterosaur jaw from the Upper Jurassic of Tendaguru (Tanzania)." *Mitteilungen aus dem Museum für Naturkunde in Berlin, Geowissenschaftliche Reihe* 2: 121–34.

Upchurch, P. 1995. "The evolutionary history of sauropod dinosaurs." *Philosophical Transactions of the Royal Society of London B*, 349: 365–90.

Upchurch, P., and P. M. Barrett. 2000. "The evolution of sauropod feeding mechanisms." In *Evolution of Herbivory in Terrestrial Vertebrates: Perspectives from the Fossil Record*, ed. H.-D. Sues, 79–122. Cambridge: Cambridge University Press.

"Urvogel und Saurier." 1984. *Frankfurter Allgemeine Zeitung*, July 19, p. 21.

"40,000 Tokioter sahen bisher Saurier aus Berlin: Lehr- und Forschungsschau findet grossen Anklang." 1984. *Berliner Zeitung*, July 22–23, p. 21.

Von Branca, W. 1914a. "Allgemeines über die Tendaguru-Expedition." *Archiv für Biontologie* 3, no. 1: 3–13.

———. 1914b. "Kurzer Bericht über die von Dr. Reck erzielten Ergebnisse im vierten Grabungsjahre 1912." *Archiv für Biontologie* 3, no. 1: 61–63.

Von Eckardt, W., and S. L. Gilman. 1975. *Bertolt Brecht's Berlin: A Scrapbook of the Twenties*. Garden City, N.Y.: Anchor Press.

Von Eckenbrecher, M. 1912. *Im Dichten Pori: Reise- und Jagdbilder aus Deutsch-Ostafrika*. Berlin: Mittler und Sohn.

Von Huene, F. 1927. "Short review of the present knowledge of the Sauropoda." *Memoirs of the Queensland Museum* 8: 121–26.

———. 1932. "Die fossile Reptil-Ordnung Saurischia, ihre Entwicklung und Geschichte." *Monographie zur Geologie und Paläontologie*, series 1, 4: 1–361.

Weaver, J. C. 1983. "The improbable endotherm: The energetics of the sauropod dinosaur *Brachiosaurus.*" *Paleobiology* 9, no. 2: 173–82.

Wedel, M. J., R. L. Cifelli, and R. K. Sanders. 2000. "Osteology, paleobiology, and relationships of the sauropod dinosaur *Sauroposeidon.*" *Acta Palaeontologica Polonica* 45, no. 4: 343–88.

Wellnhofer, P., ed. 1978. *Handbuch der Paläoherpetologie*. Teil 19, *Pterosauria*. Stuttgart: Gustav Fischer Verlag.

Werner, C., and W. Zils. 1997. "Tendaguru-Expedition Dezember 1996." *Terra Nostra* 7, no. 97: 28–32.

Whiting, C. 1982. *The Home Front: Germany*. Chicago: Time-Life Books.

Wild, R. 1991. "*Janenschia* n. g. *robusta* (E. Fraas 1908) pro *Tornieria robusta* (E. Fraas 1908) (Reptilia, Saurischia, Sauropodomorpha)." *Stuttgarter Beiträge zur Naturkunde*, series B, no. 173, pp. 1–4.

Wilson, J. A., and P. C. Sereno. 1998. "Early evolution and higher-level phylogeny of sauropod dinosaurs." Society of Vertebrate Paleontology Memoir 5, *Journal of Vertebrate Paleontology* 18, suppl. to no. 2.

Wings. O. 2002. "Early diagenetic processes and cementation in the fossilization of tetrapod bones." <http://www.Paleontology.uni-bonn.de/mitarbeiter/wings/bristol.htm>, accessed September 7.

Wyman, M. 1998. *DPs: Europe's Displaced Persons, 1945–1951*. Ithaca: Cornell University Press.

Zils, W., C. Werner, A. Moritz, and C. Saanane. 1995. "Orientierende Tendaguru-Expedition 1994." *Berliner Geowissenschaftliche Abhandlungen* 16, no. 2: 483–531.

INDEX

Bold figures indicate illustrations.

Abdallah (servant), 257
Aberhan, Martin, 304, 305, 306
airship, 113
Aitken, William G., 282
Alberti (Reck's deputy), 91, 103
Albrecht, Duke of Mecklenburg, Johann, 18, 68, 84, 198
Alexander, R. McNeill, 323, 326, 327
Ali (cook), 103
Ali (Hennig's attendant), 23
Ali (worker), 237
Am Tendaguru (Hennig), 104, 123, 286
Amani Biological-Agricultural Research Institute, 52, 208
Ambrosius, Pater, 78
Amin Dada, Idi, 292
ammonites, dating of, 310
amphibians, 315
Anderson, J. F., 326
Anderson (officer), 138, 159, 172
Arambourg, Camille, 275
Archiv für Biontologie, 130, 187, 206
Arkell, W. J., 282
Arning, Wilhelm, 1–2, 7, 11, 69, 94, 106
Auracher, Dr., 71, 81, 108
Awusi (worker), 237

Baedeker, Fritz, 18
Bamberg, Paul, 68, 104, 196
Bampfylde, District Officer, 228
Barlow, F. O., 219
Barthel, Manfred, 295, 296
Bate, Raymond, 283
Bather (British Museum's keeper of geology), 156, 161, 179
  aids expedition, 187, 189, 191, 199
  conflict with Smith, 210
  correspondence with Parkinson, 205, 208
  relations with Migeod, 165, 187, 188
Bavarian State Collection for Paleontology and Historical Geology, 250, 271
Bedford, Duke of, 140
Béland, Pierre, 291, 292, 321, 323, 332
Berckhemer, Fritz, 271
Berlin
  after WWI, 114–115, 150–151, 195
  during and after WWII, 272, 273–274, 277, 281
Berlin Society for Earth Sciences, 14

Berlin-Tanzania expedition
  1993, 300–301
  1996, 302
Bernardo (cook), 254
Berno, Pater, 286
Besser, C. W., 24, 28, 34
bin Ali, Saidi, 88, 103, 254
bin Amrani, Boheti
  character, 40
  field work with British, 138, 160, 173, 178, 180, 184, 185, 204, 215, 217, 221 241 passim
  field work with Germans, 25, 27, 28, 30, 42, 47, 56, 57, 59, 67, 69, 74, 81, 87, 88, 90, 100, 102
  picture, 4
  relations with Cutler, 130, 139, 140, 143, 159
  after Tendaguru, 254
  wages in 1925–1926, 175–176, 187
bin Manjonga, Saidi, 101–102, 103
bin Namanorow, Issa, 103
bin Salim, Issa, 58
Bingham, Mike, 283
biogeography, 318–321
biomechanics
  jaw movement, 331
  locomotion, 324, 326–327
  neck movement, 327–329
Blackwood, Dr., 132, 143, 162
blood pressure and circulation, 329–330, 332–333
Böck, Max, 14, 270, 271, 275, 2
Bohari (camp helper), 178
Böhme, Gottfried, 295
Bokari (quarry leader), 185
Boli, Isa (quarry leader), 242
Bonaparte, Jose, 298, 319
Bond, Geoffrey, 283
bone pathologies, 179, 324
Borchert, Gustav, 58, 90, 150, 247, 250, 262, 1
Bornhardt, Wilhelm, 144
Bötzow (planter), 45
Bowring, Charles, 189
Brachiosaurus mount
  casts of, in world's museums, 297–298
  construction, 262–267
  dismantling and shipping, 279, 295, 296
  pictures, 6–11
  response to, 268, 273

Gerhard Maier has spent ten years working in archaeology and vertebrate paleontology. Formerly a technician at the Royal Tyrrell Museum of Palaeontology, he now works for a major oil company. A long-standing interest in dinosaurs, travel, and history has culminated in this volume.

9 780253 342140